Electrical Power System Protection

Electrical Power System Protection

A. Wright

Emeritus Professor of Electrical Engineering,
University of Nottingham,
Nottingham, UK

and

C. Christopoulos

Professor of Electrical Engineering,
University of Nottingham,
Nottingham, UK

 CHAPMAN & HALL
London · Glasgow · New York · Tokyo · Melbourne · Madras

Published by Chapman & Hall, 2-6 Boundary Row, London SE1 8HN

Chapman & Hall, 2–6 Boundary Row, London SE1 8HN, UK

Blackie Academic & Professional, Wester Cleddens Road, Bishopbriggs, Glasgow G64 2NZ, UK

Chapman & Hall Inc., 29 West 35th Street, New York NY10001, USA

Chapman & Hall Japan, Thomson Publishing Japan, Hirakawacho Nemoto Building, 6F, 1–7–11 Hirakawa-cho, Chiyoda-ku, Tokyo 102, Japan

Chapman & Hall Australia, Thomas Nelson Australia, 102 Dodds Street, South Melbourne, Victoria 3205, Australia

Chapman & Hall India, R. Seshadri, 32 Second Main Road, CIT East, Madras 600 035, India

First edition 1993

© 1993 A. Wright and C. Christopoulos

Typeset in 10/12 pt Times by Pure Tech Corporation, Pondicherry, India

Printed and bound in Great Britain by Hartnolls Ltd, Bodmin, Cornwall

ISBN 0 412 39200 3 0 442 31648 8 (USA)

A catalogue record for this book is available from the British Library

Library of Congress Cataloging-in-Publication data available

Contents

Acknowledgements

Figure 4.26 is reproduced from BS 142: Section 3.2: 1990 with the permission of BSI. Complete copies of the standard can be obtained by post from BSI sales, Linford Wood, Milton Keynes, MK14 6LE.

Preface

Several books have been produced over the years about the protective equipment which is incorporated in electrical power systems and manufacturers continually produce detailed literature describing their products. Recognizing this situation and accepting that it is no longer possible in a single volume to provide a complete coverage of the protective equipment now available and the many factors which have to be considered when it is being developed and applied, we have concentrated on basic principles and given examples of modern relays and schemes in this work.

Chapter 1 deals with electric fuses, which were the earliest protective devices. The chapter begins with a historical introduction, as do all the chapters, and then information is provided on the construction and behaviour of fuses and finally the factors which must be taken into account when they are to be applied to circuits are examined.

Chapters 2 and 3 deal respectively with conventional current and voltage transformers and modern transducers. In each case, details are given of the constructions and behaviours of theses devices, which play important roles in supplying protective equipment.

Chapter 4 deals with relays which have constant operating times and those which have inverse time/current characteristics. After tracing their development, modern relays are described and then the factors which must be considered when applying them are considered in some detail.

The principles of current-differential schemes are set out in Chapter 5 and the causes of the imbalances which can arise in them when protected units are healthy are examined. The biasing features provided to enable satisfactory performance to be obtained are outlined.

The later chapters are devoted to the protection of the main components of the networks, namely transformers, busbars, rotating machines and transmission and distribution lines and cables. The presentation is similar to that in the earlier chapters. In each case information is provided about the construction and behaviour of the plant being protected and then the appropriate protective schemes, including current-differential, phase comparison, distance and travelling wave, are described and examined.

Appendices dealing with per-unit quantities, symmetrical components and other modal quantities are included.

We express our appreciation of the assistance given to us by Dr D.W.P. Thomas during the preparation of this book and during research into travelling-wave protective schemes. We also wish to thank Miss S E Hollingsworth for typing the manuscript.

We hope that this book will prove of value to those involved in the study, development, production and application of protective equipment and that they will enjoy working in a challenging field in which new problems continuously arise.

Arthur Wright and
Christos Christopoulos

List of symbols

A	cross-sectional area (m^2)
B	magnetic flux density (T)
C	capacitance (F)
e	instantaneous e.m.f. (V)
E	r.m.s. or constant e.m.f. (V)
i	instantaneous current (A)
I	r.m.s. or constant current (A)
L	inductance (H)
M	mutual inductance (H)
N	turns in a winding
R	resistance (Ω)
t	time (s)
v	instantaneous voltage (V)
V	r.m.s. or constant voltage (V)
φ	magnetic flux (Wb) or phase angle
μ	permeability of magnetic material (Wb/Am)

SUFFIXES

a, b, c	phases of three-phase system
p	primary circuit or winding
pk	peak value of alternating current or voltage
s	secondary circuit or winding
t	tertiary circuit or winding
1, 2, 0	positive, negative and zero-sequence quantities

1

Fuses

Fuses, which were introduced over one hundred years ago, were the first form of protection used on electrical networks. Extremely large numbers of them have been produced since that time and they are still used extensively in the lower voltage sections of power systems around the world. Fuselinks are simple and therefore relatively cheap devices, their cost being very low relative to that of the plant being protected by them. They thus satisfy a basic requirement which applies to all protective equipment.

The underlying principle associated with fuses is that a relatively short piece of conducting material, with a cross-sectional area incapable of carrying currents quite as high as those which may be permitted to flow in the protected circuit, is sacrificed, when necessary, to prevent healthy parts of the circuit being damaged and also to limit damage to faulty sections of the circuit to the lowest level possible.

Fuses incorporate one or more current-carrying elements, depending on their current ratings. Melting of the elements, followed by arcing across the breaks, occurs when overcurrents flow through them. They can interrupt very high fault currents and because of the rapidity of their operation in these circumstances, they limit the energy dissipated during fault conditions.

1.1 HISTORICAL BACKGROUND

The earliest reference to the use of fuses which the authors have been able to trace occurred during the discussion which followed the presentation of a paper by Cockburn [1] to the Society of Telegraph Engineers in 1887 when W. H. Preece stated that platinum wires had been used as fuses to protect submarine cables since 1864 and Sir David Solomons referred to the use of fuses in 1874.

Fuses must have been in use in significant numbers by 1879 and there must have been situations where a simple wire or strip element was not suitable, because in that year Professor S. P. Thompson produced what he described as an improved form of fuse. It consisted of two wires connected together by a ball of conducting material as shown in Fig. 1.1. It was stated that the ball could be an alloy of lead and tin or some other conducting material with a low melting point. When a high current flowed through the fuse for a sufficient

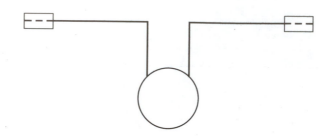

Fig. 1.1 Fuse developed by Professor S. P. Thompson. (Reproduced from Wright and Newbery, 1982, *Electric Fuses* with the permission of IEE).

period of time, the ball melted and the wires swung apart to form an adequate break in the circuit.

A variation on the above construction was patented in 1883 by C. V. Boys and H. H. Cunyngham. In their arrangement the element consisted of two leaf springs which were soldered together at their tips as shown in Fig. 1.2. Again the passage of an overcurrent for a sufficient period of time caused the solder to melt, after which the springs flexed away from each other to provide the quick and adequate separation needed to ensure current interruption.

Demonstrations of incandescent filament lamps were given by J. Swan (later to become Sir Joseph Swan) in Britain in 1878 and at about the same time in the USA by T. A. Edison. These led to a great demand for the installation of electric lighting in both public and private buildings. Initially individual consumers had their own generating plants, but shortly afterwards small central generating stations were provided to supply their own surrounding areas.

Some interesting and detailed information about early installations is given in letters written by J. H. Holmes to H. W. Clothier in 1932. An excerpt from one of these letters, which was included in Clothier's book entitled *Switchgear Stages* [2], is reproduced below. It clearly indicates that the identity of the person who first introduced fuses is not known.

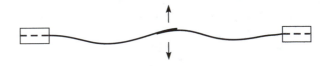

Fig. 1.2 Fuse patented by Boys and Cunyngham. (Reproduced from Wright and Newbery, 1982, *Electric Fuses* with the permission of IEE).

Letter from Mr J. H. Holmes:

Regarding the origin of fuses, I have always been uncertain as to who is entitled to the credit of being the first inventor, and am of the opinion that this is very clearly the case of 'Necessity is the Mother of Invention'.

I have been looking up some records of what was known about fuses in the early 'eighties' (1880's), and in the first volume of 'Electric Illumination', compiled by J. Dredge and published in August 1882 at the Offices of 'Engineering', on page 630 it is stated that Edison's British Patent of April 1881 appears to have been the first notification of lead safety wire. It also appears that Edison's device was called a 'safety guard'.

I think however that Swan used a device for the same purpose and before April 1881, because 'Cragside' near here, the seat of Sir W. G. (afterwards Lord) Armstrong was lighted with Swan lamps by the middle of December 1880. Swan used tinfoil for the fuse and a strip of this was jammed between two brass blocks, so as to form part of the circuit, by a plug of wood and later of steatite, and I have samples of a combined switch and fuse, and a fuse only made in this way, and which were in use at Cragside. In a Swan United Electric Light Co's catalogue dated 1883, I found such fuses illustrated and called 'safety fusing bridges'.

In the description of the Electric Lighting, on the Swan system, of the Savoy Theatre in 'Engineering', March 3rd 1882, fusible safety shunts are referred to as 'not intended so much to guard against a danger which is next to impossible to occur in practical working, but to protect the lamps themselves from destruction from too powerful a current being transmitted through them. This seems to confirm what Campbell Swinton says about the Drawing Office at Elswick in 1882*, which you quote, and I note he also says that at the Paris Exhibition of 1881 there was 'a vast array of switches, fuses, cut outs and other apparatus'.

Factors, including the concern for public safety, the cost and fragility of the lamps, referred to earlier, and the increasing level of available volt-amperes under fault conditions, made evident the need for protective equipment. As there were no obvious alternatives to fuses at that time, a number of workers endeavoured to develop reliable fuses.

Much work was done to understand the processes involved during the melting of fuse elements, a particularly significant contribution being made by Cockburn [1]. He attempted to put the design of fuses on a sound engineering basis. He studied the effects of the heat conducted away from fuse elements by their terminals and connecting cables and investigated the properties of conductors in an attempt to select the materials most suitable for use as fuse elements. He recognized that materials which oxidize significantly would be

---

* It is perhaps not generally known that fuses, as originally introduced by Swan, were designed not as a safeguard to protect the wires against overloading on short circuits, but in order to prevent the lamps from over-running. When I went to the Armstrong works at Elswick in 1882, part of the drawing office had been electrically lighted by the Swan Company, and each incandescent lamp was fitted with a separate tinfoil fuse for this purpose. The precaution was, perhaps a necessary one, as the lamps then cost 25 s [£1.25] each and were very fragile, while the arrangements for keeping a constant voltage were very crude.' (Campbell Swinton at the IEE Commemoration Meetings, February 1922, *IEE Journal*, 1922, Vol 60, p. 494).

unsuitable because the characteristics of fuses containing them would change with time. Tests which he did on a range of fuses showed that they were not being applied satisfactorily. He found instances where the minimum fusing currents were many times the rated currents of the circuits and pieces of equipment being protected. He suggested that fuses should operate at 150–200% of the rated current of the circuit being protected.

Most of the early fuses were mounted in wooden boxes, but the individual elements were not separately enclosed. As early as May 1880, however, T. A. Edison patented a fuse in which the element was enclosed in a glass tube. This was done to protect the surroundings from the effects of the rupturing of the element rather than to affect or control the fuse performance. Undoubtedly the credit for developing the filled cartridge fuse must go to W. M. Mordey, who patented the device in 1890. His patent described a fuselink with a fusible copper conductor, of either thin foil or one or more small diameter wires, enclosed in a glass tube or similar vessel. It was stated that the tube should be wholly or partially filled with finely divided, semi-conducting or badly-conducting material, which should preferably be incombustible or non-flammable. The fuse produced by Mordey is illustrated in Fig. 1.3.

Fuses were the only form of protective equipment available during the final decade of the nineteenth century. That they were produced in large ratings is evident from Clothier's paper entitled 'The construction of high-tension control-station switchgears, with a comparison of British and foreign methods' [3]. A relevant extract from this paper reads as follows:

> The High-Tension Fuse most extensively used in Germany is not unlike the well-known Bates fuse consisting of an open-ended tube of porcelain, ambroin,

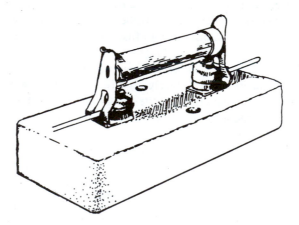

Fig. 1.3 Cartridge fuse patented by Mordey. (Reproduced from Wright and Newbery, 1982, *Electric Fuses* with the permission of IEE).

stabilit, or similar insulating material with plug terminals at each end. The fuse wire of copper or alloy is threaded through the tube and clamped by screws and plates or soldered to the terminals. For potentials of 2,000 to 10,000 volts, the length of these tubes varies between 8 inches and 15 inches. Several fine wires are connected in parallel for the higher voltages, each wire being enclosed in a separate internal tube or otherwise partitioned by insulating materials, so that each wire has a column of air to itself. Unlike the Bates fuse, there is no handle moulded with the tube, but flanges are provided at the ends, and moreover, in most cases, it is customary to have a long pair of tongs close by the switchgear with which any fuse can be clutched while the operator is at a safe distance from it. Considering the massive tube fuses—from four to five feet long—used at Deptford and Willesden, and also expensive designs such as the oil-break fuses of home manufacture, it would appear that either we overestimate the destructive effects caused in breaking high tension circuits or else the necessity of blowing a fuse without destroying the fuse holder is not considered a matter of importance in Germany.

During this century, relay-based protective schemes have been produced. These together with the circuit-breakers controlled by them are now used in conjunction with all the items of plant on the major generation and transmission networks. Fuses are still used, however, in large numbers to protect parts of the lower voltage distribution networks as well as the many electrical appliances and items in use today. As a result, fuses have been continuously developed to meet new needs and investigators have and are still studying basic phenomena, such as the arcing process, to enable fuses with the characteristics needed for new applications to be produced.

1.2 BASIC REQUIREMENTS

In 1882 an Electric Lighting Act was passed by the British Parliament and it was amended six years later to form what was known as the Electric Lighting Acts 1882 and 1888. A feature of these acts was that they required the UK Board of Trade to introduce regulations to secure the safety of the public and ensure a proper and sufficient supply of electrical energy. The early regulations included clauses stating that a suitable fuse or circuit-breaker must be present in each service line within a consumer's premises. In 1919, British Standard Specification 88, which covered fuses for rated currents up to 100 A at voltages not exceeding 250 V, was introduced. It included definitions of terms such as 'fuse carrier' and 'fusing current' and specified the maximum short-circuit currents that fuses of various rated currents should be able to interrupt and also the corresponding minimum currents at which they should operate.

Over the years there have been revisions of BS88 and other standards have been produced in Britain, the United States and European countries. To clarify the situation, the International Electrotechnical Commission (IEC) has produced standards which require that:

1. each fuse is so designed and produced that it will, throughout its life, allow the circuit in which it is included to operate continuously at currents up to its rated value;
2. each fuse will operate in a sufficiently short time when any current above a certain level flows, because of an overload, to prevent damage to the equipment being protected;
3. in the event of a fault developing on a network or piece of equipment, fuses will operate to limit the damage to a minimum and confine it to the faulted item.

These requirements, which equally apply to all other types of protective equipment, infer that fuses must have inverse time/current characteristics and that they must be applied so that discrimination is achieved during fault conditions.

1.3 FUSE TYPES AND CONSTRUCTIONS

Fuses are classifed into three categories, namely high voltage for use in circuits above 1000 V a.c., low voltage and miniature.

There are a number of basic constructional forms of fuses, namely cartridge, semi-enclosed, liquid and expulsion.

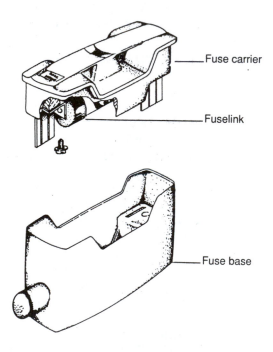

Fuse carrier

Fuselink

Fuse base

Fig. 1.4 Low-voltage cartridge fuse. (Reproduced from Wright and Newbery, 1982, *Electric Fuses* with the permission of IEE).

1.3.1 Cartridge fuses

These are designed for high voltage, low voltage and miniature applications. The fuselink, which is replaceable, is often fitted into a fuse holder that consists of a fuse carrier and fuse base, an example being shown in Fig. 1.4.

Fuselinks for low current ratings contain a single element, while those for higher ratings have a number of parallel-connected elements. The elements are usually of silver or silver-plated copper, it being necessary to ensure that oxidation will not occur, as stated earlier. Wire elements are used in fuses with ratings below 10 A, but for higher ratings strips with one or more sections of reduced cross-sectional area are used. The elements are attached to plated copper or brass end caps which together with the body form an enclosure or cartridge. The bodies, which must be good insulators, must also be robust and able to withstand the conditions which occur during interruption. They were

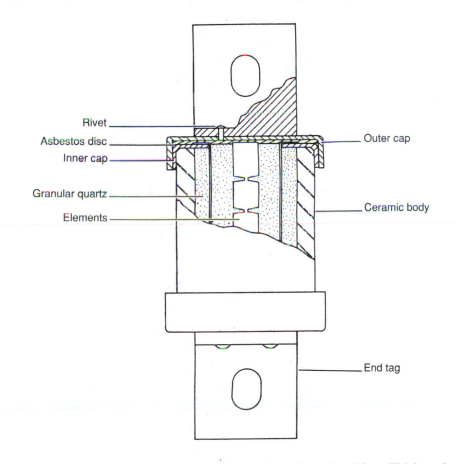

Fig. 1.5 Cross-sectional view of a cartridge fuselink. (Reproduced from Wright and Newbery, 1982, *Electric Fuses* with the permission of IEE).

almost exclusively of ceramic or glass in the past but glass-reinforced plastics have been introduced in recent years.

For high-breaking capacity fuselinks, the space within the body is usually filled with quartz of controlled grain size and chemical purity. A cross-sectional view through a typical fuselink is shown in Fig. 1.5.

1.3.2 Semi-enclosed fuses

These fuses are in the low voltage category. They also consist of a fuse base and carrier, the latter containing the replaceable wire element. They are of low breaking capacity, being capable of interrupting currents up to about 4000 A. An example of this type of fuse is shown in Fig. 1.6.

1.3.3 Expulsion fuses

These fuses, which are used for high voltage applications, contain a mechanism to withdraw the fuselink away from one of its contacts when the element melts. As a result, a long air gap is introduced into the circuit. Satisfactory interruption is achievable at current levels up to about 8000 A.

1.3.4 Liquid fuses

These fuses are also in the high voltage category. They have a liquid-filled glass body within which is a short element of wire or notched strip, held in tension by a spring. During operation, the element melts and a rapid separation occurs, as the spring contracts. As a result the arc extends and is extinguished in the liquid filling.

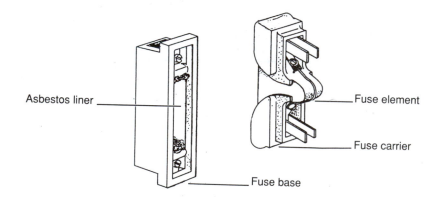

Asbestos liner

Fuse element

Fuse carrier

Fuse base

Fig. 1.6 Semi-enclosed fuse. (Reproduced from Wright and Newbery, 1982, *Electric Fuses* with the permission of IEE).

1.4 THE BEHAVIOUR OF CARTRIDGE FUSELINKS

These fuselinks incorporate one or more elements which melt and then vaporize when currents above a certain level flow through them for a certain time. Thereafter the arc or arcs which result have to be extinguished to complete the interruption process. The operating time is therefore made up of two periods, designated the pre-arcing and arcing periods. The behaviour during these periods is considered in the following sections.

1.4.1 The pre-arcing period

The element or elements and the other conducting parts of a fuselink possess resistance and as a result a fuselink must absorb electrical power when it carries current. While the current is constant and below a particular level, the temperatures of the various parts of the fuse must be at levels above the ambient value so that power, equal to the input power, is dissipated to the surroundings and a state of equilibrium exists.

Should the current be increased and maintained above a certain level, a new state of equilibrium will not be achieved because, although the temperatures of the fuselink parts will rise, the power dissipated from the fuselink will not become equal to the power input by the time that parts of the element or elements melt. Disruption of the element or elements will result and circuit interruption will take place after a period of arcing. The time from the instant when the current exceeds the critical value until the melting and initial vaporization of the elements occurs is known as the pre-arcing period. The time taken thereafter to achieve interruption is known as the arcing period.

The more the current through a fuselink exceeds the maximum value at which equilibrium can be achieved, the shorter is the time taken before melting of the element or elements occurs. This is because the power available to cause the temperatures to rise is equal to the difference between the input power, which is proportional to the square of the current, and the power dissipated from the fuselink. The latter quantity is limited because the temperatures of the fuselink parts cannot exceed the melting point of the element material.

As a result, all fuses have inverse pre-arcing time/current characteristics in the range of currents above that at which thermal equilibrium can be established. A current marginally above this latter level would theoretically cause operation after an infinite time. Clearly this current could not be determined experimentally and therefore in practice tests are done on a number of similar fuselinks to determine a current level at which each of them will operate in a particular time, typically 1–4 hours depending on the fuse rating. This current is termed the 'conventional fusing current'. Further tests are done on the same number of fuselinks to determine the current level, termed the 'conventional non-fusing current', at which none of them will operate in the same time as that used in the earlier tests. A current between the above two current values is then designated as the minimum-fusing current of the fuselink and for simplicity only this value is referred to in the remainder of this chapter.

It will be appreciated that a fuselink cannot be operated near the minimum fusing current continuously and that the rated current of the circuit and the fuse must be at a lower level. The ratio of the minimum fusing current to the rated current of a fuselink is called the fusing factor.

After a part or parts of a fuse element have melted, the current must continue to flow in the liquid metal. The situation which then arises is not fully understood but there are a number of conditions which may exist during the initial stages of vaporization. During the transition period when current levels are very high, it is possible that the material in the sections of reduced cross-sectional area boils. There would then be bubbles present in the liquid metal which would cause the resistance to rise, thus increasing the power input. This would lead to rapid vaporization of the notched sections. Alternatively, it could be postulated that the temperatures must be highest at the centres of restricted sections, causing vaporization to commence at these points. The vapour escaping through the surrounding liquid could produce hairline cracks or a gap across the notch.

Whatever process takes place, the vapour in the gaps will not be ionized initially and capacitance will be present across them. Because of the small cross-sectional areas of fuselink elements, these capacitances must be small. They will charge rapidly because of the current flow which will be maintained through them by the circuit inductance. The resulting voltage will cause the gaps to break down and thus initiate arcs. This transition period between melting and arcing must always be of very short duration.

1.4.2 The arcing period

When an arc or arcs have been established they must persist until the current reaches zero, at which time extinction will occur and it is clearly desirable that the arc or arcs should not then restrike. This process must always occur and it is desirable that satisfactory clearance should be obtained at all current levels. At relatively low currents, the duration of arcing is very short relative to the pre-arcing period and the effect of arcing on the energy let-through to the protected equipment is not very significant. This is not the case at very high current levels however, when the arcing durations may be comparable to or even greater than the pre-arcing periods.

During the pre-arcing period associated with short-circuit conditions, i.e. high currents, the current varies with time in a manner determined by the circuit itself and the resistance of the fuse. It will be marginally lower than it would have been had the fuse not been present. The current which would have flowed in these latter circumstances is known as the prospective current and typical variations which would be obtained in a circuit containing significant inductance and some resistance are illustrated in Fig. 1.7.

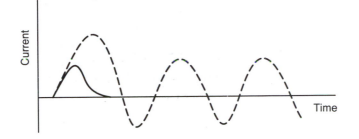

Fig. 1.7 Schematic of prospective (broken line) and actual (solid line) currents.

At the instant when arcs are initiated in a fuselink there is a rapid increase in the voltage drop across it. This voltage then rises as the arcs lengthen due to more material being vaporized from the element.

Consideration of the circuit shown in Fig. 1.8, assumed to apply for a fault condition, shows that the following relationship applies:

$$e = iR + \frac{\mathrm{d}}{\mathrm{d}t}(L \cdot i) + v_{\mathrm{f}}$$

in which e is the source e.m.f., R and L are the resistance and inductance of the circuit, v_{f} is the voltage across the fuselink and i is the current.

When the current is positive as arcing commences, it is necessary that the rate of change of current ($\mathrm{d}i/\mathrm{d}t$) should become negative so that the current falls to zero. Clearly this situation will obtain more quickly and more rapid extinction will be obtained, the greater the voltage across the fuse arcs. These conditions are illustrated in Fig. 1.9, from which it can be seen that the current which flows during fault conditions is lower than that which would have flowed had the fuse not been present. Fuses therefore have the beneficial effect

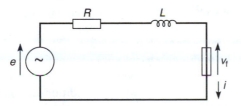

Fig. 1.8 Equivalent circuit.

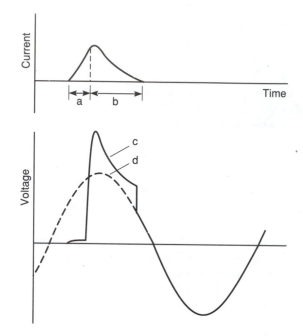

Fig. 1.9 Electrical conditions during a short circuit. (a) pre-arcing period; (b) arcing period; (c) fuselink voltage; (d) source e.m.f. (Reproduced from Wright and Newbery, 1982, *Electric Fuses* with the permission of IEE).

of providing current limiting, which assists in reducing damage to the circuits they protect.

Fuselinks containing notched strip elements may be made to reduce the arcing period by increasing the number of restrictions and thus the number of arcs in series. Care must be taken however to ensure that the rate of current reduction ($\mathrm{d}i/\mathrm{d}t$) is not so great that excessive voltages will be induced in inductive components. Permissible upper limits for fuselink voltages are included in specifications.

The ideal situation, which cannot be achieved in practice, would be for the fuselink voltage to be at the maximum allowable value throughout the arcing period. This would give the fastest possible operation.

The behaviour of fuselinks with cylindrical wire elements is less well controlled because such elements do not have well defined restrictions at which arcs occur during operation. In practice, however, conditions are not uniform throughout a wire element and distortions occur during the pre-arcing period. As a result, the cross-sectional area does vary along the element length, producing thinner and fatter sections, known as unduloids. Gaps ultimately form at the centres of thin sections, their number not being fixed however, as in notched elements. In some cases, sufficient arcs may be formed to cause excessive voltages to be produced in the protected circuits.

1.4.3 Determination of fuselink performance

The performances of fuselinks may be determined by testing in circuits such as that shown in Fig. 1.10, in which the voltage, current level and waveform may be controlled. The ability to interrupt very high currents may be ascertained as well as the operating times at lower currents. The minimum fusing current may also be determined. Because of the possible variability of fuses, each individual test must be performed on several fuselinks.

When fuselinks are being developed for new applications for which the required characteristics are known, designers rely on their experience. Prototypes are produced in sufficient quantities to enable all the necessary tests to be done. Should the performance not meet the requirements, design modifications must be made, further prototypes produced and then the tests must be repeated. This process can be both lengthy and expensive.

There has long been a desire to calculate the performance of fuselinks so that acceptable and optimum designs can be produced without recourse to trial and error methods and lengthy testing procedures. In 1941 Gibson [4] published the results of investigations, done by Professor R. Rüdenberg, in which the high-current pre-arcing behaviour of fuselinks with wire elements was determined. Because the times to reach the melting point of the elements was very short, heat transfer from the elements was neglected. The current flow was axial and the current density constant.

With notched elements, the current flow is not axial or of constant density but nevertheless with various assumptions applicable over certain current ranges and by employing finite difference or other methods, various workers [5][6] have modelled pre-arcing performance with high accuracy in recent years. It is now possible to determine the effects of changing the materials in fuselinks as well as changing their dimensions and geometry.

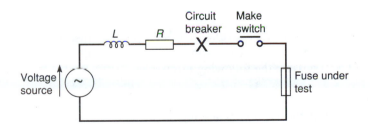

Fig. 1.10 Schematic diagram of test circuit.

The arcing process is more complex than that which takes place during the pre-arcing period and accurate modelling is therefore more difficult. Some workers have sought from experimental studies to establish empirical relationships between the current through a fuselink and the voltage across it when arcing occurs. Gnanalingam and Wilkins [7] were able to predict the arcing performances of certain fuses with reasonable accuracy using the relationship:

$$v_f \propto i^n$$

They claimed that the simulation technique is useful for screening preliminary designs and investigating the effects of system parameters such as frequency.

Such methods are not based on the phenomena taking place and do not assist in understanding the behaviour. They may not be applicable if the use of other element or filling materials are being investigated. For this reason Wright and Beaumont [8] developed a model based on a number of simplifying assumptions. They considered it to be superior to earlier models because it was based on the processes occurring during arcing and took account of energy changes. Work of this type is still proceeding to enable the amount of testing currently required to be reduced greatly in the future. Details of work done recently in this field were given in papers presented at the Fourth International Conference of Electrical Fuses and their Applications [9]. It must be accepted, however, that it will be necessary to determine fuselink characteristics experimentally for many years.

1.5 THE CONSTRUCTION OF CARTRIDGE FUSES

As indicated earlier in section 1.3, cartridge fuses are produced in a number of forms and categories for use in low voltage and high voltage circuits. Factors to be considered in all cases when such fuselinks are to be designed or used are dealt with below.

1.5.1 Fuse elements

From theoretical and other studies it is evident that fuse elements should generally be made from materials with low resistivities. If possible the materials should also possess the following properties:

(a)   low specific heat;
(b)   low melting and vaporization temperatures;
(c)   low latent heats;
(d)   low density;
(e)   high thermal conductivity.

Properties (a) to (d) are desirable when rapid operation at high currents is necessary. At the rated current level, steady state equilibrium conditions must be established in which the electrical power input to an element is conducted away from it and dissipated by the connecting cables and the outer surfaces of the fuselink. This is the reason for property (e). Of course, not only must the element material be of high thermal conductivity but so must the surrounding filling material and the body. In addition, because the power which can be dissipated from a particular fuselink is limited, it is necessary to limit the power input to the element by using elements of adequate cross-sectional area.

After extensive studies of the available materials and taking the above factors into account, it has been found that silver and copper are the most suitable materials for elements and they are used in the majority of fuselinks. Notched elements are produced in various forms, a few being shown in Fig. 1.11.

The strip thicknesses are usually in the range 0.05–0.5 mm and the widths seldom exceed 10 mm.

Occasions do arise when the necessary performance cannot be met using notched elements of a single material with a relatively high melting point. In these circumstances, advantage may be taken of the so-called 'M' effect in which a low-melting point material is deposited adjacent to one or more of

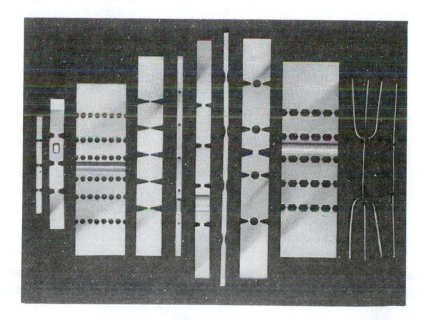

Fig. 1.11 Various fuse element designs. (Reproduced from Wright and Newbery, 1982, *Electric Fuses* with the permission of IEE).

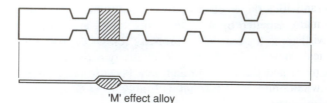

'M' effect alloy

Fig. 1.12 Fuselink element with 'M' effect alloy. (Reproduced from Wright and Newbery, 1982, *Electric Fuses* with the permission of IEE).

the restrictions on the notched strip, an example being shown in Fig. 1.12. This method resulted from work done by Metcalf who published an article entitled 'A new fuse phenomenon' [10] in 1939.

1.5.2 Fuselink bodies

Bodies must possess good electrical insulating properties and should not allow the ingress of moisture. They should be reasonably good thermal conductors and have an adequate emissivity constant so that they can transmit and radiate heat energy emanating from the elements within them. They should also be physically robust.

1.5.3 Filling material

The filling material, invariably quartz of high chemical purity and grain sizes in the region of 300 microns, conducts some of the heat energy away from the fuse element. To ensure consistency of performance it is necessary that the packing density of the filling material be maintained at a high constant level during production. It has also been found that this factor may have a very significant effect on the fuse performance at high current levels. A low packing density is undesirable as it may allow arcs to expand more rapidly and thus adversely affect the extinction process.

1.5.4 Mountings and ratings

Low voltage cartridge fuselinks of standardized performance and dimensions are available with rated currents ranging from 2 to 1250 A to provide overcurrent protection on three-phase, 415 V (line), a.c. systems and up to 250 V d.c. systems. Designs are also available at certain ratings for a.c. circuits up to 660 V and d.c. circuits up to 500 V.

Fuselinks with rated currents up to 200 A are usually fitted into fully-shrouded fuse holders comprising a carrier and base as described earlier in section 1.3.1 and illustrated in Fig. 1.4. Such fuses are produced in large numbers and many are fitted in distribution fuse boards.

Fuselinks with current ratings above about 200 A are large and produced in small quantities. They are normally installed directly in convenient positions without their own special or standardized housings. They are often incorporated in fuse-switch units, an example being shown in Fig. 1.13.

High voltage cartridge fuselinks are produced for use in a.c. systems operating at frequencies of 50 Hz and 60 Hz with rated voltages exceeding 1000 V. They are of the same basic design as low voltage fuses. Those with low rated currents have silver wire elements whilst those for higher current applications have elements of silver strip, with restricted sections along their lengths. All elements have low melting point metals applied to them to enable the 'M' effect, referred to in section 1.5.1, to be obtained. The elements are long, because to achieve satisfactory current interruption, the total voltage across the fuselink must be high and this is best achieved by having many restrictions and therefore many arcs in series. To accommodate such elements in a body of acceptable length, the elements are accommodated in a helical form on insulating formers with a star-shaped cross section.

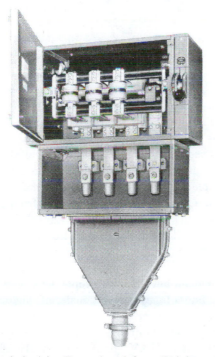

Fig. 1.13 Fuse-switch unit. (Reproduced from Wright and Newbery, 1982, *Electric Fuses* with the permission of IEE).

Whilst some high voltage fuselinks are mounted in air, it is nevertheless common practice for them to be immersed in oil in the pieces of equipment they are protecting. This is advantageous because the cooling effect of the oil allows a given current rating to be achieved in a smaller fuselink than that required for use in air. The smaller fuselink also operates more rapidly at very high current levels.

Fuselinks suitable for use in systems operating at voltages up to 33 kV (line) at a range of rated current values are produced. As an example, voltage ratings of 3.6–7.2 kV with current ratings up to 500 A are available.

1.6 SEMI-ENCLOSED FUSES

These fuses are produced for three-phase industrial applications where the system voltage does not exceed 240 V to earth and are available with current ratings up to 100 A. Their breaking capacity is low, being limited to a maximum of about 4000 A.

They are also used for domestic applications where their breaking capacity is 1 kA at 240 V a.c..

The construction of these fuses was described earlier in section 1.3.2 and illustrated in Fig. 1.6. Because of their cheapness they are likely to remain in demand.

1.7 EXPULSION FUSES

These fuses, which are used in high voltage circuits, contain a short element of tin or tinned copper wire in series with a flexible braid. These items are mounted in a fuse carrier incorporating a tube of organic material, usually closed at the top with a frangible diaphragm, and containing a liner of gas-evolving material. The fuse is also surrounded by a close-fitting sleeve of gas-generating fabric. The flexible braid is brought out of the lower end of the tube and held in tension by a spring attached to the lower end of the fuse base. Figure 1.14 shows a cross-sectional view of a typical fuselink.

The fuse carrier has pins at the lower end which act as a hinge when it is positioned in the lower contact of the mount. In the normal service position, the fuse carrier is tilted from the vertical as can be seen from Fig. 1.15.

When the fuse element melts during operation, the release of the spring tension disengages a latch, so allowing the fuse carrier to swing down due to gravity. This provides electrical isolation.

When interrupting relatively low currents the arc extinguishes within the fabric sleeve around the element, but at high currents the sleeve bursts and the arc extinguishes within the liner of gas-evolving material. These fuses are thus able to break a large range of currents.

Fuses of this type are used outdoors in three-phase circuits with current and voltage ratings up to 100 A and 72 kV (line) respectively. Their breaking capacity is limited to about 150 MVA.

The previous sections have dealt with the fuses used to protect plant and

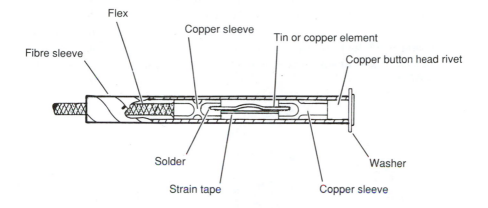

Fig. 1.14 Sectional view of an expulsion fuselink. (Reproduced from Wright and Newbery, 1982, *Electric Fuses* with the permission of IEE).

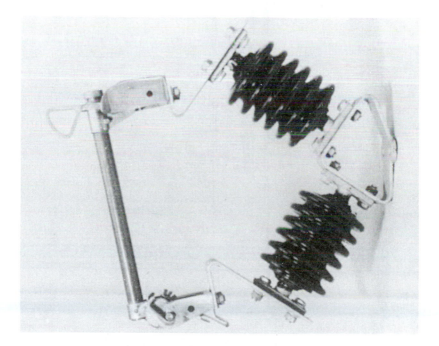

Fig. 1.15 Expulsion fuse assembly. (Reproduced from Wright and Newbery, 1982, *Electric Fuses* with the permission of IEE).

connections in power supply networks in Britain and other countries. Variations do occur in the designs produced in different countries but nevertheless the practices are broadly similar. Some specialized designs such as the boric-acid fuses produced in the United States have not been described nor have miniature and domestic fuses which are used in great quantities. Information about these and other fuses is available in other works devoted entirely to fuses [11, 12].

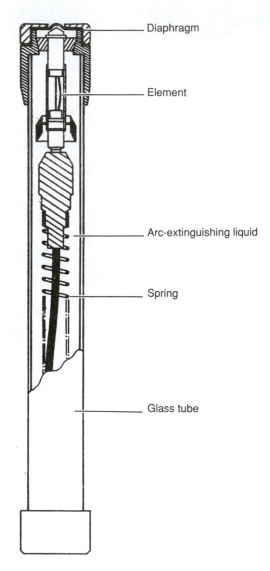

Fig. 1.16 Sectional view of a liquid fuse. (Reproduced from Wright and Newbery, 1982, *Electric Fuses* with the permission of IEE).

1.8 LIQUID FUSES

In many early fuses the arcs were quenched in liquid and this principle is used in present-day liquid fuses. In all designs, these fuses have a glass tubular body which is mounted vertically. The short element of silver wire or strip is positioned near the top of the tube. The element is held in tension by a spring anchored to the lower end of the fuse, the tube being filled with an arc-extinguishing liquid, usually a hydrocarbon. When the element melts during operation, the spring collapses and the arc is extinguished in the liquid. A cross-sectional view of a typical fuse of this type is shown in Fig. 1.16. These fuses are only used outdoors and provision is made for removing them from and replacing them into their mountings from the ground. They are mostly used to protect 11 or 33 kV pole or pad-mounted transformers in rural systems and also for spurs feeding several transformers.

The breaking capacity of these fuses is lower than that of expulsion fuses and although they have been produced in significant numbers and are still giving good service, they are no longer recommended for new installations.

1.9 THE APPLICATION OF FUSES

Fuses are used for so many different applications in power systems that it is impossible to deal with all of them and only the more common and important are considered in this section. There are however some general aims and requirements which always apply and these are dealt with initially.

A fuselink which is to protect a piece of equipment and/or a circuit should ideally satisfy several criteria. This may be illustrated by considering an example based on the simple circuit shown in Fig. 1.17.

The criteria are:

1. The minimum-fusing current of the fuse should be slightly less than the current which can be carried continuously by the source, the supply and connecting cables and the piece of equipment.

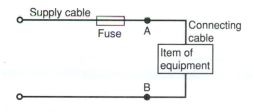

Fig. 1.17 Circuit protected by a fuse.

2. The item of equipment should be able to carry currents in excess of its rated current for limited periods and the fuse should operate, when carrying these overload currents, in times slightly shorter than those which the equipment can withstand.
3. The cables should also be able to cope with the above overcurrents without being damaged.
4. High currents may flow due to faults within the item of equipment and in these circumstances consequential damage should not occur to the remainder of the circuit. The extreme case would occur if the input terminals of the piece of equipment became short-circuited. In this situation clearance should be effected quickly enough to prevent damage to the cables.
5. A further possibility is a short circuit between the conductors of the connecting cables. The most severe situation would arise if the fault was at the input end, i.e. between points A and B. For such faults the fuse should operate rapidly enough to prevent damage to the source and the supply cables.

1.9.1 Time/current relationships

To achieve the above criteria, fuselinks should have time/current characteristics which lie close to the withstand curves of their associated circuits as shown in Fig. 1.18. Obviously the operating time of a fuse should always be less, at any current level, than the period for which its associated circuit can withstand the condition.

A very important factor which must be taken into account is that fuse operating times include the arcing period and a change of state occurs before clearance is effected. Once arcing has commenced it is clearly impossible for

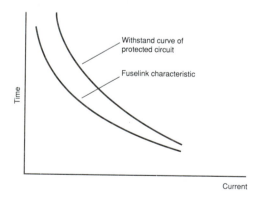

Fig. 1.18 Time/current characteristics of circuit and fuse.

an element to return to its original form and even when melting begins, distortion may result and an element may not return to its original shape on cooling. Fuselinks with low melting-point materials on their elements will be subjected to unacceptable and irreversible changes if overcurrents pass through them long enough to initiate the 'M' effect diffusion process.

There is thus a band below its time current characteristic in which a fuse should, if possible, not be called upon to perform. While in practice it is not likely that a high current will flow through a fuselink for a period just less than its operating time, the possibility nevertheless exists. If such a condition should arise, then the probable result, which must be accepted, is that the fuse may operate more quickly than expected on a future occasion.

Difficulties would certainly arise if a fuselink carried a current just below its minimum fusing value for prolonged periods. To ensure that this situation is avoided, fuselinks are assigned rated currents somewhat below their minimum fusing levels. The ratio of the minimum-fusing current to the rated value, which is defined as the fusing factor, usually has a value in the range 1.2 to 2. The significance of this factor is that circuits must be able to operate continuously at levels significantly above the rated current of the fuse protecting them to satisfy criterion 1 above, namely that the continuous rating of the circuit should exceed the minimum fusing current. This situation arises with other protective equipment in which current settings above the full load level are used. Clearly circuits and equipment are not normally designed without some overload capacity, indeed factors of safety are usual and difficulties are not encountered in providing adequate fuse protection. Clearly, relatively low fusing factors are nevertheless desirable on economic grounds.

A further factor which must be recognized is that not all fuselinks are capable of operating satisfactorily at all current levels between their maximum breaking capacities and minimum-fusing values. Satisfactory arc extinction may not take place at certain current levels, usually those just above the minimum fusing values. Care must be taken to ensure that such fuses are only used in circuits where currents of these levels will not be encountered or, if this cannot be guaranteed, then some other protective scheme must be provided to open the circuit before the fuse attempts to operate. In the USA the term 'full range' has been introduced to designate those fuses which can clear all levels of current satisfactorily.

1.9.2 I^2t

Fuses operate very rapidly under short-circuit conditions when the current levels are extremely high, clearance times being typically only a few milliseconds. Under these conditions, the current waveshapes depend on the parameters of the protected circuit, the instant in the voltage cycle at which the fault occurs and the current-limiting effect of the fuselink itself. Only a small portion of a cycle of power-frequency current flows before interruption occurs and transient components are present. Such currents cannot be assigned a single value which may be used to determine the corresponding operating time from the time/current characteristic. Because of this, use is made of a quantity, termed $I^2 t$, which is the time integral of the square of the instantaneous current which passes through a fuselink between the incidence of a circuit fault and the instant at which the fuse arc is extinguished, i.e.

$$I^2 t = \int_0^t i^2 \, dt$$

If it could be assumed that the resistances of the components of the protected circuit were constant throughout the operating period of the fuselink, then the value of $I^2 t$ would be proportional to the energy fed to the circuit and also the heating produced by it. In practice, however, the resistances must rise, some significantly. Nevertheless, the $I^2 t$ value is often described as the let-through energy, a term which is not strictly correct because it does not contain a resistance value and is not therefore in energy units (J) but in A^2 s.

To assist in the application of fuses, many manufacturers of components determine the $I^2 t$ withstand limits of their products and publish them. The $I^2 t$ values needed to operate fuselinks at very high currents tend to be independent of the waveshapes and manufacturers publish these values as well as providing time/current characteristics to enable appropriate fuselinks to be selected by users.

In practice, fuselink manufacturers provide two sets of $I^2 t$ values, one set associated with the total clearance time and the other set, with lower values, for the $I^2 t$ let-through during the pre-arcing period. Both sets are needed to enable coordination to be achieved between fuselinks which are effectively in series in networks, as is explained later in section 1.9.7.

1.9.3 Virtual time

The term virtual time is no longer used to any great extent but it may be encountered in older publications. It was introduced to assist the selection of fuselinks which would have the required performance under very high current conditions. It was defined as the $I^2 t$ value divided by the square of the r.m.s. value of the prospective current, the unit clearly being time. Again as with the quantity $I^2 t$, two values were produced for each fuselink, one associated with the total clearance period and the other with the pre-arcing period.

1.9.4 Published time/current characteristics

The characteristics published by fuse manufacturers are derived from tests done on fuselinks which are at a temperature in the range 15°–25°C when current flow through them is commenced. It must be recognized that slightly faster operation will be obtained in practice if a fuselink is in an environment with a high temperature or if a significant load current has been flowing through it for a period before an overcurrent condition occurs. This behaviour is somewhat counterbalanced by the fact that the times for which the protected circuit can carry given overcurrents will be reduced if load currents have been flowing in it and/or if it is operating in an environment with a relatively high temperature.

1.9.5 Cut-off characteristics

Manufacturers produce characteristics to show the highest possible values of current which given current-limiting fuselinks will carry for varying values of prospective current. They are used when checking that the mechanical forces to which items in the protected circuit will be subjected during short-circuit conditions are within the permissible limits. A typical characteristic is shown in Fig. 1.19.

1.9.6 Operating frequency

The characteristics provided for fuselinks are usually based on operation at 50 or 60 Hz. In general, the higher the system frequency the shorter are the operating times at very high current levels, because of the shorter duration of the half cycles of the supply voltage. Care must therefore be taken in selecting fuselinks for use in circuits operating at frequencies below 50 Hz. It must be realized that their operating times in such circumstances will be longer than those obtained when operating at 50 Hz. As a result higher arc energies will be released in the fuselinks, and to ensure that satisfactory clearance will be obtained, their ratings in terms of operating voltage may have to be lowered.

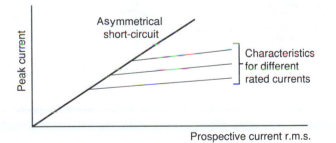

Fig. 1.19 Fuse cut-off characteristics.

For d.c. applications the voltage rating must usually be reduced very significantly, particularly if the circuit is highly inductive.

1.9.7 Discrimination and coordination

Most power circuits contain many items of equipment and these are both in series and parallel paths. To ensure a high continuity of supply, several protective devices must be included to ensure that only the minimum of interruption will occur to clear any fault condition. The protective devices must therefore be so coordinated that correct discrimination is achieved.

In practice, some networks are protected by fuses only and in other cases fuses are used near the loads while feeder connections and equipment are protected by relays. Methods of achieving correct discrimination in these two situations are considered separately below using the simple network shown in Fig. 1.20 as an example.

Networks protected by fuses

It is very common to use the above radial system and to have an upstream fuse (PD1) in the supply connection and downstream fuses (PD2, 3 and 4) in the individual load circuits. Each of the fuses PD2, 3 and 4 must have time/current characteristics which will protect their associated loads. A fault on a particular load should be cleared by its associated fuse, and although the upstream fuse (PD1) will carry the fault current and the other load currents it should not operate or be impaired.

At medium and low fault-current levels, discrimination will be ensured provided that the time/current curves of the downstream fuses are to the left of the curve for the upstream fuse. The time (vertical) displacements between the curves should nevertheless be sufficient to provide an adequate margin to

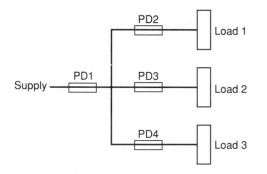

Fig. 1.20 A simple network containing fuses.

allow for the manufacturing tolerances of the fuselinks and because the upstream fuselink will be carrying current to healthy circuits in addition to that in the faulted circuit.

At levels of fault current which will result in a downstream fuselink melting in less than 100 ms, I^2t values must be compared. To ensure correct discrimination, the pre-arcing I^2t of the upstream fuselink should exceed the total operating I^2t of the downstream fuselinks by a significant margin, say 40%. Clearly the effects of the load currents of the healthy circuits flowing through the upstream fuselink are negligible when a very high fault current exists. Pre-fault conditions must also be considered and the I^2t margin should be increased if a downstream fuselink may at times be much less loaded relative to its rated current than the upstream fuselink. A suggested, but rough guide, for the extreme case where a downstream fuselink may be on no load for periods during which the upstream fuselink is nevertheless carrying a current near its rated value is that the upstream fuse rating should be increased by a further 25%.

Networks protected by fuses and other devices

In such networks, when fuses are used as the downstream devices (PD2, 3 and 4), they must be chosen in the same way as above, their time/current characteristics being such that they will adequately protect their associated circuits. Thereafter the upstream device (PD1) must have a characteristic which will provide correct discrimination and there is usually little difficulty in obtaining satisfactory performance.

In some cases, miniature circuit-breakers incorporating overcurrent protective features are used as the downstream devices and a fuselink is used as the upstream device. With this arrangement there is always an upper limit to the fault current at which discrimination can be achieved. This is because mechanical devices have a definite minimum operating time irrespective of the current level. The upstream fuse, however, will have an operating time which decreases continually as the current increases. As a result, the curves of the various devices must cross at certain current levels, above which correct discrimination will not be achieved.

The use of two co-ordinated fuses

As stated earlier, some current-limiting fuselinks are unable to satisfactorily interrupt currents across the whole range from their maximum breaking capacities to their minimum fusing levels. When this is the case, it is currents somewhat above the minimum fusing level where operation is unsatisfactory. In these circumstances, a back-up fuse should be provided, the fuse characteristics being chosen to provide an overall composite characteristic of the form shown in Fig. 1.21. In this way, the desirable current-limiting property

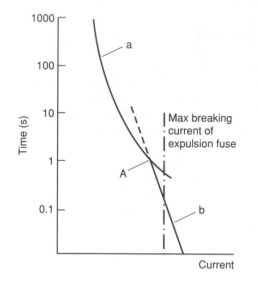

Fig. 1.21 Combined characteristic of two fuses. (a) back-up fuse characteristic; (b) current-limiting fuse characteristic. (Reproduced from Wright and Newbery, 1982, *Electric Fuses* with the permission of IEE).

may be obtained under short-circuit conditions as well as satisfactory clearance being achieved over the full operating current range.

1.9.8 The protection of power-system equipment

The preceding treatment has been general, but in practice special requirements arise when particular items in power systems are to be protected and some of these are considered below.

Cables

International Electrotechnical Commission (IEC) Publication 364 deals with Electrical Installations in Buildings. In the UK, the Institution of Electrical Engineers (IEE) produced the 15th Edition of its Regulations for Electrical Installations in 1981, this being based on IEC Publication 364. In these regulations, the term overcurrent covers both short-circuit currents and overloads, an overload being defined as an overcurrent which flows in a circuit which is perfectly sound electrically.

The current-carrying capacities of cables, which are dependent on many factors including the environments in which they are to operate, have been determined under a range of conditions and tabulated in the regulations referred to above. A cable can, of course, carry currents above its continuous current-carrying capacity for limited periods and they therefore have inverse withstand time/current characteristics. Clearly, the characteristic of each fuse should be such that fault clearance at any current level will be effected within the period for which the protected cable can carry that current. In addition, the minimum fusing current should be below the continuous current-carrying capacity of the cable. To comply with these conditions and taking account of the fusing factor (say 1.5) would require a cable to operate at only about 66% of its current capacity. This practice could be unacceptably expensive and regulations, which are intended to ensure that the life of cable insulation is not significantly shortened as a result of running the conductors at high temperatures, permit the minimum operating current of protective devices to be up to 1.45 times the current-carrying capacity of the cable.

To provide adequate protection under short-circuit conditions, fuselinks should have let-through I^2t values lower than those which can be withstood by the cables. These latter values can be determined from the recent edition of the IEE Regulations. In practice it is satisfactory to check whether satisfactory coordination will be achieved using the I^2t value of the fuselink needed to provide clearance in 5 s. Further information is available on the above topics in reference [11].

Motors

It is well known that motors draw surges of current on starting, the magnitude and duration of the surges depending on the motor, the starting method in use and the load on the motor. As an example, three-phase a.c. induction motors up to 2 MW rating operating at line voltages up to 11 kV and employing direct-on starters may draw 5 to 6 times their full-load currents on starting. In such applications, current-limiting cartridge fuselinks with rated currents up to twice the full-load current of the motors to be protected are used, to ensure that fuse operation does not occur during the starting period. Such highly-rated fuses provide protection against the very high currents associated with short circuits, but either air-break or vacuum contactors must also be included with settings such that they will clear the lower currents associated with overloading. The fuses must be able to carry three times their rated currents during starting periods and they will reach higher temperatures at these times than they do when carrying their rated current continuously. The expansion and contraction which results could cause long elements, such as those used for high voltage applications, to fail mechanically after a number of starts and therefore special fuselinks with corrugated elements are produced for motor protection in these cases. Normal low voltage fuselinks are satisfactory for this duty, however, because their elements are relatively short and sufficiently robust.

When other methods of starting which cause the surges to be smaller are employed, fuselinks with lower current ratings may be used. Again, these levels are exceeded during motor starts and fuselinks with special elements are needed for high voltage applications. It is also necessary for such fuselinks to have current ratings of at least 125% of the rated current of the protected motor to limit the temperatures of the elements under normal running conditions to acceptable levels.

To completely protect a motor may require the use of fuses and also several other devices including inverse-definite-minimum time (IDMT) overcurrent relays, instantaneous overcurrent and earth fault relays, all the relays operating a contactor. Clearly these various items must be correctly coordinated.

Power transformers

Many step-down transformers in distribution networks and industrial premises utilize fuses for their protection on both their high voltage and low voltage sides.

In Britain high voltage fuses are used in the primary circuits of three-phase, 11 kV/415 V transformers rated up to 1.5 MVA and low voltage fuses are included in the secondary circuits. High voltage fuses are also included on both sides of three-phase, 33/11 kV transformers rated up to 5 MVA.

Factors which influence the selection of these fuses are considered below.

1. Fuses in the secondary circuit of a transformer have to be chosen to protect the loads and connecting cables while those in the primary circuit must isolate the transformer if a fault occurs within it.
2. Because the magnetic core of a transformer may go into saturation repeatedly over a significant period after the primary winding is connected to a supply, so-called exciting-current surges of the form shown in Fig. 1.22 may be experienced. These surges may reach peak values many times the rated current of a transformer even though the normal exciting current may only be 2 or 3% of the rated current. Clearly they should not cause the fuses on the primary side to operate. When possible, the magnitudes and durations of the surges, which depend not only on the transformer to be protected but also on the high voltage circuit parameters, should be determined for each application so that fuses with the suitable characteristics may be chosen. In practice it has been found for distribution transformers, when actual surge-current information is not available, that the surge may be taken to be equivalent to 12 times the transformer full-load current for a duration of 100 ms. This rule, which has been in use for some time and found to be satisfactory, is specified in the British Electricity Supply Industry (ESI) Specification 12–8. It is sometimes supplemented by the requirement that melting of the fuselink elements

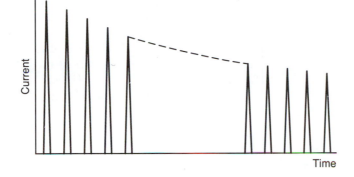

Fig. 1.22 Exciting current of a power transformer. (Reproduced from Wright and Newbery, 1982, *Electric Fuses* with the permission of IEE).

should not begin in less than 10 ms when they are carrying 25 times the rated current of the protected transformer.

3. A further factor which should be considered is that power transformers may be deliberately operated above their rated-current levels for predetermined periods, often of several hours, this being allowable because of their long thermal time constants. To allow this practice, the fuses should have ratings based on the maximum currents which may flow.

4. In those applications where a transformer is supplied by overhead lines, the possibility of overvoltages being impressed on its windings in the event of lightning strikes should be considered. The fuses in the primary circuit should ideally be able to withstand the resulting high currents without operating, but this often requires such highly-rated fuselinks that adequate protection is not provided for other conditions. A compromise has to be struck and some degree of risk must be accepted in most cases.

5. To enable faults within a protected transformer to be cleared rapidly, the currents needed for operation in the 10 s region should be as low as possible.

6. The minimum-fusing current of the primary-circuit fuselinks should be as low as possible to enable them to clear as many internal faults as possible. It must be accepted, however, that some conditions, such as interturn faults, may cause primary currents of less than the full-load value to flow and clearly these will not cause fuse operation.

7. Correct discrimination between the fuselinks on the high and low voltage sides of a transformer and between the transformer fuses and other protective devices on a network should be achieved under all conditions. Account must be taken of the fact that the transformation ratio is not fixed if a transformer is fitted with tap-changing equipment and that various

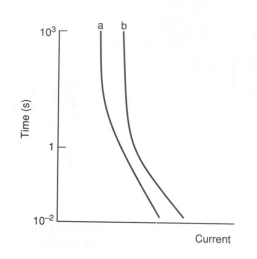

Fig. 1.23 Fuselink characteristics. (a) low-voltage fuse; (b) high-voltage fuse. (Reproduced from Wright and Newbery, 1982, *Electric Fuses* with the permission of IEE).

current distributions are possible in the phase windings when unbalanced faults occur if the high and low voltage winding connections are different, for example star-delta.

In addition to the above, account must be taken of the fact that the characteristics of the high and low voltage fuselinks will differ because their designs will be dissimilar. In general, the high voltage fuselinks will tend, at high current levels, to have smaller operating-time changes for a given percentage current change than the low voltage fuselinks and the time/current characteristics will converge at intermediate current levels before diverging again at lower currents. This situation is illustrated in Fig. 1.23.

To assist users, low voltage fuselinks complying with BS88 : Part 5 have standardized characteristics which provide discrimination with the standardized characteristics of high voltage fuselinks. This is explained in ESI Standard 12–8.

Voltage transformers

Electromagnetic voltage transformers are used on systems operating at voltages up to about 66 kV (line) to provide outputs up to a few hundred volt-amperes to measuring equipment and relays at low voltages, now standardized at 63.5 V per phase (110 V line) in Britain and many other countries.

Their step-down ratios are high and therefore the primary windings carry very small currents under normal conditions. Because the transformers are small physically, compared with power transformers, they have magnetic cores of small cross-sectional area and consequently they have primary windings of many turns of fine wire. They are thus more vulnerable than most other transformers.

Providing protection against all possible faults within a voltage transformer would necessitate the use of fuselinks, in the high voltage side, with rated currents so low that their elements would be extremely fragile and liable to break because of factors such as vibration. This would be completely unacceptable because it would cause vital protective and control equipment to be de-energized and therefore either inoperative or even worse, causing incorrect tripping or control of power circuits.

A compromise must therefore be accepted and it is now standard practice for high voltage fuselinks to have minimum rated currents of 2–3 A at all voltage levels. Such fuselinks only operate when relatively large faults occur within a transformer. They do not prevent damage occurring and it is expected that small internal faults will develop until fuse operation is initiated. Some users feel that the measure of protection provided by such fuses is not sufficient to warrant their installation and they prefer to rely solely on other protection in the system.

Fuses are included in the secondary circuits of voltage transformers to provide protection in the event of faults in the burdens. There is no difficulty in selecting suitable fuselinks for this duty because the secondary circuits usually carry currents of a few amperes. Coordination with the protection on the high voltage side does not cause any problems because the referred rated-current levels of the secondary fuselinks are very low.

Power-system capacitors

Capacitors are used quite extensively to achieve power factor improvement. In low voltage situations it is usual to have a single capacitor in each phase. Fuse manufacturers are usually prepared to provide simple recommendations in these cases. These are based on past experience and take into account the high transient inrush currents which may flow when the capacitors are switched into circuit and also the significant harmonics which may be present in the capacitor currents because of their reduced reactances at higher frequencies. As an example, when protecting capacitors of ratings greater than 25 kVA, using fuselinks satisfying BS88 : Part 2, it is recommended that the fuse ratings be at least 1.5 times the full-load current of the capacitors.

In large installations for operation at higher voltages, each phase contains several units made up of individual capacitors connected in series and parallel groups. The units may also be connected in series and parallel groups. Both star- and delta-connected three-phase capacitor banks are produced for ratings up to 1 MVA and for voltages up to 11 kV (line). For higher voltage systems, however, banks are usually connected in star.

In most European countries it is the practice to provide a fuse for each of the individual capacitors forming the units, the fuses having the appropriate relatively low current-rating and breaking capacity. In Britain however it is the practice to provide only one fuse for each unit, rather than for each capacitor, and frequently a line fuse for each phase is also included. Clearly either an individual capacitor fuse or a unit fuse in British installations should operate when necessary to isolate only part of the complete capacitor bank so that the remainder may be left in service.

Fuselinks, associated with capacitors, must meet the normal requirements of fast operation in the event of faults and also be able to carry the rated currents of the capacitors they protect as well as transient currents. The latter will flow if the voltage across a bank changes suddenly, a condition that may arise at instants when the supply is connected to a bank or when system faults affect the network voltages. The magnitudes of these transient charging or discharging currents and their durations are dependent on the capacitances and the parameters of their associated circuits. To prevent undesired fuselink operation under these conditions it is usually necessary to select fuselinks with rated currents significantly greater than the normal current through them.

In the event of a breakdown within an individual capacitor or a short circuit across a unit, current will flow into it from other healthy capacitors and units. Considering, as an example, a bank of four units connected in series-parallel as shown in Fig. 1.24, it can be seen that the discharge current (i_{df}) of the short-circuited unit would not flow through its associated fuselink. Each of the two lower fuselinks would carry the current (i_{ch}) needed to double the voltage across its associated unit and the upper healthy unit would carry the current (i_{dh}) required to reduce the charge on it to zero. In practice, if each of the units were of the same capacitance, then the surge currents i_{df}, i_{ch} and i_{dh} would be of approximately equal magnitudes and the fuselink associated with the faulted unit would carry three times the currents flowing in each of the other fuses. Following the transient surges, the fuselink protecting the faulted unit would carry four times its normal current to feed twice normal current to each of the lower units which would then have their voltages doubled. In these circumstances, correct discrimination would be readily achieved.

A further factor for which allowance must be made is that when the fuselink associated with the faulted unit operates, the voltage across the other upper unit will rise, for this example to 133%, of its earlier value, and the current through it will also rise similarly. The VAR input would thus rise by 77%. Clearly similar situations would arise whichever unit in the bank became short-circuited and all the units would have to be capable of operating above their normal VAR level. Actual arrangements with other series/parallel combinations of the capacitors should be considered in a similar way to determine the conditions which will arise during and after fuse operation.

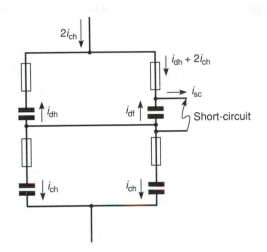

Fig. 1.24 Current flows due to the short-circuiting of a capacitor unit.
i_{sc} = short-circuit current
i_{ch} = charging current of healthy unit
i_{dh} = discharge current of healthy unit
i_{df} = discharge current of faulty unit
$i_{sc} = 2i_{ch} + i_{dh} + i_{df}$.
(Reproduced from Wright and Newbery, 1982, *Electric Fuses* with the permission of IEE).

Other equipment

Fuses are used to protect many items in the load circuits of power systems, very large numbers being associated with semiconductors employed in the expanding field of power-electronic equipment. As in all applications, the characteristics and behaviour of each fuselink should be such that costly items will not be damaged in the event of faults. In this connection semiconductors tend to require more rapid fuse operation at high currents than items such as motors, because their thermal capacity is relatively low. Information about the withstand abilities of their components is provided by manufacturers and it should be used to enable the necessary fuselinks to be selected for particular applications.

Special requirements arise in some applications because of mechanical or environmental conditions. As an example, the rectifiers in the field-winding circuits of brushless alternators and their associated fuselinks are mounted on the rotor, which may revolve at speeds up to 50 or 60 rev/s. Special fuselinks which can withstand the vibration and the large forces encountered in this situation are necessary.

Because of the variations in the requirements which must be met by fuses to be used in the many applications which exist, the IEC has produced application guides. These guides explain the basis of the ratings and how to relate them to practical situations.

Other detailed information on topics such as special applications, testing, manufacture, quality control and International and National Standards, which cannot be included in this work, is given in references listed under Further Reading at the end of this chapter.

1.10 THE FUTURE

The pre-arcing phenomena which occur in fuselinks are now well understood as a result of studies undertaken in recent times. The arcing behaviour is less clearly understood however, and it is still being investigated both theoretically and experimentally. The aim is to obtain information on the effects which may be produced on time/current characteristics and interrupting performance by changes to the following parameters:

(a) element material
(b) element geometry
(c) filling material
(d) filler grain size
(e) body material and geometry

Such work should eliminate much of the testing associated with trial and error methods of producing fuselinks for new applications. It should also enable better protection to be provided for existing equipment as well as enabling products such as new semi-conductor devices, to be adequately protected.

The production and use of fuselinks has risen dramatically over the last hundred years and it seems likely, because of their attractive qualities, that they will continue to be used for many years to protect items of equipment in the lower voltage sections of power systems.

REFERENCES

1. Cockburn, A. C. (1887) On safety fuses for electric light circuits and on the behaviour of the various metals usually employed in their construction. *J. Soc Teleg Eng*, **16**, 650–665.
2. Clothier, H. W. (1933) *Switchgear Stages*, (printed by G. F. Laybourne and Unwin Brothers).
3. Clothier, H. W. (1902) The construction of high-tension central-station switchgears, with a comparison of British and foreign methods, Paper presented to the Manchester Local Section of the IEE, 18 February, 1902. Printed in reference [2], pp 1–19.

4. Gibson, J. W. (1941) The high-rupturing-capacity cartridge fuse, with special reference to short-circuit performance. *J. IEE*, **88**, (1), 2–24.

5. Leach, J. G., Newbery, P. G. and Wright, A. (1973) Analysis of high-rupturing-capacity fuselink prearcing phenomena by a finite-difference method, *Proc. IEE*, **120**, (9), 987–993.

6. Wilkins, R. and McEwan, P. M. (1975) A.C. short-circuit performance of uniform section fuse elements, *Proc. IEE*, **123**, (3), 85–293.

7. Gnanalingan, S. and Wilkins, R. (1980) Digital simulation of fuse breaking tests, *Proc. IEE*, **127**, (6), 434–440.

8. Wright, A. and Beaumont, K. J. (1976) Analysis of high-breaking-capacity fuse-link arcing phenomena, *Proc. IEE*, **122**, (3), 252–260.

9. *Proc. of Fourth Int Conf on Electric Fuses and their Applications*, 22–25 Sept 1991, Nottingham, England, ISBN 0 95 148 28 15

10. Metcalf, A. W. (1939) A new fuse phenomenon, *BEAMA. J.*, **44**, 109–151.

11. Wright, A. and Newbery, P. G. (1982) *Electric Fuses*, Peter Peregrinus.

12. Wright, A. (1990) Construction, behaviour and application of electric fuses, *Power Eng. J.*, **4**, 141–148, (1990) Application of fuses to power networks, ibid., **4**, 298–296; (1991) Application of fuses to power-system equipment, ibid., **5**, 129–134.

FURTHER READING

Wright, A. and Newbery, P. G. (1982) *Electric Fuses*, Peter Peregrinus, ISBN 0 906048 78 8.

Miniature fuses

IEC Publications

127 (1974), 127A (1980), 127–3 (1984); 257 (1968), Amendment No 1 (1980).

British Standards

2950 (1958); 4265 (1977).

Low voltage fuses

IEC Publications

269, Parts 1 (1986), 2 (1986), 2A (1975), 3 (1987), 3A (1978), 4 (1980).

British Standards

88, Parts 1 (1975), 2 (1975), 4 (1976), 5 (1980); 646 (1977); 714 (1978); 1361 (1971); 1362 (1973); 3036 (1978).

British Electricity Supply Industry Standards (ESI)

12–10 (1978).

British Ministry of Defence Standards (DEF STAN):

59–96 Part 1 Fuselinks; 59–100 Part 1 Fuseholders.

High voltage fuses

IEC Publications

282, Parts 1 (1985), 2 (1970), 3 (1976); 291 (1969), 291A (1975); 549 (1976); 644 (1979); 691 (1980); 787 (1983), Amendment No 1 (1985).

British Standards

2692, Parts 1 (1975), 2 (1956).

British Electricity Supply Industry Standards

12–8 (1976).

2

Current transformers

The term current transformer (commonly abbreviated to ct) is used to describe a piece of equipment consisting of a pair of mutually-coupled windings mounted around a core, usually of magnetic material. Such transformers are normally used to step down high currents flowing in their primary windings to lower levels suitable for feeding to measuring and protective equipment. Isolation of the latter equipment from the primary circuits is ensured by the insulation normally associated with the primary windings.

Current transformers have been produced and used in very large numbers at all voltage and current levels in power systems around the world for most of this century.

2.1 HISTORICAL BACKGROUND

During the last two decades of the nineteenth century, electricity supply systems were being set up to meet the demands caused by the rapidly expanding use of electric lighting.

Some of the initial installations provided direct voltage supplies, but it was soon recognized that alternating voltage systems were preferable because higher transmission voltages could be used and greater efficiency was thereby achievable. This development, of course, depended on the availability of suitable power transformers.

The principle of electromagnetic induction on which all transformers are dependent had been demonstrated by Michael Faraday in 1831 and indeed it is recorded [1] that by 1838 Joseph Henry was able to change an 'intensity' current, as he called a current in a high voltage circuit, to a 'quantity' current, by which he meant a current in a low voltage circuit. He had effectively built a step-down voltage transformer, which he called a voltage changer, but it was only used subsequently in laboratory demonstrations.

In 1882 Goulard and Gibbs took out British patents on a transformer designed to, as they said, isolate consumers from the power lines and for reducing the mains voltage to a safe value. Systems were then developed using these transformers but they proved unsatisfactory because the primary windings of the transformers supplying individual consumers were connected in series with each other. In 1885, three Hungarians, Zipernowsky, Beri and Blathy, realized that parallel connections should be used. In the same year it is recorded [2] that Dr S. Z. de Ferranti, who was working on the Grosvenor Gallery Station in New Bond Street, London, had parallel-connected transformers installed to step down from the 2400 V output voltage of the alternators then in use.

Thereafter, alternator and transmission voltages rose quite rapidly, transformers being used at first, however, only to step down the voltages supplied to the consumers.

In 1891, at the Electrical Exhibition in Frankfurt, a transmission line 109 miles long was energized at 25 kV from a transformer to demonstrate that power could be transmitted over long distances, and by 1897 machines operating at 13 kV were installed and operating at Paderno in Italy.

At this time, current and voltage transformers could readily have been produced because of their basic similarity to power transformers and it seems probable that they were installed to feed measuring instruments associated with the high voltage circuits then in existence.

Another matter of significance is that doubts were being expressed around the turn of the century about the use of fuses as protective devices, possibly because their manufacture was not adequately controlled. Apparently they were causing unnecessary interruptions of supplies to consumers. This was referred to by Andrews [3] in a paper presented to the IEE in 1898. The following two quotations are from that paper:

'Some engineers attribute the immunity of their systems from failures (i.e. interruptions of supply), to the fact that they use fuses made from copper of the same cross-sectional area as the mains'.

'If any fuses that it is customary to be used can be omitted, everyone will admit that they are a source of danger, and consequently better omitted'.

Because of this situation and the fact that systems could not really be left unprotected, so-called cut-outs, which had contacts capable of interrupting fault currents, were developed. They incorporated magnetic devices which operated when overcurrents above certain levels flowed. Subsequently, circuit-breakers with trip coils were produced. By this time power system voltages and currents had reached levels at which existing fuses could not be used. One method of dealing with this situation was to connect a fuse, of low current rating, across the trip coil of a circuit-breaker and to feed this combination from a current transformer, as shown in Fig. 2.1. The alternative of using a current transformer to feed a relay, which in turn operated the trip coil, was also possible. Current transformers therefore became essential components in power systems, feeding both measuring and protective equipment.

Although current transformers were probably used several years earlier, as suggested above, the first indication of their use which the authors have been able to trace was in a paper presented to the IEE in 1902 by Clothier [4]. Figure

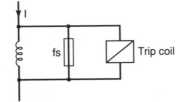

Fig. 2.1 Fused trip coil.

16 in that paper showed a device labelled as an ammeter transformer. The term current transformer was used in 1904 in a paper, also presented to the IEE, by Edgcumbe and Punga [5]. Certainly, current transformers were in wide use by 1910 when Young [6] presented a paper to the IEE on their theory and design. He described transformers produced by several companies, including a type suitable for use at voltages and currents up to 7 kV and 500 A respectively, which was manufactured at that time by British Thomson Houston Ltd of Rugby, England. It is shown in Fig. 2.2.

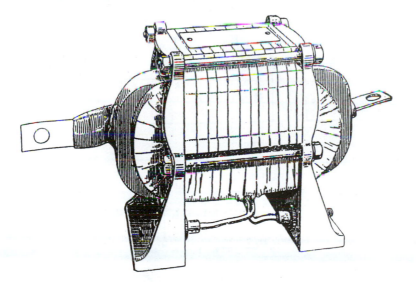

Fig. 2.2 Current transformer manufactured by British Thomson Houston Co Ltd. (Reproduced from Young, 1910, *Journal of the IEE*, 45, with the permission of IEE).

2.2 CONVENTIONAL CURRENT TRANSFORMERS

Information about the behaviour, construction and use of current transformers
is provided in this section.

2.2.1 Equivalent circuits

Current transformers behave similarly to all two-winding transformers and
may be represented by the well-known circuit shown in Fig 2.3(a) which is
based on the winding directions and current and voltage polarities shown in
Fig. 2.3(b).

Because the connected burdens are normally of low impedance, the secondary output VA and voltage are relatively low, typically a maximum of 20 VA
and 20 V for a current transformer with a secondary winding rated at 1 A. The
voltage across the primary winding, because of the turns ratio, is not therefore
likely to exceed a fraction of one volt, a negligible value relative to the rated
voltage of the primary circuit. For this reason, the equivalent circuit of a
current transformer can be simplified to that shown in Fig. 2.4. This circuit
can be made to represent any current transformer operating with any burden
and primary current under either steady state or transient conditions.

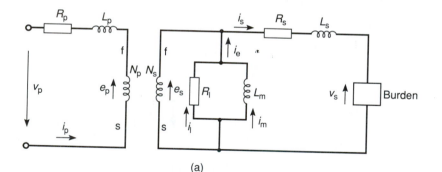

(a)

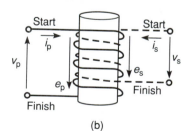

(b)

Fig. 2.3 Equivalent circuit of a current transformer.

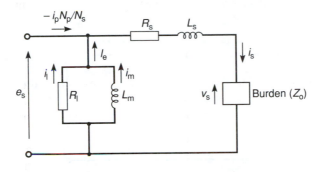

Fig. 2.4 Equivalent circuit of a current transformer referred to the secondary.

It will be seen that the exciting current (i_e) is dependent on the exciting impedance, presented by R_ℓ in parallel with L_m and the secondary e.m.f. (e_s) needed to drive the secondary current (i_s) through the total secondary-circuit impedance. Because the secondary current of a current transformer may vary over a wide range, i.e. from zero under no-load conditions to very large values when there is a fault on the primary circuit, the secondary e.m.f. and excitation current may also vary greatly and in this respect the behaviour is very different from that of voltage transformers.

Because of the non-linearity of the excitation characteristics of the magnetic materials used for transformer cores, the exciting impedance of a given current transformer is not constant, both the magnetizing inductance (L_m) and loss resistance (R_ℓ) varying with the core flux (φ), needed to provide the secondary e.m.f. (e_s). Allowance may be made for this non-linearity, if necessary, when determining the behaviour of a particular transformer under specified conditions, calculation then being done using step by step or other methods. If great accuracy is not required, however, simplifications can be effected by assigning constant values, the averages over a cycle, to R_ℓ and L_m.

An alternative method of representing a current transformer is to employ the concept of mutual inductance (M_{ps}). By definition and based on the conventions used above:

$$e_s = - M_{ps} \, di_p/dt$$

when there is zero current in the secondary winding. It has been shown [7] that the circuit shown in Fig. 2.5 is equally as satisfactory a model as that shown in Fig. 2.4.

The self-inductance, L_{ss} of a secondary winding is given by:

$$L_{ss} = L_s - M_{ps} \, N_s/N_p$$

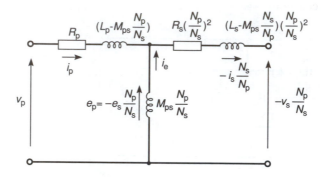

Fig. 2.5 Alternative equivalent circuit of a current transformer.

in which L_s is the leakage inductance used in the circuit of Fig. 2.4. It is clear therefore that the value of L_s for a given transformer may be determined experimentally by measuring its self (L_{ss}) and mutual (M_{ps}) inductances.

2.2.2 Behaviour under normal steady state conditions

Provided that the impedance of the burden connected to a current transformer does not exceed the value corresponding to the rated VA output of the transformer, and the primary current is sinusoidal and of the rated level or less, then the e.m.f. induced in the secondary winding will not be great enough to cause the core to saturate. The exciting m.m.f. of a well-designed transformer operating under these conditions will be very small compared with the primary winding ampere turns ($i_p N_p$). The exciting current (i_e) will nevertheless in practice be non-sinusoidal.

A fundamental, which applies to all two-winding transformers, with the conventions adopted in the previous section, is that the sum, at every instant, of the ampere-turns provided by the primary and secondary windings ($i_p N_p + i_s N_s$) must be equal to the ampere-turns needed to excite the core ($i_e N_s$). Consequently, if the primary current wave is a pure sinusoid, then the secondary current must contain harmonics of the same size as those present in the exciting current. The resulting degree of distortion of the secondary current waveform will nevertheless be very much less than that of the exciting current because the secondary current will contain a much greater fundamental component than the exciting current, maybe 100 times. Secondary current harmonics are therefore negligible and for all practical purposes the secondary waveform can be regarded as sinusoidal.

On this basis, if the burden and secondary winding impedances are linear and equal to $Z_s \angle \beta$ then an e.m.f. of:

$$e_s = I_{spk} Z_s \sin(\omega t + \alpha + \beta)$$

must be induced in the secondary winding, if the secondary current is given by:

$$i_s = I_{spk} \sin(\omega t + \alpha)$$

To provide the necessary secondary e.m.f. (e_s), the core flux density must vary as follows:

$$\Phi = -\frac{1}{N_s} \int e_s dt = \frac{I_{spk} Z_2}{\omega N_s} \cos(\omega t + \alpha + \beta) + k_1 \tag{2.1}$$

It is normal when determining the accuracy of transformation to assume that the core flux alternates symmetrically about zero, in which event the constant k_1 in equation (2.1) is zero. The exciting m.m.f. needed for a particular condition could be determined from the excitation characteristic of the core, which will be of the form shown in Fig. 2.6. It is this need for an exciting m.m.f. that causes errors in practical transformers and at any instant the secondary current is given by:

$$i_s = -i_p \frac{N_p}{N_s} - i_e \tag{2.2}$$

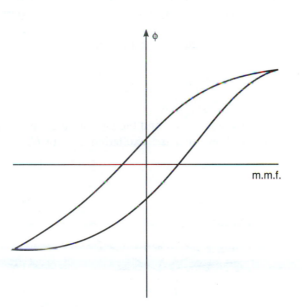

Fig. 2.6 Core-excitation characteristic.

in which $i_e = \dfrac{\text{exciting m.m.f.}}{N_s}$

It is difficult to express the error of a transformer in simple terms if the excitation non-linearity is taken into account and it is accepted practice to express errors in current and phase form by considering only the fundamental components of the exciting current and secondary current. Doing so is equivalent to assuming that the B ~ H characteristic of a core is elliptical. This is also the case when the components R_ℓ and L_m in the equivalent circuits, described in the previous section, are assumed to be constant. A steady state phasor diagram of the form shown in Fig. 2.7 may then be produced.

Current error is defined as the amount by which the actual transformation ratio departs from the desired ratio (K_n). It is therefore given by:

$$\text{Current error} = \frac{\dfrac{|I_p|}{|I_s|} - K_n}{K_n} \qquad \text{per unit} \qquad (2.3)$$

The phase error is defined as the angle (θ) between the phasor representing the primary m.m.f. $I_p N_p$ and the secondary m.m.f. reversed $(-I_s N_s)$. It is regarded as positive when the secondary m.m.f. reversed leads the primary m.m.f. It has been shown [7] that current transformers which have a turns ratio equal to the desired ratio (i.e. $N_s/N_p = K_n$), have positive current errors, i.e. $|I_s| < |I_p| N_p/N_s$, unless the secondary circuit has a leading power factor of less than $\cos \delta$, in which δ is $\tan^{-1} |I_\ell/I_m|$ and their phase errors are also positive unless the secondary circuits have a lagging power factor of less than $\cos (\pi/2 - \delta)$.

Turns compensation

Consideration of the equivalent circuit and the phasor diagram shown in Figs. 2.4 and 2.7 respectively shows that the excitation current (I_e) would, for a

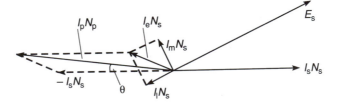

Fig. 2.7 Phasor diagram for a current transformer.

fixed burden impedance, be proportional to the secondary circuit (I_s) if the excitation characteristics of its core were linear. If this were so, both the current and phase errors would be constant and independent of the primary and secondary current levels. In these circumstances, the current error could be reduced to its minimum possible level by using the appropriate number of secondary winding turns. Clearly if the current error would have been positive, i.e. $|I_p|N_p > |I_s|N_s$, with a turns ratio equal to the desired transformation ratio, then more accurate behaviour could be produced by decreasing the number of secondary winding turns.

In practice, the current error does vary with secondary current level and burden impedance, but nevertheless, secondary winding turns may be adjusted to take advantage of the full range of current errors allowable in the various classes of accuracy quoted in standard specifications. As an example, Class 1 accuracy in BS 3938 permits current errors up to $\pm 1\%$ and the use of extra secondary winding turns enables the permissible negative error band to be used.

This adjustment of secondary winding turns from the nominal value is known as turns compensation and it can be done with considerable precision on transformers with a large number of secondary turns. For instance, an adjustment of one secondary turn in the secondary winding of a transformer with a ratio of 500/1 and a single-turn primary winding would change the current error by 0.2%, whereas in a transformer of ratio 500/5 the change would be 1%. It will be evident that turns compensation does not affect the phase error significantly.

The magnitudes of transformation errors

It is clear from the above that the accuracy obtained in any given operating condition can only be improved by reducing the ratio of the excitation m.m.f. needed to support the core flux, relative to the m.m.f. provided by the secondary winding. This ratio is affected by the core material and dimensions, the number of primary and secondary winding turns, the configuration of the secondary winding and the size of its conductors. These factors are considered separately below.

Core materials and dimensions The core-flux variation in a given transformer supplying a particular burden is proportional to the e.m.f. which must be induced in the secondary winding and thus the secondary current. For any value of core flux, the exciting current is proportional to the length of the path traversed by the flux, and inversely proportional to the cross-sectional area of the core. Ideally, therefore, a ring core of the smallest possible diameter and greatest possible cross-sectional area should be used. In practice, however, the core dimensions are limited by factors such as the size of the primary winding conductor and the insulation around it.

When reduction of errors is the only important criterion, the core material should have the highest possible permeability.

The number of primary and secondary winding turns   A considerable reduction of errors can be obtained by providing transformers with multi-turn, rather than bar or single-turn primary windings. For a given transformation ratio, an increase in primary winding turns from 1 to n implies a corresponding increase in secondary winding turns (to nN_s). For a given secondary-circuit impedance, the secondary e.m.f. for a particular condition is not changed, but the core flux (φ) is reduced because of the extra turns. Further, the exciting current is reduced to i_e/n^2. As a result, the errors are reduced to $1/n^2$ times the original value, so, for example, the use of a two-turn primary winding would reduce the errors to one quarter of those which would have been obtained with a transformer having a single-turn primary winding.

Admittedly, the secondary winding impedance of a transformer of a given size must rise if the numbers of turns in its windings are increased, but unless the external burden is negligible and the secondary winding impedance is increased in proportion to the square of the number of turns, then an improvement in accuracy must be obtained.

Secondary windings   Whatever burden is to be connected to a transformer, it is desirable that the secondary winding impedance should be as low as possible, as this reduces the required driving e.m.f. and core flux to the minimum attainable values.

To achieve a low-leakage reactance (ωL_s), the secondary winding should be toroidal and uniformly wound around the whole of the circumference of the core. Conductors of the largest possible cross-sectional area should be used for this winding to keep its resistance low. It is for this latter reason that secondary winding conductors of current-carrying capacities much in excess of those needed for transformer current ratings are usually employed, it being not uncommon to use even 0.055 cm^2 conductor in secondary windings rated at 1 ampere.

Secondary current ratings   Current transformers used in power systems are usually housed in switchgear or other power equipment, which may, because of the site layout, be a considerable distance from the control and relay rooms in which the protective and measuring equipment fed by the transformers is mounted. The interconnecting cables, which typically have conductors of 0.036 cm^2 cross-sectional area, may thus be of appreciable length, typically 100 m. Such cables would introduce a resistance of about 0.5 Ω into the secondary circuit of each current transformer. In the case of a transformer with a secondary winding rated at 5 A, an e.m.f. of 2.5 V would be required to drive the current through the cables, and therefore an added burden of 12.5 VA would be introduced. If, however, a transformer with a rated secondary current of 1 A were to be used, an e.m.f. of only 0.5 V, representing a burden of 0.5 VA, would be added, i.e. the lead burden could be reduced to 4% of the level of that with a 5 A rated transformer.

A disadvantage of utilizing transformers with low secondary current ratings is that they are more costly to produce because they have to be wound with more turns of smaller conductor. It is therefore advantageous to use a higher current rating (5 A) in applications where the burdens are near the transformers but low ratings (1 A or 0.5 A) for other applications.

2.2.3 Behaviour under abnormal conditions

Current transformers feeding measuring equipment are required to perform with high accuracy under normal operating conditions, that is up to a level somewhat above the rated current of their circuits, to allow for possible sustained overloads. Their performances during the brief periods when very high currents are flowing because of system faults are not usually of importance.

A transformer associated with protective and control equipment must however perform acceptably, often in conjunction with others, over the whole range of currents which may flow in it. These currents are dependent on the network in which the protective equipment is used and on factors such as the fault conditions encountered and the instants at which they occur. The magnitudes of these currents may therefore vary greatly and their waveforms will not always be sinusoidal but may contain harmonics and transient components when faults occur.

The behaviour of current transformers carrying two common types of transient waves which arise in power networks are considered below.

Networks with negligible shunt admittance

When the shunt admittances of power lines and other equipment are neglected most power networks may be reduced to a number of sources of sinusoidal e.m.f. interconnected via impedances consisting of series resistance and inductance, as shown in Fig. 2.8. Whenever a circuit-breaker in such a network is closed, the phase currents which will subsequently flow through it will be of the form:

$$i_p = I_{ppk} \{\sin(\omega t + \alpha) - \sin(\alpha) \cdot \exp(-t/\tau)\} \qquad (2.4)$$

in which α is dependent on the instant of switch closure and τ is the ratio of the inductance to the resistance in the circuit, i.e. $\tau = L_p/R_p$.

Normally the currents which flow are relatively small, but very high currents may flow if a circuit-breaker is closed on to a faulted piece of equipment. The worst possible condition, illustrated in Fig. 2.9, would be that of closing a circuit-breaker on to a close-up fault when the busbars were connected to a

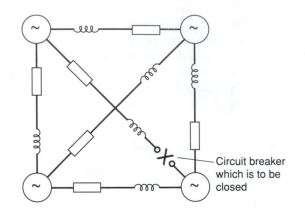

Note: the circuit is closed in single-line form for simplicity

Fig. 2.8 Single-line diagram of a power network.

large (low impedance) source. In this event the value of I_{ppk} in equation (2.4) would be many times the full-load rating of the circuit and the value of the time constant of the transient term (τ) could be of the order of 0.1 s or more.

Clearly similar currents would flow in circuits after the incidence of a short circuit. The behaviour of a current transformer carrying such current waves is now considered.

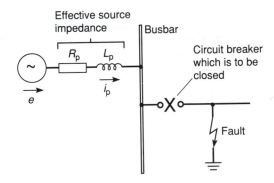

Fig. 2.9 Closure of a circuit breaker when a fault is present.

Current transformer with a purely-resistive burden To simplify the situation it will be assumed that the transformer has a uniformly-distributed toroidal secondary winding of negligible leakage inductance and that its core is so highly permeable that the exciting current is insignificant.

In these circumstances, the transformation would be ideal, i.e.

$$i_s = -\frac{N_p}{N_s} i_p$$

and the secondary e.m.f. needed would be given by:

$$e_s = i_s R_s$$

in which R_s is the resistance of the secondary circuit.

The core-flux variation would be:

$$\Phi = -\frac{1}{N_s} \int e_s \, dt = -\frac{R_s}{N_s} \int i_s \, dt$$

Thus for the current given by equation (2.4), Φ would be:

$$\Phi = \frac{I_{ppk} N_p R_s}{\omega N_s^2} \{-\cos(\omega t + \alpha) + \omega\tau \sin\alpha \exp(-t/\tau)\} + k \qquad (2.5)$$

in which k is a constant of integration.

A limiting case occurs when $\alpha = \pi/2$. For this situation:

$$\Phi = \frac{I_{ppk} N_p R_s}{\omega N_s^2} \{\sin\omega t + \omega\tau \exp(-t/\tau) + k\}$$

If the core flux is assumed to have been zero at the incidence of the fault, a reasonable assumption even if load current had been flowing before the fault, because the associated flux variations would have been small, then the core flux would be:

$$\Phi = \frac{I_{ppk} N_p R_s}{\omega N_s^2} \{\sin\omega t - \omega\tau (1 - \exp - t/\tau)\} \qquad (2.6)$$

Equations (2.5) and (2.6) show that the core flux has three components, the sinusoidal one being the steady state variation. The resultant of the constant and exponentially-decaying components rises and in the limiting case reaches a value equal to $\omega\tau$ times the peak steady state excursion. This is illustrated in Fig. 2.10 using a time constant (τ) of 0.1 s.

The variations of the ratios of the maximum flux level reached, relative to the steady state peak values with the point on wave of switching or fault incidence (α), for various values of $\omega\tau$, are shown in Fig. 2.11. It will be seen that the maximum flux level may lie between twice and $(1 + \omega\tau)$ times the peak steady state value.

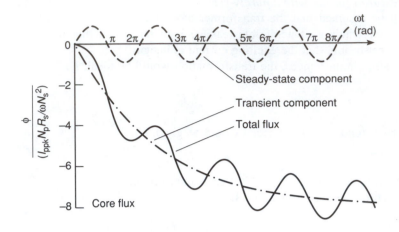

Note: the above curve is for a circuit in which $\omega L_p/R_p = 8$

Fig. 2.10 Core-flux variation.

This result applies, as stated earlier, for cores of infinite permeability. In practice the differences produced by taking account of the finite permeabilities normally encountered are extremely small and may be neglected.

It will be clear that the flux levels reached during normal conditions have to be very low, when resistive burdens are connected, if saturation of the core is to be avoided when large transient currents are flowing. For example, if the incidence of a fault causes the steady state component of the current to be 20 times the rated value, and the $\omega\tau$ value is also 20, then the peak-flux density under normal conditions should not exceed $1/(20 \times 21)$, i.e. $1/420$ of the saturation flux density of the core, only about 0.003 T, if saturation is not to occur.

Current transformer with an inductive burden To simplify this treatment it will again be assumed that the transformer has a core of infinite permeability and also that the impedance in its secondary circuit is purely inductive.

In these circumstances the transformation would be ideal at every instant, i.e.

$$i_s = -\frac{N_p}{N_s} i_p$$

and the secondary e.m.f. needed to drive this current through the secondary circuit inductance (L_s) would be:

$$e_s = L_s \, di_s/dt = -L_s \frac{N_p}{N_s} \, di_p/dt$$

To induce this e.m.f. the core flux would be:

$$\Phi = -\frac{1}{N_s} \int e_s \, dt = \frac{L_s N_p}{N_s^2} i_p + k$$

Again, as in the previous section, the flux levels during normal conditions will be very low, in which event the constant of integration may be taken to be zero and the required flux variation would be proportional to the primary current i_p, i.e.

$$\Phi = \frac{L_s N_p}{N_s^2} i_p \qquad (2.7)$$

As a result, much smaller flux variations are needed with inductive burdens than with a resistive burden of the same VA, i.e. of the same impedance, and

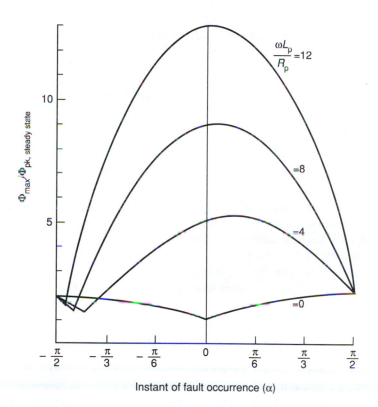

Fig. 2.11 Variation of maximum flux with time of incidence of faults and circuit $\omega L/R$ ratio.

the maximum flux density will not normally exceed double that of the peak value of its steady state component.

When secondary circuits contain both inductance and resistance the flux variations may be determined by summing the fluxes given in equations (2.5) and (2.7) in the correct proportions. The ratio of maximum flux density to the peak value of its steady state component will never exceed that which would be needed for a purely-resistive burden of the same VA.

Current transformer with a capacitive burden   As in the previous sections, it will again be assumed that the transformer has a core of infinite permeability. In addition it will be taken for simplicity that the secondary circuit is purely capacitive.

The transformation would again be ideal at every instant, i.e.

$$i_s = -\frac{N_p}{N_s} i_p$$

and the secondary e.m.f. needed to drive the secondary current i_s through the capacitance (C_s) would be:

$$e_s = \frac{1}{C_s} \int i_s \, dt = -\frac{N_p}{N_s C_s} \int i_p \, dt$$

This e.m.f. would require a core flux variation given by:

$$\Phi = -\frac{1}{N_s} \int e_s \, dt = \frac{N_p}{N_s^2 C_s} \int (i_p \, dt) dt \tag{2.8}$$

In the first integration of i_p a constant of integration (k_1), dependent on earlier conditions, could be introduced, thereby causing the presence of the terms $k_1 t + k_2$ in the second integration. The first of these terms ($k_1 t$) represents a constant rate of flux change in one direction and if it persisted then the onset of saturation would be unavoidable.

Careful consideration is therefore necessary if capacitors are to be connected in secondary circuits.

Networks containing components with distributed parameters

As shown on page 49, the currents which flow in power circuits which contain only lumped, linear impedances usually contain a transient component which decays exponentially in addition to a sinusoidal component. Should both series capacitance and inductance be present then a transient component with an exponentially-decaying sinusoidal waveform can be present.

When a power system contains transmission lines or cables, however, travelling waves are set up whenever there is a disturbance, and these cause the current waveforms which the current transformers are called upon to carry, to be very different to those considered in the previous section. Because the subject of power-line transients is much too vast to be dealt with in this book, only a brief indication will be given below of some of the currents which may flow.

Line energized from a source providing a direct voltage   It is well known that the input current to a healthy, single-phase, loss-free line when connected to an infinite source by the closing of a circuit-breaker would be as shown in Fig. 2.12(a) if the remote end of the line was open-circuited. Should the same

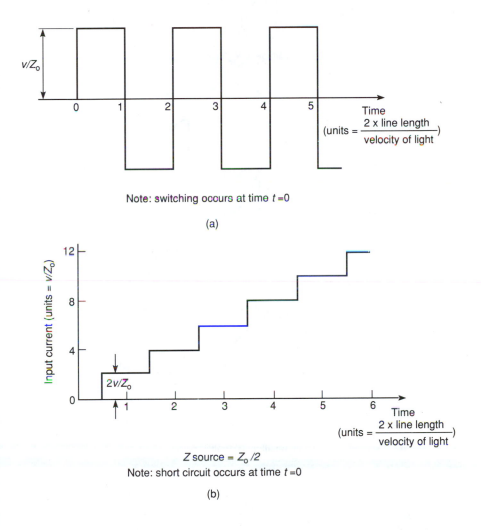

Fig. 2.12 Input currents to an ideal line energized from a source of direct voltage.

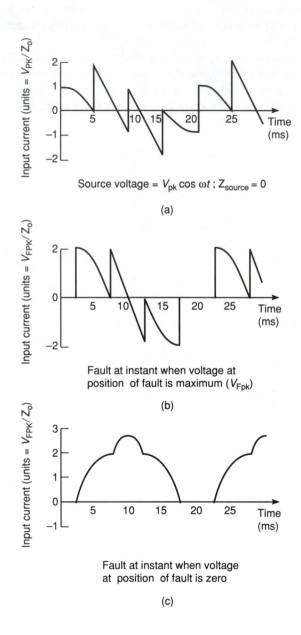

Source voltage = $V_{pk} \cos \omega t$; $Z_{source} = 0$

(a)

Fault at instant when voltage at
position of fault is maximum (V_{Fpk})

(b)

Fault at instant when voltage
at position of fault is zero

(c)

Fig. 2.13 Input currents to an ideal line energized from a source of alternating voltage.

line be suddenly short-circuited at its remote end, however, the input current would be as shown in Fig. 2.12(b).

Line energized from a source providing an alternating voltage The variations of the input currents to the above line which would occur under three different conditions, when connected to an infinite source of alternating voltage, are shown in Fig. 2.13. Two important features of these current waves are that step changes and a unidirectional component may be present.

In practice, the presence of resistance or losses in a line causes the step changes to attenuate and further effects are produced by the presence of impedance in the source. These effects are, however, quite small in the case of modern extremely high voltage transmission lines which have relatively low resistances and which tend to be connected to large sources.

More details of the current waveforms which may be encountered were given by one of the authors in an earlier work [7]. The responses of current transformers carrying such primary currents are now considered. To simplify the treatment, the behaviour obtained when primary currents of single step and alternating rectangular waveforms are examined, assuming a current transformer with an infinitely permeable core, for which no exciting current is required.

Under the above conditions, ideal transformation would always be obtained, i.e. the secondary current would be given by:

$$i_s = -\frac{N_p}{N_s} i_p$$

With a secondary circuit containing only resistance (R_s), the secondary e.m.f. would then be:

$$e_s = i_s R_s = -\frac{N_p}{N_s} R_s i_p$$

and the core flux would be given by:

$$\Phi = -\frac{1}{N_s} \int e_s \, dt = \frac{N_p}{N_s^2} \cdot R_s \int i_p \, dt \qquad (2.9)$$

When carrying a step function current, the flux would increase linearly and clearly the core would inevitably become saturated. To sustain the rectangular primary current wave shown in Fig. 2.14(a), the flux wave would be triangular as shown in Fig. 2.14(b) and this clearly indicates that the step-current changes would be correctly reproduced in the secondary circuit.

In the more practical situations encountered in systems energized from sources providing sinusoidal e.m.f.s, the current waves tend to contain sinusoidally-varying sections and steps. Effectively unidirectional components may be present during fault conditions and these tend to decrease slowly in

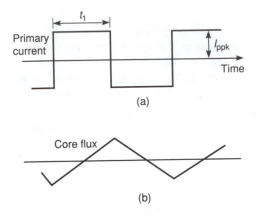

Fig. 2.14 Core-flux variation in an ideal transformer carrying a rectangular primary-current wave.

circuits in which the resistances are low. In these circumstances, the core fluxes of current transformers would be driven towards saturation. This is clear from the limiting case, for a loss-free circuit, illustrated earlier in Fig. 2.13(c). In this instance, the current would contain a constant direct component requiring a linearly-changing flux component which would ultimately cause saturation to occur.

It was shown earlier (page 53) that the flux variation needed in the core of an ideal current transformer with a purely-inductive secondary circuit (L_s) would be:

$$\Phi = \frac{L_s N_p}{N_s^2} \cdot i_p + k$$

This is clearly a much less onerous condition than that encountered with resistive secondary circuits because the flux will not tend in practice to contain a linearly-changing component.

The above treatment has been presented to indicate the core-flux variations which may be needed in current transformer cores to ensure accurate transformation.

When determining the suitability of current transformers which are to operate with particular protective schemes, modern computing techniques should be used to determine the currents which will actually flow during abnormal conditions and then, using the appropriate secondary-circuit parameters, the core flux and secondary current variations should be determined, allowance being made for the core excitation characteristics [8, 9].

2.2.4 The effects of core saturation on transformation behaviour

In many applications, as indicated above, impractically large core-flux variations might be needed under certain circumstances to approach ideal transformation, i.e.

$$i_s \simeq -\frac{N_p}{N_s} i_p$$

As a result, saturation of transformers must, on occasion, be accepted and the associated protective equipment must be so designed that it will nevertheless function correctly. To enable this to be done it is necessary to determine the secondary current errors or distortions which will be introduced when core saturation occurs.

Current transformer with a resistive secondary circuit To simplify the presentation, the approximate excitation characteristic shown in Fig. 2.15 is assumed. At flux values below the saturation level ($\Phi < \Phi_{sat}$) the permeability is infinite and the exciting current is zero. For flux variations in this zone, ($-\Phi_{sat} < \Phi < \Phi_{sat}$), ideal transformation ($i_s = -N_p i_p/N_s$) would therefore be achieved.

When a core is saturated, however, ($\Phi = \Phi_{sat}$ or $-\Phi_{sat}$), then the rate of change of flux will be zero and the secondary e.m.f. (e_s) must then also be zero. With zero e.m.f., no current will flow in the resistive secondary circuit

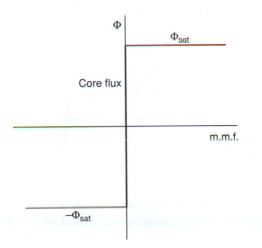

Fig. 2.15 Approximate excitation characteristic.

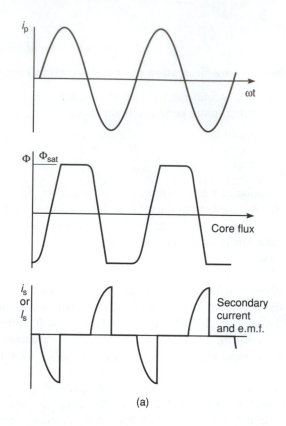

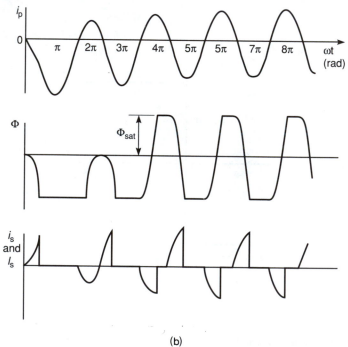

Fig. 2.16 Behaviour when core saturation occurs in a transformer with a resistive secondary circuit.

($i_s = 0$) and the primary m.m.f. will maintain saturation until the magnitude of the flux needs to reduce again and this will not occur until the polarity of the required e.m.f. reverses. When the secondary circuit is resistive this will be at the next current zero. As an example, if saturation occurs during a positive half cycle of the secondary current, the e.m.f. will not become negative until after the next current zero. Considering sinusoidal primary current conditions, the behaviour would be as shown in Fig. 2.16(a) from which it will be seen that the tails of the half cycles of secondary current are chopped off, and the greater the desired flux excursion, the greater the portions of wave lost in each half cycle.

This pattern of behaviour is repeated whatever the primary current wave-form including situations where transient components are present, an illustration being provided in Fig. 2.16(b).

Clearly the effect of saturation is to distort the waveform of the secondary current and also to reduce its r.m.s. value.

Current transformer with a purely inductive secondary circuit Under ideal conditions, the core flux needed when a secondary circuit is inductive is given by:

$$\Phi = \frac{L_s N_p}{N_s^2} i_p + k$$

If symmetrical conditions obtain, then the constant of integration term (k) will be zero and the flux will vary in phase with the primary current.

Saturation, if it is to occur, must therefore take place in the first half of each half cycle. Thereafter, the secondary e.m.f. (e_s) will be zero which will prevent the secondary current from changing, i.e. $L_s\, d i_s/dt = 0$. This condition will then be maintained because the primary m.m.f. will exceed that provided by the secondary winding. Saturation will end when the primary and secondary m.m.f.s become equal again. Thereafter the magnitudes of the core flux and secondary current will reduce. This behaviour is shown in Fig. 2.17 and the general effect is always to cut off the tops of the secondary current waves and, as with resistive secondary circuits, the r.m.s. values are lower than the ideal values, i.e. $I_s < N_p I_p/N_s$.

Current transformer with a secondary circuit containing resistance and inductance in series As in the above two situations, ideal behaviour would be obtained whenever the magnitude of the core flux is below the saturation level. Once saturation occurs, the flux remains constant and no secondary e.m.f. is produced. In these circumstances:

$$e_s = L_s\, d i_s/dt + R_s\, i_s = 0$$

The secondary current during a saturated period is therefore given by:

$$i_s = A \exp(-R_s t/L_s)$$

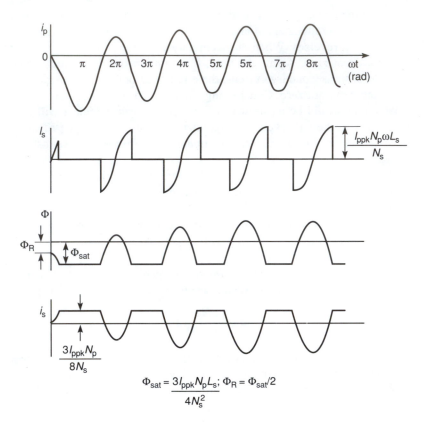

$$\Phi_{sat} = \frac{3I_{ppk}N_pL_s}{4N_s^2}; \quad \Phi_R = \Phi_{sat}/2$$

Fig. 2.17 Behaviour when core saturation occurs in a transformer with an inductive secondary circuit.

in which A is the value of the secondary current at the instant at which saturation occurs. Again saturation is maintained because the magnitude of the primary m.m.f. exceeds that of the secondary m.m.f. Saturation will end once again when the secondary current reaches its ideal value. An example of a secondary current waveshape which would be obtained is shown in Fig. 2.18.

The general effects which result from saturation of the cores of current transformers may be summarized as reductions of the r.m.s. values of their secondary currents below the ideal values and distortion of the current waveforms, the actual form of the distortion depending on the particular secondary circuit parameters. The importance of these effects in protective applications and methods of allowing for them will be considered in later chapters.

2.2.5 Remanent core flux

In the previous sections idealized excitation characteristics have been assumed to simplify the presentation. With practical transformers, however, the characteristics are in the form of symmetrical loops, one of which is shown in Fig. 2.19, and it is therefore possible to have flux in a core when there is zero exciting current. During a particular normal steady state condition requiring a sinusoidal flux variation, operation could be around a minor hysteresis loop displaced along the Φ axis of the Φ ~ m.m.f. diagram, as shown dotted in Fig. 2.19, because this would not require the exciting current to contain a unidirectional component. This situation could arise due to remanent flux left in the core at an earlier time.

The amounts of flux which may be retained and their persistence are of interest to protective-equipment engineers as the presence of remanent flux in the core of a transformer reduces the available flux swing in one direction, and makes the avoidance of saturation during fault conditions more difficult.

There are a number of conditions which may leave remanent fluxes in current transformer cores. The largest levels tend to be left after the clearance of faults and two possible situations are examined below.

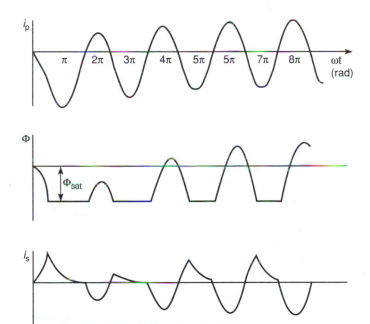

Note: this figure is based on the following parameters:
$\omega L_p/R_p = 8$; $\omega L_s/R_s = 1$; $\Phi_{sat} = (I_{ppk}N_pZ_s)/(2\omega N_s^2)$; $\Phi_R = 0$; $\alpha-\gamma_1 = 0$

Fig. 2.18 Behaviour when core saturation occurs in a transformer with a resistive and inductive secondary circuit.

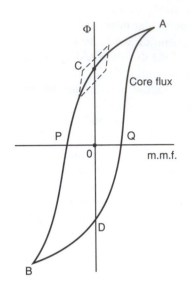

Fig. 2.19 Core-excitation characteristics.

The clearance of a symmetrical fault current

The passage of a symmetrical sinusoidal fault current through a primary winding may cause the core flux to vary around a symmetrically-positioned major hysteresis loop such as that shown in Fig. 2.19. Since circuit-breakers normally interrupt a fault at a current zero, the operation of the core iron at the instant of interruption would be at one of two fixed points on the hysteresis loop, the points depending on the impedance present in the secondary circuit. With a purely-resistive circuit the flux would have its maximum value at the instant of clearance, i.e. A or B in Fig. 2.19, whereas with a purely-reactive circuit, it would be zero, i.e. at P or Q.

For the working point to be at A, say, a direct magnetising m.m.f. would be required and this could only be provided by a direct current in the secondary winding after the primary circuit had been opened. To drive this secondary current, the core flux would have to decrease to induce the necessary e.m.f. The operating point would therefore move along the hysteresis loop until equilibrium was established at point C where no secondary current or e.m.f. would be required.

The maximum possible remanent flux (OC) will be left when interruption occurs at points along the sections AC or BD of the hysteresis loop, conditions which will occur with inductive secondary circuits of sufficiently high power factor.

Protective current transformers, because of their need to operate satisfactorily during faults, usually operate at very small flux levels and the steady state swings which occur even when large fault currents are flowing may be so small that only rather insignificant levels of remanent flux may be left after clearance.

The clearance of asymmetrical fault currents

The flux variations needed in transformer cores when their primary windings carry currents containing exponentially-decaying components were examined on page 51. It was shown that large flux swings are necessary when secondary circuits are highly resistive. A typical flux excursion for a transformer in a primary circuit with an $\omega L_p/R_p$ ratio of 8 and a current with a sinusoidal component starting at zero is shown in Fig. 2.20. This is based on a purely-resistive secondary circuit and a ratio of exciting reactance (ωL_m) to secondary resistance (R_s) of 8. It is the presence of the exciting reactance which is responsible for the unidirectional flux component falling after $\omega t = 3\pi$ to eventually reach a constant value.

If clearance of this fault current occurred within a few cycles, the core flux would be at a high value at the instant of interruption and thereafter the flux would decay, as explained in the preceding section, until the working point was on the vertical axis of the Φ – m.m.f. diagram, a considerable remanent flux being left in the core.

The persistence of remanent flux

Because of the non-linear nature of the excitation characteristics of magnetic materials, it is not possible to determine, simply by quantitive analytical means, the effects of remanent flux on subsequent behaviour or on the variations

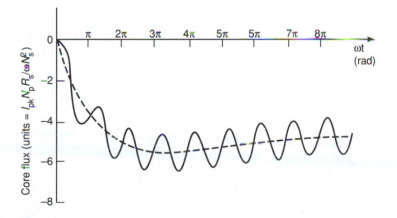

Fig. 2.20 Flux excursion when a fault is present.

which will occur in the remanent flux. In practice, numerical step-by-step or other methods are used to obtain accurate predictions of behaviour.

To illustrate the behaviour which may be obtained, however, a qualitative presentation of the effects of the passage of sinusoidal primary currents on remanent flux is provided below and some experimentally-obtained results are included to illustrate the changes which occur.

When steady state primary current flows again in a transformer after the clearance of a fault, sinusoidal flux variations are needed to provide the secondary winding e.m.f. These would start from the remanent value (OR in Fig. 2.21) and could progress around a minor loop such as RS. Operation around such a loop could not, however, continue because its offset nature would require the exciting current to have a direct component. Operation would therefore continue around minor loops moving progressively to the left and downwards until the exciting current had no direct component. Equilibrium would be established at this position, shown as loop TU in Fig. 2.21. Experimentally-obtained loops of this type for four different current levels are shown in Fig. 2.22. If the primary current ceased after reaching an equilibrium state, the remanent flux would be somewhat lower than the earlier value, the reduction being dependent on the magnitude of the primary current which has flowed and thus the size of the minor hysteresis loop. For normal current levels the loops in protective current transformers tend to be small and the resulting reductions of remanent flux are usually small. It must therefore be accepted that a large part of the remanent flux left by a fault will tend to remain until another large fault occurs. The possible effects resulting from this situation are shown in Fig. 2.23 which includes experimentally-obtained records of flux density against both time and m.m.f. for two fault conditions. Those shown in Fig. 2.23(a), which are for a fault requiring a flux swing in the direction of

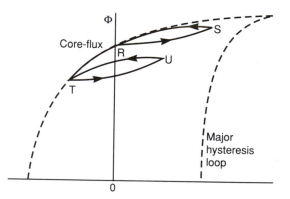

Fig. 2.21 Change in remanent flux in a core.

Fig. 2.22 Minor hysteresis loops.

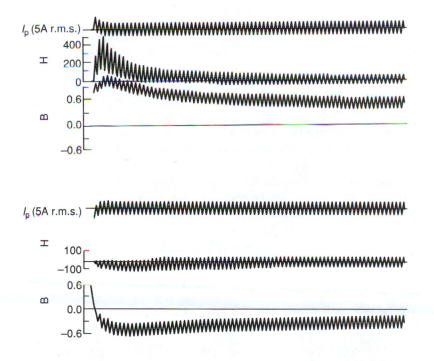

Fig. 2.23 Flux variation when a fault occurs.

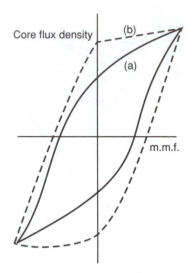

Fig. 2.24 Core-excitation characteristics.

the remanent flux, cause the core to saturate. Those shown in Fig. 2.23(b) are for a fault of the opposite polarity which cause the flux to swing in the opposite direction to the initial remanent flux.

The deviations of the secondary currents of current transformers from their ideal values ($i_s = - N_p/N_s\, i_p$) as a result of saturation of their cores would cause some protective equipments fed by them to maloperate. This is clearly unacceptable and therefore, in these applications, the flux range between the maximum possible remanent flux value (OR) in Fig. 2.21 and the saturation value must be great enough to allow the flux swings needed when large asymmetric fault currents are flowing. Clearly core materials with high saturation flux densities and excitation loops with relatively low flux densities at exciting current zeros, as shown in Fig. 2.24 (curve *a*) are desirable rather than almost rectangular loops of the form shown in Fig. 2.24 (curve *b*). The ratio of maximum-possible remanent-flux density to the saturation density is referred to as remanence factor.

2.2.6 Operation with a secondary circuit which is open or of a high impedance

Current transformers are unlikely to be operated with their secondary windings on open circuit, because of the widespread knowledge of the high voltages which may be produced under such conditions. There are, however, cases of near open-circuiting which occur at times when certain protective schemes are used. Such situations are possible, for example, when circulating-current schemes incorporating high-impedance relays are used. An examination of the way in which high voltages are produced is therefore of interest.

As in earlier sections, a transformer with a particular excitation characteristic will be assumed, to simplify the treatment. In this instance, the characteristic shown in Fig. 2.25(a) will be used. Over the portion BC, the current

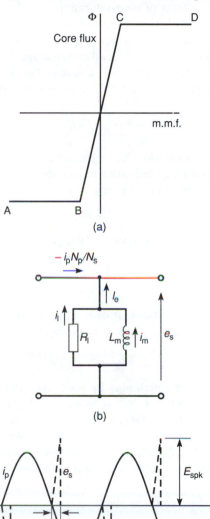

Fig. 2.25 Behaviour when a secondary winding is an open circuit.

is proportional to the flux and therefore the associated behaviour may be represented by a constant magnetizing inductance (L_m) in the equivalent circuit of Fig. 2.25(b). Over the portions AB and CD, no flux change occurs as the current changes and this behaviour may be simulated by zero magnetizing inductance. A further assumption can be made that a constant resistance (R_1) may be included in the equivalent circuit to take account of the hysteresis and eddy-current effects in the core. This condition is similar to that considered on page 59 when the effects of core saturation were being examined. It was shown that saturation ends at primary current zeros and that secondary current then flows until saturation occurs again. In the case of an open-circuited secondary winding, the current through the high magnetizing inductance (L_m) during unsaturated periods is very small and thus the referred primary current ($-N_p i_p / N_s$) flows through the resistor R_ℓ requiring an e.m.f. (e_s) given by:

$$e_s = -\frac{N_p}{N_s} R_\ell\, i_p$$

This variation is shown in Fig. 2.25(c) assuming a sinusoidal primary current. In practice saturation occurs after a very short interval of time (t_s) and over this period the primary current may be expressed as:

$$i_p = I_{ppk}\,\omega t$$

Over the period t_s the flux changes by $2\Phi_{sat}$ and therefore:

$$2\Phi_{sat} = -\frac{1}{N_s} \int_0^{t_s} e_s\, dt$$

From the above three equations, it can be shown that the peak secondary e.m.f. (E_{spk}) which occurs at the time t_s is given by:

$$E_{spk} = 2\ \left[\Phi_{sat}\,\omega N_p\, I_{ppk}\, R_\ell \right]^{\frac{1}{2}} \tag{2.10}$$

An alternative treatment which may be used when the hysteresis and eddy current losses in a core are very low is to assume that the resistor R_ℓ has an infinite value. In these circumstances, the primary referred current must always flow through the magnetizing inductance (L_m). Whenever the current is above a certain value, the core will be saturated and as a result the incremental inductance L_m will then be zero. Near current zeros, however, the magnetizing inductance, as before, may be taken to have a constant non-zero value. At these times the secondary e.m.f. is given by:

$$e_s = -L_m \frac{d}{dt}\left(-\frac{N_p}{N_s} i_p\right)$$

For a primary current of $i_p = I_{ppk} \sin \omega t$, the secondary e.m.f. would have a maximum value at the current zeros of:

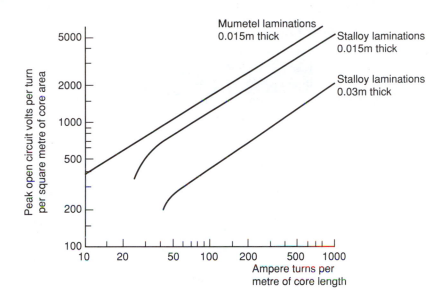

Fig. 2.26 Voltages during open-circuit conditions.

$$E_{spk} = \frac{N_p}{N_s} \omega L_m I_{ppk} \qquad (2.11)$$

In practice, the behaviour will lie between the two limiting conditions described above. More accurate modelling can be performed by using non-linear components with values related to the core flux and its rate of change.

The normal reason for determining the peak voltages which may be produced is to ensure that the insulation of a transformer and its burden, including the connecting cables, is adequate or suitably protected by voltage-limiting devices. As it is usual to allow a significant factor of safety on insulation levels, highly-accurate determination of the voltages which may be produced by current transformers is not justified. It is probably better to make use of experimentally-derived results such as those shown in Fig. 2.26 which were obtained by Wright [7].

2.2.7 The construction of current transformers

Each current transformer, as stated earlier, has a core of magnetic material which is linked with the primary and secondary windings. These three components are considered separately below.

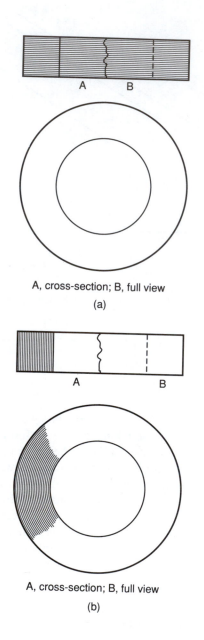

A, cross-section; B, full view

(a)

A, cross-section; B, full view

(b)

Fig. 2.27 Current transformer cores.

Magnetic cores

Transformation errors are caused by the need to excite the core of a transformer and thereby induce the necessary e.m.f. in its secondary winding. Clearly, therefore, it is desirable that secondary-circuit impedances should be kept as low as possible and for this reason the secondary winding leakage inductances should be reduced to the minimum attainable levels. This may be done by using ring-shaped cores around which toroidal secondary windings of one or more layers are uniformly distributed. Cores usually consist of a cylindrical stack of laminations as shown in Fig. 2.27(a) or are made up of strip wound in spiral form, like a clockspring, as shown in Fig. 2.27(b). The latter method is much to be preferred when grain-orientated magnetic materials are being used as this ensures that the flux is able to follow the path of minimum reluctance.

The core material should ideally be of very high permeability, have a high saturation flux density and, as shown in the previous section, not be capable of retaining large levels of remanent flux. Although nickel-iron alloys have very high permeabilities, they do not have the other desirable properties and therefore protective current transformer cores are usually made from either hot-rolled or cold-rolled silicon-iron alloys.

Because it is relatively expensive to produce toroidal secondary windings on ring cores, cores made up of rectangular strips or C-shaped sections are sometimes employed when a high standard of performance is not required. This reduces manufacturing costs because the cores can be assembled into secondary windings which have been produced earlier, in coil form on conventional coil-winding machines. Such an arrangement is shown in Fig. 2.28.

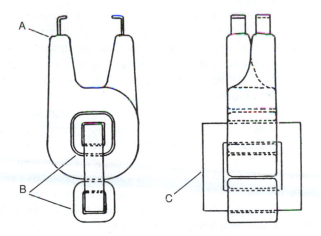

A, primary winding; B, secondary windings (connected in series); C, built-up core

Fig. 2.28 Current transformer with a built-up core.

Secondary windings

As explained above, secondary windings are usually toroidal and uniformly spaced around the circumferences of the cores. In applications where the transformer ratio is low, e.g. 50/5 A, the secondary winding turns will be relatively small in number and they are then spaced out to occupy a single layer. When the ratios are high, the numbers of secondary turns will be great enough to require windings of two or more layers but again the turns will be spaced to fully occupy each of the layers.

As stated earlier in section 2.2.2, it is usual for secondary windings to be made with a conductor of a cross-sectional area considerably greater than that needed to carry the rated secondary current continuously. This practice is adopted to keep secondary winding resistances low and thus reduce the e.m.f.s which must be induced.

The windings should be insulated to withstand the voltages which may appear across them under all normal and abnormal conditions.

Primary windings and positioning

Transformers used in circuits with high rated currents, say 500 A and above, usually have a single-turn primary winding. For lower current levels, however, multi-turn windings are commonly used to enable greater transformation accuracies to be obtained. This improved performance, as explained earlier, is obtained because, for a given ratio, the number of secondary winding turns is proportional to the primary winding turns, and the greater the number of secondary turns, the smaller is the flux variation and thus the exciting current needed to provide a given secondary e.m.f.

Many modern protective schemes monitor the zones between the sets of current transformers associated with them, and as any fault within a protected zone is cleared by opening circuit-breakers to isolate the unhealthy section of the system, discrimination approaching the ideal can only be achieved when the current transformers are near the contacts of the circuit-breakers. This topic was treated in detail in reference [8].

When very high voltage switchgear incorporates bushings through which the main conductors enter and leave, the above desired proximity is achieved by mounting the secondary wound current transformer cores over these bushings. In addition to the transformers which are needed to feed the main protective equipment, others are needed to feed back-up protective relays and indicating instruments and possibly accurate measuring equipment. In practice, therefore, several wound cores may be stacked together on each phase bushing, a typical arrangement being shown in Fig. 2.29.

This arrangement is not only ideal operationally, but it is very economic as it reduces the total amount of costly high voltage dielectric material to a minimum. It does however, limit the primary winding to only one turn.

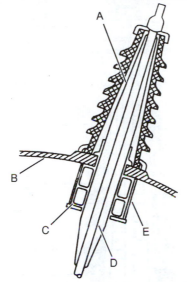

A, capacitor bushing; B, tank top; C, current transformer cores and windings;
D, primary conductor; E, clamps and supports

Fig. 2.29 Current transformer mounted around a bushing.

Some forms of high voltage switchgear, such as small-oil-volume and air-blast designs, do not have suitable bushings over which transformers may be mounted. Separate current transformers must then be provided adjacent to the switchgear, unless the latter is associated with and near to power equipment with bushings on which the current-transformer wound cores may be mounted. Separate high voltage current transformers are constructed in the post form shown in Fig. 2.30. Oil-impregnated paper is used to insulate the primary conductor, which is a U-shaped tube. Metallic foils are inserted in the insulation at a number of levels to control the electric stress distribution and means must be provided to prevent moisture from entering the housing. The external diameter of the primary tube must be large enough to keep the electrical stress in the dielectric to an acceptable value and this usually causes the internal diameter of the tube to be great enough to permit a lightly insulated cable to be passed through it. By making suitable connections a two-turn primary winding can thereby be provided if desired. The secondary wound cores are fitted around the lower part of the primary tube as shown.

Gas-insulated current transformers have been produced for use in high voltage systems. In one arrangement the cores and secondary windings are housed in a tank which is at line potential. It is supported on a porcelain insulator inside which the secondary circuit connections pass. A 'straight-through' primary conductor passes centrally through the cores and secondary windings, from which it is insulated by a gas, usually sulphur hexa-fluoride, under pressure.

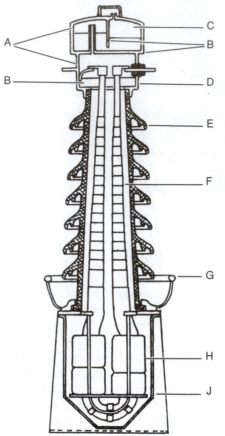

A, nitrogen; B, Oil; C, Air; D, terminal chamber; E, porcelain insulator;
F, insulated primary conductor; G, arcing ring; H, cores and secondary winding; J, tank

Fig. 2.30 Post-type current transformer.

In lower voltage metalclad and metal-enclosed switchgear, one of the most satisfactory arrangements is to stack the secondary wound cores and use a common primary winding with the necessary number of turns. The completed assembly is then housed in a special transformer chamber, which forms part of the switchgear. A typical arrangement is shown in Fig. 2.31. In some cases however, when the necessary performance can be obtained, transformers with single-turn primary windings (usually called bar primaries) are used. These transformers are produced in two forms, in one the core and secondary winding are fully insulated and in the other the main insulation is on the primary conductor. For these lower voltage applications, the insulation may be varnished tape and cloth or synthetic resin.

In all designs, primary windings must be sufficiently robust and so supported that they will not be mechanically damaged by the forces which may

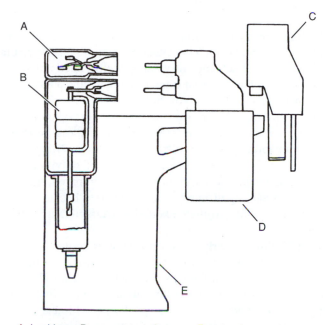

A, bushbars; B, current transformers; C, operating mechanism;
D, circuit breaker (racked out); E, main frame

Fig. 2.31 Current transformer housed in switchgear.

act on them during short-circuit conditions. In addition they must comply with regulations concerned with their current-carrying capability. As an example, BS 3938 requires that a current transformer with an integral primary winding must be capable of carrying its rated short time current, for its rated time, with 50% of the rated burden connected to the secondary terminals, without suffering any damage such as would adversely affect it either electrically or mechanically.

The rated short-time current of a transformer is the rupturing capacity of the circuit-breaker with which it is associated and the time rating is the maximum period for which the transformer may be subjected to the full short-circuit current, and this depends on the protective equipment fitted to the circuit.

2.3 LINEAR COUPLERS

Although iron-cored current transformers are and have been used in conjunction with power-system protective equipment for almost a hundred years, they are not ideal because of the non-linearity of their excitation characteristics and their ability to retain large flux levels in their cores.

Protective schemes, in which the outputs of non-ideal current transformers at different points in a power system are compared, could operate correctly provided that the current transformers each had the same percentage transformation errors at each instant. This situation cannot be achieved in practice, however, with protective schemes protecting multi-ended units, such as busbars, because the output of a current transformer carrying a large current might be required to balance with the sum of the outputs of a number of transformers carrying small currents. The transformers would be operating at different points on their excitation characteristics and when these are non-linear the errors of the transformers would differ, preventing perfect balance being achieved. Even in those cases where the outputs of only two current transformers carrying the same primary currents are compared, unbalance would occur if they were retaining different remanent fluxes because they would not then be at the same points on their excitation characteristics and would not have the same errors.

The above difficulties clearly would not arise if protective schemes were energized from transformers with cores of unit relative permeability, i.e. of wood or air, because such transformers are unable to retain flux in their cores when their excitation is removed and they always behave in a proportional or linear manner. These devices, which are usually called linear couplers, do not produce secondary currents approximating to those which would be provided by an ideal current transformer ($i_s = - i_p N_p/N_s$) and consequently protective equipment must be specially designed to operate from them. This may be done by inserting interface equipment, such as amplifiers, so that an infinite impedance is presented to the outputs of each of the couplers, causing them to operate as mutual inductors, thus providing output e.m.f.s proportional to the rates of change of their primary or input currents. Alternatively, the input impedance of the protective equipment could be matched to that of the couplers, to enable the maximum possible outputs to be obtained from them.

The procedure for designing linear couplers required to supply particular outputs was dealt with fully by Mathews and Nellist [10] and therefore this section will be confined to analysing the general behaviour of linear couplers during steady state and transient conditions.

2.3.1 The output of a linear coupler with a burden of infinite impedance

In the limiting case where an infinite impedance is presented to the secondary winding of a linear coupler, the device operates, as stated above, as a mutual inductor and the output is in the form of an e.m.f. given by:

$$e_s = - M_{ps} \, di_p/dt$$

in which M_{ps} is the mutual inductance between the windings. Its value is given by:

$$M_{ps} = \frac{N_p N_s \mu_0 A_c}{l_c}$$

in which N_p and N_s are the turns of the primary and secondary windings respectively, A_c is the cross-sectional area of the core and l_c is its magnetic path length.

During steady state conditions with a sinusoidal primary current $i_p = I_{ppk} \sin(\omega t + \alpha)$ the output e.m.f. would clearly be:

$$e_s = -\omega M_{ps} I_{ppk} \cos(\omega t + \alpha)$$

2.3.2 The output of a linear coupler with a burden of finite impedance

Ideally it should be possible to connect any type of burden to a linear coupler, and therefore the most general case in which the secondary circuit consists of resistance, inductance and capacitance in series is examined below.

The equivalent circuit of a linear coupler is the same as that of a normal current transformer, but because there are no core losses the resistance (R_ℓ) is infinite and therefore not included. The exciting inductance, designated $M_{ps}N_s/N_p$, in the circuit shown in Fig. 2.32 is constant under all conditions.

The basic equation for a coupler is therefore:

$$-\left\{ i_s R_s + L_s \frac{di_s}{dt} + \frac{1}{C_s} \int i_s \, dt \right\} = M_{ps} \cdot \frac{d}{dt}\left(i_p + \frac{N_s}{N_p} i_s \right) \tag{2.12}$$

For the condition of switching a primary circuit containing series inductance (L_p) and resistance (R_p) on to a fault, or for a fault occurring on such a circuit when unloaded, the primary current would be:

$$i_p = I_{ppk}\left\{ \cos(\omega t + \alpha) - \cos\alpha \exp\left(-\frac{R_p}{L_p} t \right) \right\}$$

The corresponding secondary current would be:

$$i_s = \frac{\omega M_{ps} I_{ppk}}{\left[R_s^2 + \left\{ \omega\left(L_s + M_{ps}\frac{N_s}{N_p} \right) - (1/\omega C_s) \right\}^2 \right]^{\frac{1}{2}}} \sin(\omega t + \alpha - \beta)$$

$$+ \frac{M_{ps} I_{ppk} \cos\alpha}{L_s + M_{ps}\frac{N_s}{N_p} - (L_p R_s/R_p) + L_p^2/R_p^2 C_s} \exp\left(-\frac{R_p}{L_p} t \right)$$

$$+ k_a \exp(-mt) + k_c \exp(-nt) \tag{2.13}$$

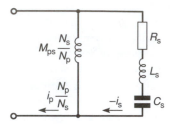

Fig. 2.32 Equivalent circuit of a linear coupler, referred to secondary level.

in which:

$$\beta = \tan^{-1}\left[\frac{\omega\left(L_s + M_{ps}\dfrac{N_s}{N_p}\right) - 1/\omega C_s}{R_s}\right]$$

and m and n are given by:

$$m = \left[R_s/2\left(L_s + M_{ps}\frac{N_s}{N_p}\right)\right] + \left\{\left[R_s^2/4\left(L_s + M_{ps}\frac{N_s}{N_p}\right)^2\right] - \left[1/C_s\left(L_s + M_{ps}\frac{N_s}{N_p}\right)\right]\right\}^{\frac{1}{2}}$$

$$n = \left[R_s/2\left(L_s + M_{ps}\frac{N_s}{N_p}\right)\right] - \left\{\left[R_s^2/4\left(L_s + M_{ps}\frac{N_s}{N_p}\right)^2\right] - \left[1/C_s\left(L_s + M_{ps}\frac{N_s}{N_p}\right)\right]\right\}^{\frac{1}{2}}$$

Examination of equation (2.13) shows, as would be expected, that as the mutual inductance approaches infinity, the secondary current approaches the value normally regarded as the ideal ($i_s = i_p N_p/N_s$). The low permeability of the core of a linear coupler prevents a high mutual inductance from being achieved, however, and there is consequently always a considerable divergence from the performance which is expected of a conventional current transformer.

The steady state sinusoidal behaviour of a linear coupler can be determined from the first term in equation (2.13). The ratio of the magnitude of this output to that of the sinusoidal component of the primary current is given by:

$$\text{Transformation ratio} = \frac{I_s}{I_p} = \cdot\frac{\omega M_{ps}}{\left[R_s^2 + \left\{\omega(L_s + M_{ps}N_s/N_p) - \dfrac{1}{\omega C_s}\right\}^2\right]^{\frac{1}{2}}}$$

This expression shows that the output of a coupler falls significantly as inductive reactance is added to its burden. It also falls, but at a reduced rate as the burden resistance is increased. The output can, however, exceed the ideal value $(I_s = -N_p/N_s I_p)$ if the secondary circuit is made sufficiently capacitive. The phase error is $(\pi/2 - \beta)$ (leading) and examination of the expression for β given above shows that increase of burden resistance or capacitive reactance causes the phase error to increase.

It can be seen that there are three transient terms in the expression for the secondary current (equation 2.13). One of these is produced by the transient component of the primary current whilst the others are produced by the coupler and its burden.

The deviations between both the steady state and transient outputs from the normal ideal value may be considerable and as shown they depend on the coupler and burden parameters. When particular protective schemes, for example differential equipments, are being supplied, satisfactory behaviour should be obtainable, however, provided that the coupler and burden parameters are matched. Clearly couplers should be as nearly identical as possible, i.e. their mutual inductances should be the same, and the burdens must be virtually equal.

2.4 CURRENT TRANSFORMERS WITH AIR-GAPPED CORES

Current transformers with air-gapped magnetic cores were introduced many years ago to provide more linear behaviour than that obtainable from transformers with homogeneous cores of highly permeable magnetic material.

Figure 2.33(a) shows the excitation characteristics of magnetic material and air and also a combination of the two which should be obtained when using an iron core with one or more air gaps as shown in Fig. 2.33(b).

The resultant characteristic is obviously more linear than that of the magnetic material and the remanent flux which may be left in the core is lower. Clearly, however, complete linearity would not be obtained until the magnetic path was completely in air or other material of unit relative permeability, i.e. as in a linear coupler.

Air-gapped core designs occupy therefore an intermediate position between conventional current transformers and linear couplers and a satisfactory compromise performance has to be achieved. The introduction of air gaps or sections of low permeability in a core increases the excitation needed. While this could be counteracted by using large cores with large cross-sectional areas, this in turn increases the lengths of the secondary winding turns and their resistances. To obtain performances comparable with those of conventional transformers, air-gapped designs have to be large and relatively expensive.

Such transformers have been produced and used to supply protective equipment which has functioned satisfactorily. As with linear couplers, care must be taken during manufacture to ensure that transformers which are to operate together in groups are correctly matched, particular attention being taken to

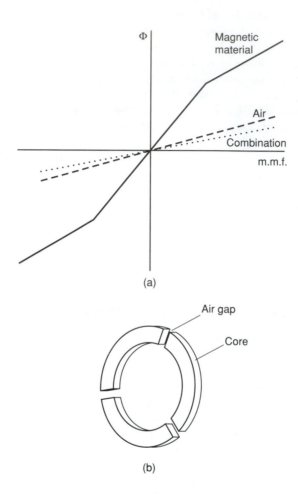

Fig. 2.33 Air-gapped core and its excitation characteristic.

ensure that the gapped sections have the correct reluctances. The so-called distributed air-gap transformer (abbreviated to DAG), which was developed many years ago, has several gaps uniformly distributed around its core, this being considered preferable to a core with only one relatively long gap.

2.5 NON-CONVENTIONAL CURRENT TRANSDUCERS

It is evident from the previous sections that the iron-cored current transformers, used for feeding measuring and protective equipment in power systems, are not ideal. In addition, those which are used on very high voltage systems in positions where they cannot be fitted over bushings in major items of equipment have the disadvantage of being costly, because of the insulation needed on their primary windings. These factors have caused several investigations to be initiated during the last ten to twenty years to produce alternative current transducers. Workers have incorporated modern signal-processing equipment and other electronic devices in protective schemes to eliminate the need for them to be provided with inputs of several volt-amperes.

Work was done to eliminate the costly insulation needed on high voltage primary windings by mounting lightly insulated conventional secondary wound cores directly around the main system conductors. The secondary outputs were processed to produce light signals conveyed to the relaying equipment via optical fibres. The latter can convey signals over a wide frequency band and being insulators they do not need bulky insulation around them. They also have the advantage that they are not affected by electromagnetic interference.

Blatt [11] described an arrangement in which a number of sensing coils were placed at different points in a switchgear site. Simultaneous outputs from these coils, monitored under microprocessor control, were analysed to provide data on the individual circuit currents. An alternative device, called a dynamic current transducer, developed by Mercure [12], used the current-carrying conductor as the primary of a coaxial plastic-core Rogowski coil. The signal from the secondary was transmitted to relaying equipment using an analog fibre-optic link.

Current and voltage transducers of an entirely new type making use of certain optical materials have been developed in recent years. When these materials are placed in magnetic or electric fields the polarization of light beams passing through them is affected by the strengths of the fields.

A current transducer of this type contains a piece of opto-magnetic material which is mounted near the conductor to be monitored and the material is thus in a magnetic field with a strength proportional to the conductor current. Light from a light-emitting diode is linearly polarized and then transmitted through the opto-magnetic material. On exiting from the material, the direction of polarization is shifted by an angle proportional to the conductor current. Measurement of this angle enables a signal proportional to the current to be generated and this may then be transmitted, via an optical fibre or otherwise, to supply continuous information about the current to measuring and/or protective equipment. More detailed information about such devices and their performance is provided in references [13–16] at the end of this chapter.

The above devices, which are clearly more complex than conventional current transformers, will not necessarily provide ideal performance at all times. As an example, variable current errors could be introduced because the behaviour of opto-magnetic materials is affected by ambient temperature variations. Nevertheless, these transducers appear to be suitable for use with modern protective schemes.

To date they have not been installed as replacements for existing current transformers because the latter are performing satisfactorily and the large capital costs involved would be too great to be justified. It does seem probable, however, that current transducers containing optical components will be used in future installations where advanced protective schemes, requiring accurate detailed information about the system currents, are to be used. Conventional current transformers will nevertheless continue to be used for the majority of protective applications and certainly this will be the case on systems operating at the lower voltage levels.

2.6 SPECIFICATIONS AND TESTING

Current transformers used with measuring instruments or accurate metering equipment must perform within given error bands during normal steady state conditions. The permissible current and phase errors for given classes of accuracy are specified in documents such as BS 3938. The classes of accuracy quoted in BS 3938 are 0.1, 0.2, 0.5, 1, 3, 5, these being the respective percentage current errors permitted at rated current. The corresponding phase-error limits are 5, 10, 30 and 60 minutes of angle for the first four classes, no phase-error values being specified for classes 3 and 5. The durations of faults and other abnormal conditions are usually so brief that their effects on performance are not considered to be significant and consequently no accuracy limits are quoted for very high current levels.

The situation is more complex for current transformers supplying protective equipment because satisfactory transformation must be achieved during fault conditions. For applications such as time-graded overcurrent protection a measure of core saturation may be accepted in the current transformers during transient conditions because the minimum operating times of these relays are achieved at ten times rated current and errors at currents somewhat below this level do not significantly affect the performance. For example, an error reducing the current by 10% at 10 times rated current increases the operating time by only 7%. Current transformers of classes 5P and 10P, which are defined in BS 3938, are therefore suitable in these situations. The P is included to denote protective applications and the 5P and 10P classes permit the exciting current to be 5% and 10% of the rated current respectively.

For other protective applications, such as high-speed differential schemes, current transformers must function to acceptable standards up to the highest current levels which may be encountered during fault conditions. BS 3938 includes a class X category for such transformers. It is stated in the specification that the requirements are so numerous and dependent on the protective gear involved that useful guidance cannot be provided. It is further stated that the transformers should be so designed that balance is maintained within the protective equipment supplied by them.

ESI Standard 35–15 [17] covers transformers for low voltage distribution systems up to 33 kV and a supplement to IEC Publication 185 (1966) [18] quotes additional requirements for current transformers associated with protection systems for which transient performance is significant.

Testing methods to be used by manufacturers, to ensure that their current transformers comply with the permissible accuracy limits, are outlined in specification documents such as BS 3938. However, for those transformers which are to be used with complex protective schemes, such as those applied to long high voltage transmission lines, conjunctive testing of the scheme and prototype transformers is usually performed on the manufacturers' premises to ensure that satisfactory performance will be obtained. Thereafter, testing must be done on production transformers to ensure that their characteristics are similar to those of the prototypes.

It is clear from the above that general guidance on the selection of current transformers to be used with protective schemes cannot be provided, but more specific guidance is provided, where possible, in the later chapters in which particular types of protective schemes are considered. In addition, references to works which may prove helpful are provided at the end of this chapter.

2.7 THE FUTURE

It is probable that conventional current transformers will, because of their simplicity and reliability, continue to be used to supply many protective equipments in the future and this will certainly be so for those applications associated with relatively low voltage networks.

It does seem probable, however, that non-conventional current transducers, such as those containing optical components, will be used in future installations where advanced protective schemes, requiring accurate and detailed information about the system currents, are to be used.

REFERENCES

1. Sharlin, H. I (1963) *The Making of the Electrical Age*, Abelard-Schuman, New York.
2. Parson, R. H. (1940) *Early Days of the Power Station Industry*, Cambridge University Press.
3. Andrews, L. (1898) The prevention of interruptions to electricity supply, *J. Inst Elec. Eng.*, **27**, 487–523.
4. Clothier, H. W. (1902) The construction of high-tension central station switchgears, with a comparison of British and foreign methods, *J. IEE*, p. 1247.
5. Edgcumbe, K. and Punga, F. (1904) Direct reading measuring instruments for switchboard use, *IEE*, **33**, 620–66.
6. Young, A. P. (1910) The theory and design of current transformers, *J. Inst Elec. Eng.*, **45**, 670–678.
7. Wright, A. (1968) *Current Transformers – their transient and steady state performance*, Chapman and Hall.
8. Gray, W. and Wright, A. (1953) Voltage transformers and current transformers associated with switchgear, *Proc. IEE*, **100**, 223.

9. Poljac, M. and Kolibas, N. (1988) Computation of current transformer transient performance, *Trans. IEEE,* **PD-3**, 1816–1822.
10. Mathews, P. and Nellist, B. D. (1962) The design of air-cored toroids or linear couplers, *Proc. IEE*, **109**, (A), 229.
11. Blatt, D. W. E. (1989) Monitoring and fault protection of high voltage switchgear by the REMMIT method, *Fourth Int Conf on Developments in Power System Protection*, 11–13 April 1989, Edinburgh, UK, *IEE Conf Publ.*, vol. 302, pp. 167–171.
12. Mercure, H. P. (1987) Development of a novel measuring device: the dynamic current transducer, *IEEE Trans. on Power Delivery*, PWRD-2, pp. 1003–1007.
13. Erickson, D. C. (1980) The use of fibre optics for communications, measurement and control within high voltage substations, *IEE Trans. on Power Apparatus and Systems*, PAS-99, pp. 1057–1065.
14. Kanoi, M., Takahashi, G., Sato, T., Higaki, M., Mori, E. and Okumura, K. (1986) Optical voltage and current measuring system for electric power systems, *IEE Trans. on Power Delivery*, PWRD-1, pp. 91–97.
15. Nojima, K., Nishiwaki, S., Okubo, H. and Yanabu, S. (1987) Measurement of surge current and voltage waveforms using optical-transmission techniques, *Proc. IEE* **134**, (C) 415–422.
16. Mitsui, T., Hosoe, K., Usami, H. and Miyamoto, S. (1987) Development of fibre-optic voltage sensors and magnetic field sensors, *IEEE Trans. on Power Delivery*, PWRD-2, pp. 87–93.
17. ESI Standard 35–15, April 1989, Protection and measurement transformers for H.V. distribution systems up to 33 kV, 58 pages.
18. Supplement to IEC Publication 185 (1966), Current transformers for protection systems for which transient performance is significant.

FURTHER READING

British Standard Specification for Current Transformers

BS 3938: 1973

IEC Publication

185 (1966), 'Current transformers'; IEEE Relaying Committee Report (1990)
Gapped core current transformer characteristics and performance *IEEE Trans. on Power Delivery*, **5**, (4), 1732–1740.

3

Voltage transformers

The term voltage transformer (commonly abbreviated to v.t.) is used when referring to a low output, step-down transformer with two or more windings per phase mounted over the limbs of a core of magnetic material.

Both single- and three-phase designs are produced. Clearly their primary windings must be rated at the voltage levels of the power circuits to which they are to be connected, whereas their secondary winding voltages are standardized in Britain at 63.5 V (phase) i.e. 110 V (line). In other countries, similar but slightly different values have been adopted, for example 120 V (line) and 100 V (line) are standards in the USA and many other countries.

The burden of a voltage transformer may consist of several parallel-connected items such as a voltmeter winding and the voltage windings of a wattmeter and directional and distance-type protective schemes. The total current to the burden does not normally exceed a few amperes, i.e. a burden of a few hundred volt-amperes at the most.

3.1 HISTORICAL BACKGROUND

It was stated in the previous chapter that the operating voltages of electrical power systems rose rapidly during the last two decades of the nineteenth century and that current and voltage transformers must have been introduced to feed measuring instruments in that period.

The production of voltage transformers presumably presented few problems because they would have been merely small versions of the power transformers then being produced. Understandably, therefore, there are few direct references to them in the literature produced around the turn of the century. However, in 1898, in a paper by Andrews [1], also referred to in the previous chapter, details were given of a discriminating cut-out which had been developed to detect reverse-current flow in machines. This device, which operated on the same principle as that used in later directional relays, needed both current and voltage inputs and from Fig. 3.1 (Fig. 1 in Andrew's paper) it can be seen that a voltage transformer was included in the circuit.

In the paper presented to the IEE in 1902 by Clothier [2], reference was made to the German practice of feeding voltmeters from transformers. Later,

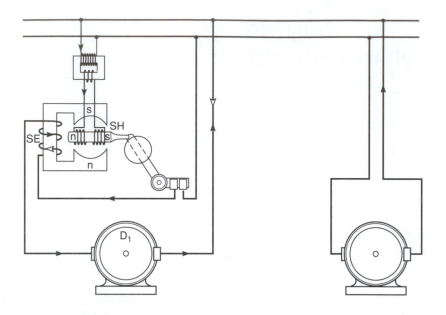

Fig. 3.1 Circuit incorporating a voltage transformer. (Reproduced from Andrews, 1898, *Journal of the IEE*, 27, with the permission of the IEE).

in 1904, a paper by Edgcumbe and Punga [3] contained a section dealing with both current and voltage transformers. It is therefore clear that voltage transformers were being used to supply measuring instruments and protective devices before that date.

Designs basically similar to those produced originally have been used in very large numbers up to the present time on circuits operating at voltages up to about 66 kV (line). For higher voltage applications, transformers incorporating capacitor dividers, known as capacitor voltage transformers, are used. The name capacitor voltage transformer was suggested by Wellings *et al.* [4] in a paper presented to the IEE in 1936.

Information about the behaviour, construction and use of both electromagnetic and capacitor voltage transformers is provided in the following sections of this chapter.

3.2 ELECTROMAGNETIC VOLTAGE TRANSFORMERS

Three-phase voltage transformers usually have both their primary and secondary windings connected in star and therefore their phases are effectively independent and their step-down ratios would ideally be the same as the ratios of the turns in their primary and secondary windings, i.e. $v_s/v_p = N_s/N_p$. The

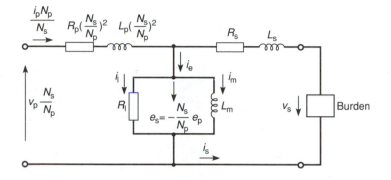

Fig. 3.2 Equivalent circuit of a voltage transformer.

behaviour of a single-phase voltage transformer or any phase of a three-phase transformer may thus be determined from the well-known equivalent circuit shown in Fig. 3.2.

3.2.1 Behaviour during steady state conditions

It is clear from the equivalent circuit that ideal transformation cannot be achieved in practice, because of the voltage drops which occur across the primary and secondary windings, i.e. $I_p Z_p$ and $I_s Z_s$.

To minimize the errors for a given secondary current, the winding impedances (Z_p and Z_s) and the exciting current (I_e) should be as low as possible. Economic and consequential dimensional constraints limit the levels to which the winding resistances and exciting currents may be reduced however and the leakage inductances (L_p and L_s) are dependent on the spacing needed between the windings for insulation.

The voltage error is defined as:

$$\text{Voltage error} = \frac{K_n \, |V_s| - V_p}{V_p} \quad \text{per unit}$$

in which K_n is the desired or rated transformation ratio. From the phasor diagram shown in Fig. 3.3 and assuming that $K_n = N_p/N_s$, the following approximate expression for voltage error can be derived:

Voltage error =

$$- \frac{I_s \, (R_s \cos \varphi_s + \omega L_s \sin \varphi_s) + I_p \dfrac{N_s}{N_p} [R_p \cos (\varphi_p - \alpha) + \omega L_p \sin (\varphi_p - \alpha)]}{V_p \dfrac{N_s}{N_p}}$$

per unit

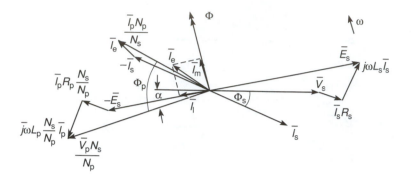

Fig. 3.3 Phasor diagram. Note: To increase clarity of the diagram, voltage drops, such as $I_s R_s$, are shown to be much greater than those which occur in practice.

The phase displacement error, shown in Fig. 3.3, is $-\alpha$ rad (V_s lags its ideal position) and for the small values normally encountered in practice it may be expressed as:

$$\alpha = \frac{I_s\,(\omega L_s \cos \varphi_s - R_s \sin \varphi_s) + I_p \dfrac{N_s}{N_p}\,[\omega L_p \cos \varphi_p - R_p \sin \varphi_p]}{V_p \dfrac{N_s}{N_p}} \quad \text{rad}$$

3.2.2 Behaviour during abnormal conditions

An extreme condition may arise on energizing a circuit to which a voltage transformer is connected. Should the switching occur at a voltage zero, the voltage impressed across the primary winding of the transformer would be:

$$v_p = V_{ppk} \sin \omega t$$

The core-flux variation required to induce the necessary e.m.f. in the primary winding (N_p turns), assuming the latter to have zero impedance, would be:

$$\Phi = \frac{1}{N_p} \int e_p \, dt = \frac{1}{N_p} \int -v_p \, dt$$

$$= \frac{V_{ppk}}{\omega N_p} \{\cos \omega t - k\}$$

in which the constant of integration (k) would be equal to unity if the initial core flux was zero.

This flux would then reach a peak value of $2V_{ppk}/\omega N_p$, i.e. twice the normal steady state peak excursion. This condition would usually cause the core to

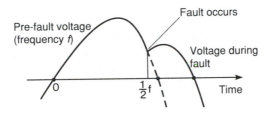

Fig. 3.4 Possible conditions caused by the incidence of a fault.

saturate and an accompanying exciting-current surge, with peak values up to several times the normal rated current of the primary winding, would flow. The related situation in current transformers was discussed in section 2.2.4. In the case of a voltage transformer, however, the surge of exciting current causes a relatively large voltage drop across the primary winding and, as a result, the secondary winding voltage is both distorted and has an r.m.s. value below the ideal value for several cycles. Such a situation could affect the behaviour of protective equipment.

Clearly, abnormal voltages are present in power circuits during fault conditions. Usually, however, the voltages are reduced below their normal levels at such times and, in these circumstances, saturation of voltage transformer cores will not tend to occur and accurate transformation should be achieved. It must nevertheless be recognized that the phase of the sinusoidal component of the voltage applied to a transformer may be changed by the presence of a fault as shown in Fig. 3.4. As a result, the period between two voltage zeros may be greater than normal and therefore a greater flux variation will be needed during this period. This situation is, however, not likely to be as severe as that considered above when a circuit is energized, and it can probably be disregarded.

3.2.3 Voltage transformer construction

Voltage transformers are basically very small versions of power transformers and therefore only those factors and features which are special to voltage transformers are discussed below.

Connections

As stated earlier, three-phase voltage transformers usually have star-connected primary and secondary windings. They may also have tertiary windings connected in star or broken-delta. Single-phase transformers are also produced and in many cases they are used on three-phase systems with the above connections. In some countries an alternative arrangement that is popular is to use two such transformers with their primary windings connected between pairs of phases so that they are each energized at a line voltage. Their secondary windings are connected in open delta and thus outputs proportional to the three-line voltages are provided. The phase voltages cannot however be determined from these outputs unless it is known that they do not have zero-sequence components.

Magnetic cores

Cores constructed from laminations of silicon-iron alloy are produced in both single-phase and three-phase forms.

While three-phase, three-limb cores would allow satisfactory operation to be obtained when the three-system phase voltages add to zero at every instant, they would not provide acceptable transformation in the event of faults to earth on the system. In these circumstances, the phase voltages do not add to zero, i.e. a zero-sequence component is present, and in consequence the fluxes in the three limbs do not add to zero. An extra flux path must therefore be provided. Often, to provide symmetry, two extra limbs of small cross-sectional area are added, the five-limb configuration commonly adopted being shown in Fig. 3.5(a)

Single-phase transformers use either the well known, core- or shell-type cores shown in Fig. 3.5(b).

Primary windings

Each primary winding is normally energized at the rated voltage of the circuit to which it is connected, i.e.

$$v_p = V_{ppk} \sin (\omega t + \alpha)$$

Neglecting the voltage drop in the winding, there must therefore be an e.m.f. induced in the winding of $e_p = - v_p$ and this must be produced by a core-flux variation of:

$$\Phi = - \frac{1}{N_p} \int e_p \, dt = \frac{1}{N_p} \int v_p \, dt$$

$$= - \frac{V_{ppk}}{\omega N_p} \cos (\omega t + \alpha)$$

The peak flux density should not exceed the saturation density of the core (B_{sat}) and therefore:

$$B_{sat} \geqslant \frac{V_{ppk}}{\omega N_p A_c}$$

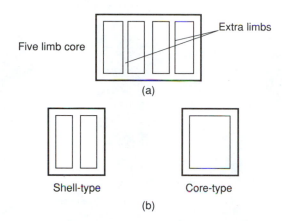

(a)

Five limb core

Extra limbs

Shell-type

Core-type

(b)

Fig. 3.5 Core configurations.

in which A_c is the cross-sectional area of the core.

In practice, transformers, depending on their duty, are required to be able to operate without saturating at voltages above the rated value of the circuits to which they are to be connected. Standards include a voltage factor (V_f) which may range in value from 1.2 to 1.9.

Taking this into account:

$$B_{sat} \geq \frac{V_f V_{ppk}}{\omega N_p A_c} \tag{3.1}$$

in which A_c is the cross-sectional area of the core.

Considering, as an example, a transformer with a core having a saturation flux density of 1.5 T and limbs of 25 cm² cross-sectional area, it can be found from equation (3.1) that its primary windings would require to have at least 55 000 turns if it were for use on a 66 kV (line), 50 Hz system and to have a voltage factor of 1.2. For an output of 200 VA/phase, the primary winding current would be less than 10 mA. It will be appreciated that the production of a winding with such a large number of turns of very fine wire is very difficult and expensive especially when it is borne in mind that it has to be insulated to withstand a high voltage. It is for this reason that electromagnetic voltage transformers are not produced for use on the very high voltage systems in use today.

Secondary windings

These windings present no difficulties because they have relatively few turns of conductor capable of carrying a few amperes.

Fig. 3.6 Typical oil-filled voltage transformer. (Reproduced from *Low oil volume voltage-transformers*, GEC High Voltage Switchgear Limited with the permission of Instrument Transformers Limited).

Tertiary windings

These may be provided and connected in open- or broken-delta to give an output under earth fault conditions which can be fed to protective equipment. This output is proportional to the sum of the three zero-sequence voltages present on the system.

Housing and insulation

Voltage transformers are often mounted in metal tanks, the cores and windings being immersed in oil, in a similar way to power transformers, a typical example being shown in Fig. 3.6. In recent years, however, resin-insulated designs have been produced for use on lower voltage circuits.

3.3 CAPACITOR-VOLTAGE TRANSFORMERS

As stated earlier these transformers, which incorporate capacitor dividers, were introduced for use on very high voltage circuits, when it was recognized that suitable electromagnetic transformers could not be produced for such applications.

3.3.1 Capacitor dividers

Capacitor dividers have the attraction of being loss-free, i.e. they do not consume power. When they are unloaded, i.e. $i_{out} = 0$, the output voltage (v_{out}), as can be seen from Fig. 3.7(a), is given by:

$$v_{out} = \frac{C_1}{C_1 + C_2} \cdot v_p \qquad (3.2)$$

Using Thevenim's Theorem, a divider supplying a burden may be represented by the circuit shown in Fig. 3.7(b) and this makes it clear that the output or burden voltage (v_{out}) is affected by the current in the burden (i_{out}) because of the voltage drop across the capacitor ($C_1 + C_2$). This drop, for a given ratio of the capacitances C_1 and C_2, reduces as both capacitances are increased in the same proportion and therefore the capacitances should be as large as is practicable.

It can be seen from equation (3.2) that a reduction of the capacitance C_2 for a fixed value of the capacitance C_1 increases the output voltage (v_{out}) and thus for a given voltage-ampere output, the output current (i_{out}) reduces. As a result the percentage effect on the output voltage reduces.

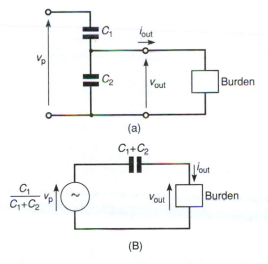

Fig. 3.7 Capacitor divider and its equivalent circuit.

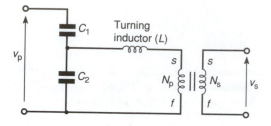

Fig. 3.8 Circuit of a capacitor-voltage transformer.

3.3.2 · Circuit of a capacitor-voltage transformer

Because of the above considerations, capacitor-voltage transformers incorporate dividers in which the capacitor connected to the system (C_1) is made as large as possible, usually 2000 pF, and the lower capacitor (C_2) has a value which will provide a voltage (v_{out}) which can be further transformed down by a conventional voltage transformer. The r.m.s. value of the voltage (V_{out}) which is termed the intermediate voltage, may be up to about 30 kV (phase).

To further reduce the errors caused by the presence of the effective capacitance ($C_1 + C_2$), shown in Fig. 3.7(b), an inductor (L) is connected in series with the primary winding of the step-down transformer. This is shown in Fig. 3.8.

3.3.3 Steady state behaviour

An equivalent circuit of a capacitor-voltage transformer is shown in Fig. 3.9.

The sum of the reactance of the inductor (L) and the leakage reactance of the primary windings and the leakage reactance of the secondary windings of

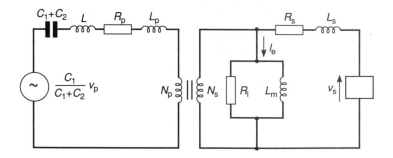

Fig. 3.9 Equivalent circuit of a capacitor-voltage transformer.

the step-down transformer, referred to the primary level is made equal to the sum of reactances of the capacitors C_1 and C_2 at the normal power-supply frequency, i.e. $\omega\{L + L_p + L_s (N_p/N_s)^2\} = 1/\omega(C_1 + C_2)$. The transformation errors, which can be calculated for any given condition using the equivalent circuit, are dependent on the system frequency and they should be determined not only at the nominal frequency but at the lower and higher limiting values likely to be encountered in service.

3.3.4 Behaviour during abnormal conditions

When a fault occurs on a power system, the voltages at different points on it may fall suddenly and then they may suddenly increase when the fault is cleared. Such a condition is illustrated in trace (i) of Fig. 3.10(a) which was

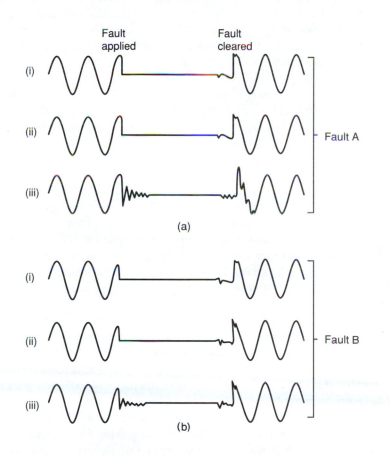

Fig. 3.10 Output voltage waveforms from voltage transformers under fault conditions. (Reproduced from Gray and Wright, 1953, *Proc. IEE*, 100 with the permission of the IEE).

obtained experimentally using an unloaded resistor divider. The outputs of an electromagnetic voltage transformer and a capacitor voltage transformer obtained at the same time are shown in traces (ii) and (iii) respectively in Fig. 3.10(a). In both cases the transformers were unloaded. A further set of traces obtained with the rated burdens connected to the transformers are shown in Fig. 3.10(b). These results clearly indicate that the response of the electromagnetic transformer is close to the ideal, whereas the capacitor voltage transformer generates undesirable damped oscillations when the sudden changes of system voltage occur. These oscillations, which arise because of the tuned nature of the transformer circuit, together with sustained oscillations which can arise because of ferro-resonance, can affect the behaviour of protective equipment. To counteract these undesirable transient errors, Harder [5] developed compensating circuitry which he patented in the United States in 1950. Subsequently, Hughes [6] studied the behaviour of distance-relaying equipment supplied from basic capacitor-voltage transformers and from those fitted with compensating equipment. All capacitor-voltage transformers are regarded, however, as having a narrow bandwidth and an alternative developed to provide a wide bandwidth was the cascade voltage transformer which was made up of several individual electromagnetic transformers, the primary windings of which were connected in series. In this way, the primary voltage was broken down into sections, each of the transformers having sufficiently low voltages across them to enable satisfactory primary windings to be produced.

3.4 RECENT DEVELOPMENTS

The acceptance in recent years of the inclusion of electronic equipment in protective schemes has opened up the possibility of using devices such as amplifiers in voltage transducers. The incorporation of an amplifier to provide the volt-amperes required by a burden enables a capacitor divider to be effectively unloaded at all times and thus to provide a constant step-down ratio. In these circumstances, the tuning inductor (L) used in a conventional capacitor voltage transformer is not necessary. Such designs have been in service since 1975.

Stalewski and Weller [7] presented a paper to the IEE in 1979 in which they described a capacitor-divider voltage sensor which incorporated both a pre-amplifier and a main amplifier, the basic circuit being shown in Fig. 3.11. Experience gained during high voltage system trials of this sensor extending over the five years prior to the publication of the paper showed that it provided faithful reproduction of the primary voltage during transient conditions. During this trial period, 27 system faults occurred in the vicinity of the substation where the prototype unit was installed, including three faults actually on the 400 kV line to which it was connected. Very satisfactory performance was obtained and further improvements were effected as a result of the information obtained. Sensors of this type, which must of course have a reliable power

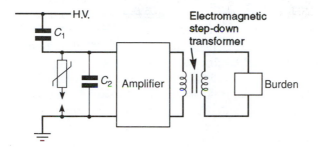

Fig. 3.11 Capacitor-divider voltage sensor.

source to supply the amplifiers, should be capable of supplying outputs of the accuracies needed by the latest protective schemes.

An alternative approach, using electro-optic materials, has been examined in recent years. These materials, which are similar to those referred to earlier in section 2.5, change the polarization of light passing through them when they are placed in electric fields. A voltage transducer may therefore be produced by placing a piece of electro-optic material near a conductor so that it is subject to an electric field proportional to the potential of the conductor. As in the current transducer, a polarized light beam is passed through the electro-optic material and the changes in polarization, which are proportional to the conductor potential, are measured to provide a signal which may be used to supply measuring or protective equipment via optical fibres. It has been shown that these transducers are capable of faithfully reproducing the primary voltages applied to them under steady state and transient conditions. Further details of this development are provided in publications by Kanoi *et al* [8], Najima *et al* [9] and Mitsui *et al*. [10].

3.5 SPECIFICATIONS AND TESTING

British Standard 3941, published in 1975, specifies the requirements for both electromagnetic- and capacitor-voltage transformers intended for use with electrical measuring instruments and electrical protective equipment. It is generally in line with the corresponding IEC Recommendations given in Publications 186, 186A and 358.

The accuracy classes for measurement purposes are 0.1, 0.2, 0.5, 1.0 and 3.0 and for protective applications they are 3P and 6P. The permissible steady state voltage and phase displacement errors for each of the above classes are given in the Table 3.1.

These limits must be complied with at any voltage between 80% and 120% of the rated value and with burdens of between 25% and 100% of the rated burden, at a power factor of 0.8 lagging.

Table 3.1

Accuracy class	Voltage errror (%)	Phase displacement error (minutes)
0.1	±0.1	±5
0.2	±0.2	±10
0.5	±0.5	±20
1.0	±1.0	±40
3.0	±3.0	Not specified
3P	±3.0	±120
6P	±6.0	±240

In addition to having to comply with the above steady state limits, capacitor voltage transformers also have to meet conditions associated with their transient response and effects produced by ferro-resonance. They are required, one cycle after a short circuit is applied across their primary terminals, to have a secondary output voltage of less than 10% of the peak value which existed before the short circuit. Also with their secondary terminals on short circuit and zero volt-ampere burden, ferro-resonance should not be sustained for more than 2 seconds after the short circuit is removed, and when 120% of rated voltage is applied the secondary voltage should revert to a value which does not differ from its normal value by more than 10% within 10 cycles after a short-circuit across the secondary winding is removed. Methods of conducting type tests, commissioning tests and routine tests are described in specification documents.

3.6 THE FUTURE

Conventional voltage transformers will continue to be used to supply many protective equipments in the future and certainly this will be so for applications associated with relatively low voltage power networks.

The needs of protective equipments, which are to be applied to transmission lines operating at extremely high voltages, for very accurate and detailed information about system voltages will almost certainly justify the use of non-conventional transducers operating in ways such as those described in section 3.4.

REFERENCES

1. Andrews, L. (1898) The prevention of interruption to electricity supply, *J. Inst. Elec Eng.*, **27**, 487–523.
2. Clothier, H. W. (1902) The construction of high-tension central station switchgears, with a comparison of British and foreign methods, *J. IEE*, p. 1247.

3. Edgcumbe, K. and Punga, F. (1904) Direct reading measuring instruments for switchboard use, *IEE*, **33**, 620–66.
4. Wellings, J. G., Mortlock, J. R. and Matthews, P. (1936) Capacitor voltage transformers, *J. IEE*, **79**, 577.
5. Harder, E. L. (1950) Transient compensation of potential-device burdens, *US Patent 2 510 631*.
6. Hughes, M. A. (1974) Distance relay performance as affected by capacitor voltage transformers, *Proc. IEE*, **121**, 1557–1566.
7. Stalewski, A. and Weller, G. C. (1979), Novel capacitor-divider voltage sensors for high voltage transmission systems, *Proc. IEE*, **126**, 1186–1195.
8. Kanoi, M. Takahashi, G., Sato, T., Higaki, M., Mori, E. and Okumura, K. (1986) Optical voltage and current measuring system for electric power systems, *Trans. IEEE*, **PWRD-1**, 91–97.
9. Najima, K., Nishiwaki, S., Okuho, H., Yanabu, S. (1987) Measurement of surge current and voltage waveforms using optical-transmission techniques, *Proc. IEE*, **134**, 415–422.
10. Mitsui, T., Hosoe, K., Usami, H., Myamoto, S. (1987), Development of fibre-optic voltage sensors and magnetic field sensors, *Trans. IEEE*, **PWRD-2**, 87–93.

FURTHER READING

British Standard Specification for Voltage Transformers

BS 3941: 1975

IEC Publication

186 (1987), *Voltage Transformers*.

British Electricity Supply Standard

35–15 (1989) *Protection and Measurement Transformers for HV Distribution Systems up to 33 kV* (58 pages).

4

Overcurrent and earth fault protection

Fuses are a very satisfactory form of protection for the lower voltage and current sections of power networks because their operating time/current characteristics are similar in form to the withstand time/current characteristics of the circuits they protect. Their use enables circuits to be kept in service until times when faulted or overloaded parts of a network must be disconnected to ensure that healthy equipment does not suffer consequential damage.

Fuses are not capable, however, of interrupting large currents in high voltage circuits and they have the disadvantage that they cannot be tested regularly after installation, a facility which is usually required in protective equipment associated with major plant. As a result, relays with inverse operating time/current characteristics, similar to those of fuses, were developed to provide overcurrent and earth fault protection. These relays, which are supplied by current transformers, initiate the opening of circuit-breakers, when necessary, by completing trip-coil circuits. They may therefore be used to protect circuits operating at voltages up to the highest levels. Over the years such relays have been installed around the world in very large numbers to provide either the main protection of circuits or to give secondary or back-up protection to other more complex protective schemes.

The current and time settings of these relays are adjustable to allow them to be applied in a similar way to fuses so that correct discrimination may be achieved during fault or overload conditions. In some applications, however, the use of time grading alone may not be sufficient to ensure correct operation under all possible system conditions, and to improve protective performance in such circumstances relays monitoring the direction of current flow, i.e. directional relays, are used in conjunction with overcurrent and earth fault relays.

Following the next section (4.1), which traces early relay developments, details are provided of modern relays used in time-graded protective schemes together with methods of applying them.

4.1 HISTORICAL BACKGROUND

Protective relays were introduced quite widely into power systems during the first two decades of this century and there are many references to them in the literature of that period.

Reverse-current relays, i.e. directional relays, were quite common. A reference was made in Chapter 1 to a design described in a paper [1] presented to the Institution of Electrical Engineers (IEE) by Andrews in 1898. An alternative design was described in the *Electrical Review* in June 1904 [2], the device then referred to being shown in Fig. 4.1.

An overcurrent inverse-time limit relay was designed in 1902 by C. E. L. Brown, one of the founders of Brown Boveri Ltd. He took out patents on the relay in several countries including Britain (Patent No 15500–03), these probably being the first world-wide patents on an a.c. relay. His original relay had an aluminium disc which rotated in the horizontal plane, the drive being provided by a shaded-pole electromagnet. It employed eddy-current braking and was basically similar to modern designs. This relay is shown in Fig. 4.2. Apparently it overran significantly if the current supplied to it was interrupted before the disc had completed its travel. Its behaviour in this respect was improved by modifications introduced in 1905. A relay of this improved type was described in an article by Frey in the November 1924 issue of the Brown Boveri Review [3]. It was claimed that experience with these relays over the previous 20 years had overcome the prejudice which had existed against somewhat indefinite time settings.

In spite of the introduction of induction relays several other types of current-operated relays and associated time-delay devices were in use in the first decade of this century, one being described in reference [2]. It was stated that

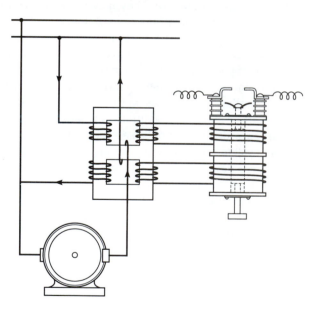

Fig. 4.1 An early reverse-current relay.

Fig. 4.2 An early over-current relay type A2 produced by Brown Boveri and Co Ltd. (Reproduced with the permission of ABB Relays AG, Baden, Switzerland).

this relay had a solenoid plunger fitted with a miniature diving bell with two superposed compartments, the lower one being bottomless. Small holes were pierced in the top and middle partitions. The upper chamber dipped into oil and the lower one into mercury. This mechanism delayed the lifting of the plunger.

Relays with both definite time lags and with inverse time/current characteristics were referred to in an article entitled 'Protective features of high-tension system' [4] which appeared in *Electrical World* in 1909. This article showed time/current characteristics which could be obtained using relays which incorporated bellows to provide the time delays, one such relay being shown in Fig. 4.3.

The use of eddy-current braking was referred to by Murphy during the discussion on a paper on protective equipment presented to the IEE by Wedmore [5] in 1915.

Reference was made to induction-type relays by Edgcumbe in a paper [6] presented to the IEE in 1920. He stated that such relays are to be preferred to fuse-shunted solenoids and went on to say that a fuse is 'at best' a capricious piece of apparatus.

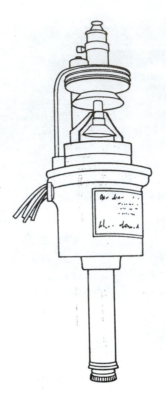

Fig. 4.3 Bellows-type overload relay.

Subsequent relay developments and their applications are described in the following sections.

4.2 RELAY CONNECTIONS AND OPERATION

The relaying arrangements needed to detect earth faults and overcurrent conditions, which may occur on three-phase networks are considered below.

4.2.1 The detection of earth faults

Because of the impedance which may be present in the ground-return path of a circuit, earth fault currents may be quite low. It is nevertheless desirable that they be detected and cleared by the opening of the appropriate circuit-breakers. Earth fault relays should therefore have settings which will enable fault currents well below the rated currents of the circuits they protect to be detected.

It is well known [7] that a single-phase, earth fault current may be represented by the set of symmetrical-component phasors shown in Fig. 4.4 and that its instantaneous value may be expressed as:

$$i_f = 3\,i_0 = i_a + i_b + i_c$$

It is therefore clear that an earth fault on any phase of a circuit may be detected by feeding a single relay with a current proportional to the sum of the three phase currents. This arrangement is illustrated in Fig. 4.5(a). Because a relay connected in this way is not affected by normal balanced currents, it may have a low current setting corresponding to 20% or less of the rated current of the protected circuit.

4.2.2 The detection of overcurrents

As with fuses, relays provided to detect overcurrents must have minimum operating currents above those corresponding to the rated currents of the circuits they protect.

When an earth fault relay is provided to protect a circuit, the associated overcurrent relays only need to operate in the event of three-phase faults or interphase faults (e.g. phase a to phase b). This may clearly be achieved by installing two overcurrent relays supplied with currents proportional to the

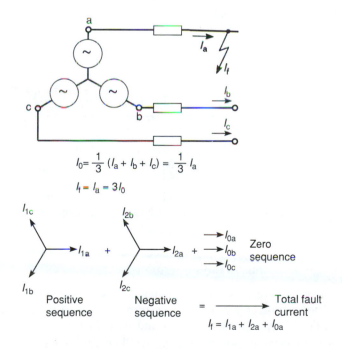

Fig. 4.4 Symmetrical components for an earth fault on phase 'a'.

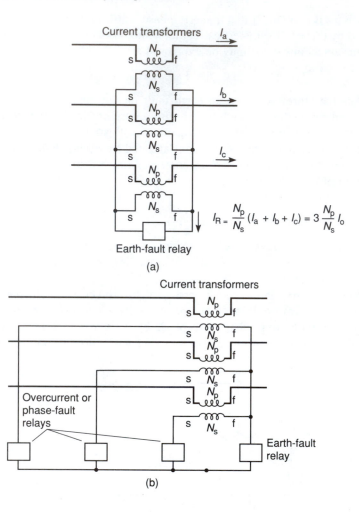

Fig. 4.5 Current transformers and relay connections.

currents in two of the phases, say a and c. This arrangement, which is shown in Fig. 4.5(b), is commonly used although other arrangements are sometimes employed.

4.2.3 Electro-mechanical relays

All the early relays were electro-mechanical, movement of armatures or discs being used to operate contacts, which in turn caused the tripping of circuit-breakers. Several types were produced and used in large quantities. Where necessary, designs were modified to improve their performances and simplify their production and this has enabled them to meet protective requirements up to the present time. In recent years, however, the use of electronic control and protective equipment has been increasingly accepted in power systems, and manufacturers now produce electronic relays which perform similarly to their electro-mechanical equivalents.

Because protective equipment practice was established when only electro-mechanical relays were available, the following sections are devoted to the behaviour of these relays and then, in section 4.3, equivalent modern electronic relays are described.

Three different basic types of relays have been produced to provide over-current and earth fault protection, namely those which operate instantaneously when the currents in them exceed set values, those which operate after a definite or fixed time delay, and those which have inverse time/current characteristics.

Instantaneous relays

These relays are relatively simple, employing, for example, a hinged armature which is attracted to the pole of an electromagnet when the current in the operating coil exceeds a certain value. The armature is coupled, via a linkage, to the contacts. The attractive force on the armature is dependent on the current and therefore the time taken for the closure of the air gap does vary with current, but it is so short, a few milliseconds, that it is regarded as instantaneous. The operating coil is often tapped, connections being brought out to a plug board so that a range of current settings is available. As an alternative, the air gaps may be varied by moving the armature back stop to provide the desired current setting.

Typical relays of this type are shown in Fig. 4.6(a-c).

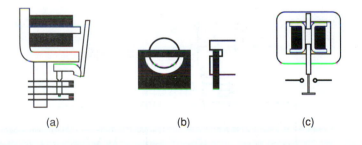

(a) (b) (c)

Fig. 4.6 Attracted-armature relays. (a) D.C. relay; (b) Shaded loop modification to pole of relay (a) for a.c. operation; (c) Solenoid relay. (Reproduced from *Protective Relays – Application Guide, 3rd edn,* GEC Measurements, 1987 with the permission of GEC Alsthom Protection and Control Ltd).

Relays with definite time-lag operation

Various forms of relay have been produced for this purpose. They clearly must monitor the current supplied by their associated current transformers and trigger a timing system whenever the current exceeds the set value. A relay of the type described in the previous section could be used, closure of its contacts causing the energization of a timing device which would in turn initiate the opening of an appropriate circuit-breaker after a fixed (definite) time.

In such an arrangement the timing device should instantly return to its initial position if the current-sensing relay resets as a result of the circuit current falling to an acceptable level before operation is completed.

To enable such relays to be applied to systems, both the current and time settings must be adjustable.

Timing has been achieved in various ways including dashpots, mechanical escapements and eddy-current braked discs.

Relays with inverse time/current characteristics (IDMT)

There are several ways in which inverse time/current characteristics could be produced and indeed, as stated earlier, relays with bellows, dashpots and thermal devices have been used. Induction-type relays based on watt-hour meters or with shaded-pole electromagnets were introduced by several manufacturers in the early years of this century. They were soon in use in vast numbers around the world and both national and international time/current standards were based on their performance.

Movements based on those of watt-hour meters have two electromagnets of the form shown in Fig. 4.7. The main electromagnet (1) has two windings, one of which is fed from a current transformer in the protected circuit whilst the other acts as a secondary which feeds the winding on the second electromagnet

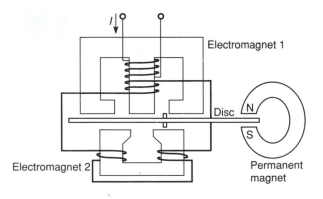

Fig. 4.7 IDMT relay movement with two electromagnets.

(2). The first electromagnet provides a magnetic flux (φ_1) and because of it an e.m.f. is induced in its second winding, the phase of this e.m.f. being displaced from that of the current in the first winding. As a result, the current driven through the winding on the second electromagnet is not of the same phase as that of the current supplied to the first electromagnet and therefore the phase of the flux (φ_2) produced by the second electromagnet differs from that of the flux φ_1.

The two fluxes both cut the aluminium or copper disc, which is mounted in jewelled bearings, and thus a torque, dependent on the input current, is produced on it.

The disc is restrained against a movable back stop by either a hair spring or, in designs in which the disc rotates in the vertical plane, by a weight mounted on an arm driven by the disc. The disc therefore only rotates when the input current exceeds a certain value, its speed of rotation being dependent on the input current and the eddy-current braking produced by a permanent magnet. As the disc rotates it drives moving contacts, either directly or via gearing, until the contacts close to energize the trip coil of the associated circuit-breaker.

Because of the air gaps associated with the electromagnets, the two fluxes (φ_1 and φ_2) cutting the disc tend to increase linearly with the input current until a level is reached at which saturation of the core materials begins. Up to this current level the driving torque on the disc increases in proportion to the square of the input current but at higher currents the torque tends to become constant. The opposing torque provided by the hair spring or weight varies with the distance travelled by the disc, having had an initial value which prevents movement below the minimum operating current. The eddy-current braking, however, is dependent on the speed of the disc. The behaviour of such relays is therefore quite complex and their time/current characteristics were expressed either in graphical form or by quoting the operating times at a number of current levels, as shown in Table 4.1.

The characteristic shown in Table 4.1, which was referred to as 3/10, i.e. the operating time is 3 s at 10 times rated current, was widely used. Other characteristics, such as 40/20 and 30/5, were also produced.

Relays with these time/current relationships were described as 'inverse definite minimum-time relays' (abbreviated to IDMT relays). The definite minimum operating time obtained at currents above a certain level is an essential feature of these relays as adequate discriminating time margins could not otherwise be obtained between relays applied to adjacent sections of networks during short-circuit conditions.

Table 4.1

Multiple of current setting	1.3	2	5	10	20 or more
time (s)	∞	10	4.3	3.0	2.2

To allow flexibility in the use of these relays, tappings were provided on the main input windings, either a current value or percentage being assigned to each tapping. Overcurrent relays for use with current transformers having secondary windings rated at 5 A were provided with tappings marked either 50% to 200% in 25% steps or 2.5 A to 10 A in 1.25 A steps. Earth fault relay tappings were marked 20% to 80% in 20% steps (1, 2, 3, 4 A) or 10% to 40% in 10% steps (0.5, 1, 1.5, 2 A). In addition, to allow time gradings to be achieved, the operating times of a relay could be adjusted using a control knob labelled 'time multiplier'. Rotation of this knob moved the back stop of the disc and varied the travel needed to cause the contacts to close. Calibration was provided in decimal form up to a maximum of 1.0 (full travel). The times quoted in Table 4.1 were for a time multiplier setting of 1.0 and therefore the times quoted would be halved if a time multiplier setting of 0.5 was used. These practices are still in use today.

As an illustration, an overcurrent relay for use with a current transformer having a ratio of 400/5 could be used to protect a circuit, required to carry a maximum current of 300 A initially, but suitable for eventual uprating to 500 A. The relay would therefore be given a 75% current setting, corresponding to 300 A, on installation, but this would be changed to 125% (500 A) later when the load developed. These settings would be acceptable because operation would not occur below about 1.3 times the settings, i.e. 390 A initially and 650 A later.

To enable time grading to be achieved, a time multiplier setting of 0.5 might be used. In these circumstances, and with a 75% current setting in the above application, operation would occur after 1.5 s for a fault current of 3000 A, i.e. 10 times the current setting.

Relays with the same ranges of current and time settings were also produced for use with current transformers having secondary windings rated at 1 A.

The alternative design of IDMT induction relays has a single shaded-pole electromagnet of the basic form shown in Fig. 4.8. It has the same time/current characteristic and time and current setting controls as the two-magnet design

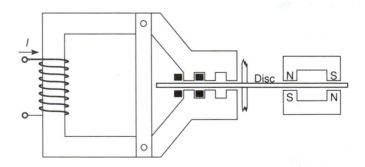

Fig. 4.8 IDMT relay movement with a shaded-pole electromagnet.

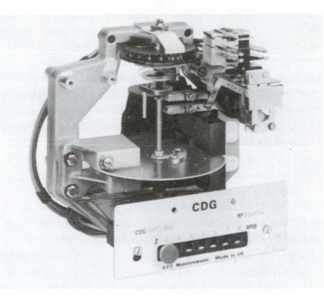

Fig. 4.9 Modern IDMT relay. (Reproduced from *Protective Relays – Application Guide, 3rd edn*, GEC Measurements, 1987 with the permission of GEC Alsthom Protection and Control Ltd).

described above. Both types of relay have been widely used up to the present day and they are interchangeable.

Since the early induction relays were produced, attempts have been made to modify their designs, one aim being to increase the driving power obtained for a given volt-ampere input. Such work has not been very successful however, because drives with greater efficiency have not allowed the earlier time/current characteristics, on which standards were based, to be obtained. Some improvements have however been made to simplify production, a modern element being shown in Fig. 4.9.

A factor of significance associated with induction-type relays is that their discs and other moving parts acquire kinetic energy during operation and in consequence they do not come to rest or reset instantly in the event of their input currents reducing to rated or lower before contact closure occurs. A relay could therefore overshoot and cause undesired tripping of its circuit-breaker when an adjacent relay operates correctly to initiate the removal of a fault. To ensure that this situation will not arise, allowance must be made for this effect when the time-grading intervals between relays are being selected.

4.2.4 Directional relays

It was recognized at the beginning of this century that protective devices whose behaviour depended only on circuit currents would not provide the necessary discrimination during some operating conditions. Polarized relays, i.e. relays with directional properties, based on wattmeter movements were therefore developed. These incorporated a disc or vane and had two electro-magnets, one having a winding fed from a current transformer and the other having a winding supplied by a voltage transformer. Under steady state sinu-soidal conditions, the flux (φ_c) provided by the electromagnet fed with the circuit current would ideally be in phase with the circuit current whereas the flux (φ_v) in the other electromagnet would ideally be in quadrature with the voltage applied to its winding. For given values of voltage and current, maxi-mum torque would be exerted on the disc in one direction when the input voltage and current were in phase and maximum torque would be exerted in the opposite direction when the voltage and current were in antiphase.

The torque of such relays is therefore proportional to power or VI cos φ, as would be expected of a wattmeter movement. A relay of this type could thus be made to close its contacts when the phase displacement between the voltage and current was less than $\pi/2$ radians, whereas for phase displacements be-tween $\pi/2$ and π radians the disc or vane would restrain against its back stop.

A relay with such a characteristic might not discriminate satisfactorily, however, because fault currents usually lag their respective voltages by quite a large angle. As an example, if a relay associated with phase 'a' were supplied with the 'a' phase voltage and current, these two quantities could be almost $\pi/2$ radians apart during an earth fault on that phase. As a result the torques on the disc would be very small for faults on either side of the relaying position. This difficulty can be overcome by modifying relay characteristics so that they produce maximum torque when the currents fed to them lag their respective voltages by an angle similar to that which is likely to be present during system faults, say $\pi/3$ radians. This behaviour, illustrated in Fig. 4.10, can be obtained by incorporating a phase-shift network between the voltage

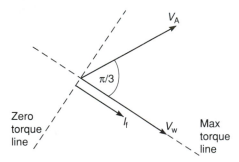

Fig. 4.10 Voltages associated with a directional relay. (V_A is the voltage provided by the *VT* and V_W is the voltage applied to the relay winding).

terminals of a relay and its voltage winding. Clearly, maximum torque can be produced in this way at any desired angle between the input voltage and current.

A further important factor which must be recognized is that the torque produced on a relay disc is proportional to the magnitude of the input voltage. As a consequence, very small torques are likely to be produced for faults near a relaying position, the voltage of a faulted phase being very depressed under such conditions.

This situation would occur for faults in either direction from the relaying position and correct discrimination might not be achieved. To minimize the zone in which a relay may not correctly detect faults, directional relays capable of operating satisfactorily at voltage levels down to 3% or less of the normal level have been produced and indeed such requirements are included in national and international standards. A practice which can improve performance in this respect is to connect the current and voltage windings of individual directional relays to different phases. As an example, a relay may have its current winding energized from phase 'a' but its voltage winding supplied with a voltage proportional to that between phases 'b' and 'c'. With such connections, the relay voltage would not be depressed in the event of an earth fault on phase 'a'. Account must clearly be taken of such connections when setting the phase angle between the voltage and current at which a relay is to produce maximum torque.

Modern induction-type directional relay elements are generally similar to those produced early this century, an example being shown in Fig. 4.11.

Fig. 4.11 A modern directional element. (Reproduced from *Overcurrent and Earth Fault IDMTL Relays*, 1989, with the permission of Reyrolle Protection).

4.2.5 Thermal inverse time/current relays

These relays, which have been produced for many years, operate as a result of the heating of elements. In this sense they are similar to fuselinks. Each relay contains an element which is fed from a current transformer in the circuit being protected, the input power therefore being proportional to the square of the current. The heat produced is transmitted to a bimetal strip, fixed at one end. The strip bends as it heats and after a certain deflection has been produced it causes contacts to close and thus energizes the trip coil of its associated circuit-breaker. The time taken for operation to occur at a particular current level can be adjusted by varying the deflection needed to cause contact closure.

It will be appreciated that the bimetal strip deflects whenever the heater carries current and therefore the deflection associated with the rated current of the circuit being protected must not cause contact closure.

Because of this factor, the operating time of a thermal relay, unlike that of an induction-type relay, depends not only on the overcurrent flowing but also on the current which was flowing earlier. In some applications, this behaviour is desirable because it matches the behaviour of the protected circuits, namely that the periods for which they can tolerate overcurrents is dependent on the currents they have been carrying earlier.

The deflections of the bimetal strips in these relays are affected by the ambient temperature and this could affect the time/current behaviour obtained. To eliminate this effect, many designs incorporate pairs of bimetal strips, only

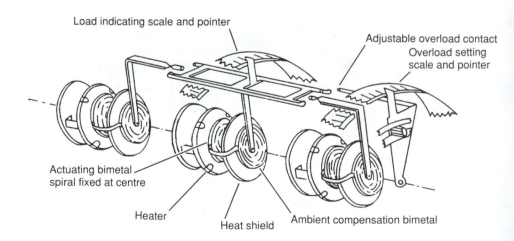

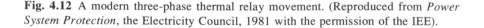

Fig. 4.12 A modern three-phase thermal relay movement. (Reproduced from *Power System Protection*, the Electricity Council, 1981 with the permission of the IEE).

one of the pair having a current-fed heater adjacent to it, the differential movement between the strips being used to control the operating time.

As with other relays, it is necessary that a range of current settings should be available. This can be achieved by including a tapped interposing current transformer to feed the heater.

A typical modern thermal relay is shown in Fig. 4.12 in which the bimetal strips are of spiral form.

A limitation of relays of this class is that they do not reset instantly when the currents supplied to them fall to the rated value or less after an overcurrent has been flowing. They are not therefore very suitable for inclusion in time-grading schemes unless they are associated only with load circuits. They are, however, particularly suitable for the protection of small motors because they can be set so that they do not operate during current surges caused by starting, but nevertheless they will operate when faults or sustained overcurrents occur.

4.3 ELECTRONIC RELAYS

Manufacturers have attempted to produce static relays to replace electro-mechanical relays for many years. Although electronic devices such as thermionic valves were readily available from the 1930s, they were not considered reliable enough to be used in protective equipment. Alternatives such as transductors were introduced in the 1950s, a paper on this development being presented to the IEE by Edgeley and Hamilton in 1952 [8]. Subsequently relays incorporating silicon planar transistors were developed.

The later rapid progress in electronic equipment including the introduction of reliable integrated circuits, logic gates and microprocessors made it possible to produce equivalents of overcurrent and earth fault relays and also directional relays using both analogue and digital processing.

Because of the inherent flexibility of the circuitry involved, it is now possible to obtain the performance required in any application very readily, for example, any desired time/current characteristic can be produced. This is a feature which is not available on electro-mechanical relays. In addition, weaknesses of the latter relays, such as slow resetting and overshooting are not present in electronic versions. It must be recognized, however, that electronic relays must have reliable power supplies and they must be made immune to electrical interference which may be produced by power equipment in their vicinity.

At present manufacturers are producing both electro-mechanical relays, which remain popular because of their simplicity and the high level of performance provided by them over many years, and also a range of equivalent electronic relays.

It is not possible in this work to give details of the circuitry used in modern electronic relays and indeed it is likely to change quite quickly in the future. Because of this, only a few examples, based on block diagrams are given below.

4.3.1 Basic electronic processing

Modern solid-state relays function by using digital circuitry to process incoming signals derived from current and voltage transducers. Usually the input signals are sampled at fixed intervals to produce a succession of discrete pulses of the form shown in Fig. 4.13(a,b). The magnitude of each pulse is stored as a binary number with a fixed number of bits. As an example, if an 8-bit coding system is used, then 2^8 (i.e. 256) signal levels can be stored. Clearly the greater the number of bits used, the greater is the resolution and accuracy achieved.

The sampling process removes some of the high-frequency content of an input signal and it is generally necessary to sample at a rate or frequency of at least twice that of the highest frequency of interest.

The circuit block which samples and codes the input signals is called an analogue to digital converter (abbreviated to A/D). The digital signals produced by such a converter can then be processed to extract relaying information, by performing a logical operation based on an algorithmic implementation of the relaying principle. The logic used for this purpose may be discrete, i.e. a combination of electronic gates and operational amplifiers, or a suitable microprocessor. Which of these alternatives is adopted depends on factors such as the complexity of the relaying principle, the speed of operation and the cost.

Discrete logic can provide rapid operation and it can be configured to deal with several tasks in parallel. It lacks flexibility, however, because its circuits

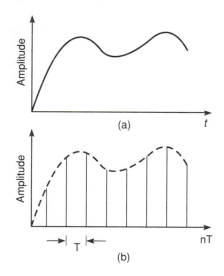

Fig. 4.13 Sampling of a continuous time-varying signal.

must be changed to implement different tasks. In contrast, microprocessors are able to perform complex sequential tasks but are not so suitable for dealing with parallel tasks. This limits their processing speeds, but continuing technological developments are providing significant improvements. Microprocessors possess the advantage of flexibility, in that they can be used for a variety of tasks without circuit alterations because their logic functions are implemented in software.

Relays employing both discrete logic and microprocessors have been produced in recent years and a few examples are described in the following sections.

4.3.2 Current-operated relay

A microprocessor-based, current-operated relay [9] manufactured by GEC-Alsthom is shown in Fig. 4.14. The output of the main current transformer is fed to a step-down interposing transformer in the relay. Its output after full-wave rectification is fed to a circuit consisting of a set of shunt-connected resistors which may be switched by the user to obtain the desired current setting. The voltage thus obtained is applied to an A/D converter, the output of which is supplied to a microprocessor based on the Motorola 6805 chip.

This relay allows users to select a current setting at which instantaneous operation will be initiated and also a current setting and associated inverse time/current characteristic, seven different curve shapes being available.

Configurations which can provide three-phase overcurrent and earth fault protection are available and directional features can be obtained using the relay described in the following section.

4.3.3 Directional relay

A directional relay employing discrete logic [10], produced by GEC-Alsthom, is shown in Fig. 4.15(a).

A voltage signal provided by the main voltage transducer is supplied to an interposing voltage transformer in the relay. The output from this transformer is phase shifted by an angle selected by setting a switch on the relay panel. After amplification, the phase-shifted signal is converted into square pulses thus clearly identifying the zero-crossing points.

The current signal obtained from the main current transducer is treated similarly except that a phase change is not introduced.

The two signals produced by the above processing are then fed to a comparator containing one 'or' and two 'and' gates, an integrator and level detector. The behaviour of this comparator is illustrated for three conditions in Fig. 4.15(b).

When the inputs to the squarers are in phase with each other then the inputs to each of the 'and' gates are always different, i.e. 1 and 0 or 0 and 1, as shown

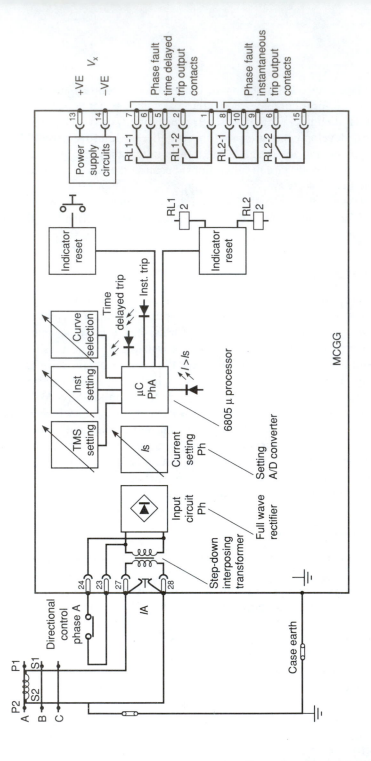

Fig. 4.14 Microprocessor-based, current-operated relay. (Reproduced from *MCGG 22*, GEC Measurements with the permission of GEC Alsthom Protection and Control Ltd).

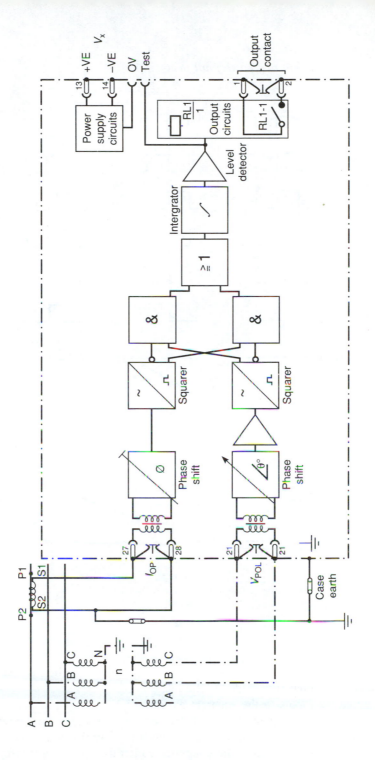

Fig. 4.15(a) GEC Alsthom directional relay METI II. (Reproduced from *METI II*, GEC Measurements with the permission of GEC Alsthom Protection and Control Ltd).

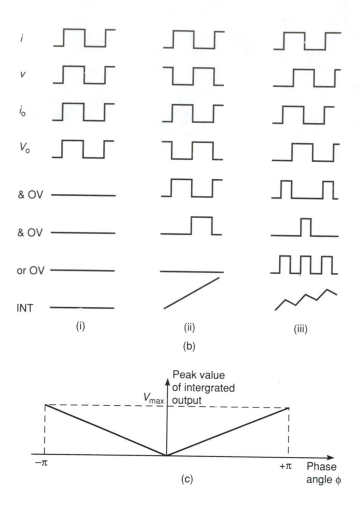

Fig. 4.15(b), (c)

in Fig. 4.15(b)(i). As a result zero inputs are always applied to the 'or' gate and a zero signal is provided to the integrator.

When the inputs to the squarers are in antiphase with each other then the inputs to the 'and' gates will alternate between 1,1 and 0,0 as shown in Fig. 4.15(b)(ii). As a result the 'or' gate inputs are either 0,1 or 1,0 and a continuous output signal is fed to the integrator which then provides a ramp output.

For the intermediate condition, shown in Fig. 4.15(b)(iii), in which the squarer outputs are displaced by $\pi/2$ rad, each 'and' gate produces an output for only a quarter of each period ($\pi/2$ rad) and as a result the 'or' gate produces an output (1) for half of each period.

It will be appreciated that the 'or' gate produces an output (1) for a fraction of $\theta/2\pi$ of each period in which θ is the phase displacement either lagging or leading between the signals supplied to the squarers and that the output from the integrator rises during a cycle to a peak value proportional to the value of the phase displacement θ. This behaviour is illustrated in Fig. 4.15(c).

The angular width of the operating zone of the relay can be controlled by adjusting the setting of the level detector. As an example, if a setting of $0.5\,V_{max}$ were used, relay operation would occur for values of θ in the ranges $\pi/2 \to \pi$ and $-\pi/2 \to -\pi$ or with a setting of $0.25\,V_{max}$, operation would occur

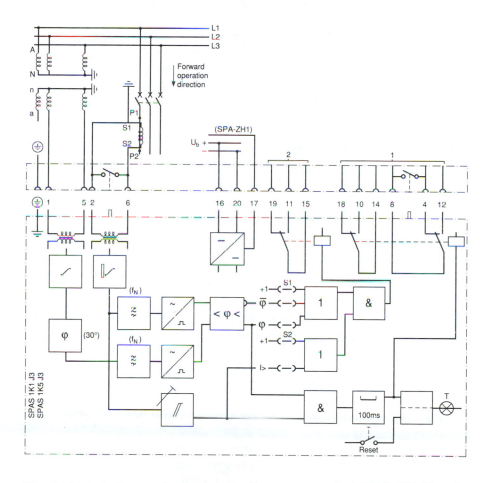

Fig. 4.16 Block diagram for the directional-overcurrent relays SPAS 1K1 J3 and SPAS 1K5 J3 manufactured by ABB. (Reproduced from *ABB Buyer's Guide 1989–90*, with the permission of ABB Relays AG, Baden, Switzerland).

for values of θ in the ranges $\pi/4 \rightarrow \pi$ and $-\pi/4 \rightarrow -\pi$. These operating zones are shown in Fig. 4.15(c).

The operating zones can then be set in the desired positions by selecting the appropriate phase shift in the voltage circuit and a shift of π radians can be obtained by reversing the connections from either the current or voltage transducers.

Whenever circuit conditions cause the phase displacements between the voltage and current to fall within the operating zone of the relay, the level detector produces an output signal which may be used directly to open a circuit-breaker or it may be used in logic circuitry which also takes account of overcurrent or earth fault signals.

A block diagram of an alternative design produced by Asea Brown Boveri [11] is shown in Fig. 4.16.

The output of the main voltage transducer is supplied to a threshold detector in this relay. It is then phase-shifted before being applied to the input of a band-pass filter, the output of which is then squared before being fed to a comparator. The second input to the comparator is obtained after similar processing of the signal provided by the main current transducer. As in the GEC-Alsthom relay described above, an output signal is provided when the phase difference between the initial voltage and current signals is within desired limits. Again the output signal may be used in conjunction with signals obtained from current-operated relay elements.

Further information on microprocessor-based relays, which it has not been possible to provide in this work, may be obtained from references [12, 13] and manufacturers' literature.

4.4 APPLICATIONS OF OVERCURRENT, EARTH FAULT AND DIRECTIONAL RELAYS

In most applications, several individual relays or elements must be associated with each circuit-breaker to enable the necessary discrimination to be achieved during fault conditions and several elements may, for the convenience of users, be mounted in a common case or housing. As an example, one earth fault IDMT and two overcurrent IDMT elements together with associated directional elements are commonly mounted in one case. In a similar way in the following treatment, a single box is shown adjacent to each circuit-breaker in circuit diagrams to indicate a relay consisting of a group of associated relay elements.

The various types of relays described in sections 4.2 and 4.3 enable discrimination to be achieved during fault conditions on many networks in the following ways:

(a) by monitoring only current levels, i.e. current grading;
(b) by introducing time-delayed operation, i.e. time grading;

(c) by combining time-delayed operation with sensing of the direction of current flow.

Applications where these methods may be used are examined in the following sections.

4.4.1 Current grading

Current grading can be used to obtain correct discrimination in circuits where there will be a large difference in the ratio of fault current to rated current in sections of a network. Such a situation can arise when there is a unit of relatively high impedance between two sections. This is illustrated in the network shown in Fig. 4.17. The current in a load circuit in the event of a fault on it (at point F_1) would be limited relative to its rated current because of the impedance of the transformer, T. The value of the current fed to the primary winding of the transformer for this condition would therefore be much lower than the current which would flow in the event of a primary-circuit fault at point F_2.

To illustrate this, the currents which would flow in the main feeder cable of the network shown in Fig. 4.17 during three-phase and phase-to-phase faults at point F_1 and F_2 would be as follows:

Three-phase fault at point F_2 $\qquad I = \dfrac{E_s}{Z_{s1}} = 20\,\text{pu}$

Three-phase fault at point F_1 $\qquad I = \dfrac{E_s}{Z_{s1} + Zt_1} = 4\,\text{pu}$

Phase-to-phase fault at point F_2 $\qquad I = \dfrac{\sqrt{3}\,E_s}{Z_{s1} + Z_{s2}} = 24.75\,\text{pu}$

Phase-to-phase fault at point F_1 $\qquad I = \dfrac{\sqrt{3}\,E_s}{Z_{s1} + Zt_1 + Z_{s2} + Zt_2} = 3.69\,\text{pu}$

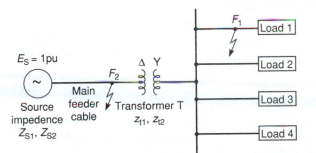

Fig. 4.17 Power network. $Z_{s1} = 0.05$ pu. $Z_{s2} = 0.02$ pu. $Z_{t1} = 0.2$ pu. $Z_t = 0.2$ pu.

The above values are based on the rating of the main feeder. If each of the loads had the same VA rating, the rating of the main feeder would be four times that of each of the load circuits and therefore, in the event of faults at point F_1, the per-unit value of the current in the load circuit would be four times that for the feeder cable, i.e. for three-phase and phase-to-phase faults at point F_1, the per-unit currents in the faulted cable would be 16 pu and 14.76 pu respectively.

In the above circumstances instantaneous overcurrent relays could be used in the load circuits with current settings of say 150% or 1.5 pu to protect against faults and sustained overloads. To obtain correct discrimination however and give rapid clearance for large faults, instantaneous relays set at 500% or 5 pu could be used on the main feeder circuit. In addition, relays with fixed or inverse time/current delays might be included to clear sustained overloads from the main feeder circuit.

With a delta-star step-down transformer as shown in Fig. 4.17, zero-sequence currents would not flow in the main feeder circuit as a result of earth faults on the load circuits. Because of this, instantaneous earth fault relays with low current settings could be used on the main feeder and load circuits without loss of discrimination.

It will be realized that current grading can only be relied upon in a limited number of applications but, because of their simplicity, instantaneous relays with suitable current settings should be used whenever possible. Certainly they can often be used on the load circuits of radially-fed networks, time grading or other methods being employed on the circuits nearer the source.

4.4.2 Time grading using relays with definite operating times

In networks where there are several sections connected in series without significant impedances at their junctions and where the source impedance is much greater than the impedances of the sections, there will be little difference between the levels of the currents which will flow for faults in different positions on the network. In these circumstances, current grading would clearly not enable satisfactory performance to be obtained.

Correct discrimination could, however, be obtained during fault conditions by using relays set to operate after different time delays. The timing differences between the relays associated with adjacent sections could clearly be made sufficient to allow the appropriate circuit-breaker to open and clear the fault on its section before the relay associated with the adjacent section nearer to the source could initiate the opening of its circuit-breaker. In practice, grading intervals of the order of 0.5 s are adequate, this being the interval used in the scheme shown in Fig. 4.18(a), for the protection of a network containing several short lines in series. Because discrimination is dependent only on time grading, the individual overcurrent relays can be set to any desired levels

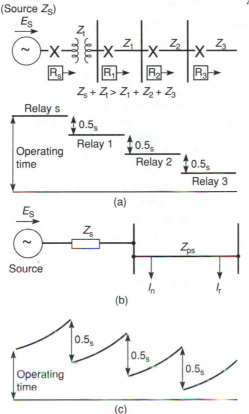

Fig. 4.18 Discrimination achieved by time-graded relays.

above the rated currents of their circuits. Similarly earth fault relays may be set independently at levels of the desired sensitivity.

It must be recognized that relays used near the sources in networks with several sections connected in series may have unacceptably high operating times for faults in the sections they protect because such faults may be of current levels which can only be allowed to persist for short periods. It is evident therefore that although it is very easy to select the operating times needed to ensure that correct discrimination will be achieved when relays with definite time lags are used, they have the weakness that their use must be restricted to networks with relatively few series-connected sections.

4.4.3 Time grading using relays with inverse time/current characteristics

In networks such as that shown in Fig. 4.18(b), where the impedance (Z_s) between a source of e.m.f. (E_s) and the input end of a protected section is small compared with the impedance (Z_{ps}) of the protected section, there will be a significant variation in the current levels of faults with their positions on the protected section. For the extreme cases of faults at the near and remote ends of a section, the currents would be:

$$I_n = \frac{E_s}{Z_s} \quad \text{and} \quad I_r = \frac{E_s}{Z_s + Z_{ps}}$$

respectively. Taking as an example, a situation where $Z_{ps} = 5Z_s$, the current (I_n) for a near-end fault would be six times that for a fault at the remote end (I_r). In these circumstances, and if relays with truly inverse time/current characteristics were used (i.e. $1/t \propto I$ or $It = \text{constant}$), the operating time for a near-end fault would be one sixth of that for a remote fault. The behaviour which would be obtained with such relays applied to a network with three-lines in series, allowing 0.5 s discriminating margins, is illustrated in Fig. 4.18(c). Clearly this behaviour is more desirable than that which would be obtained using relays with definite time delays because the large fault currents would be interrupted more quickly.

The energy let-through during a fault may be taken to be proportional to $I^2 t$ and on this basis the energy let-through in the above example would be six times greater for near-end faults than for those at the remote ends of lines. This situation might be regarded as unacceptable and it could clearly be avoided by using relays in which operation occurs at a constant I^2 value (i.e. $t \propto 1/I^2$). Relays with time characteristics of this form would also enable better discrimination to be achieved in applications where the protected sections have relatively low impedances and where, as a result, the currents during faults are not greatly affected by the fault position. In these circumstances, the greater sensitivity of such relays to such current differences would be beneficial.

For networks with sources of low volt-ampere output or in which some circuits are remote from their sources, the effective values of source impedance may be high and the behaviour obtained when using relays with inverse time/current characteristics of any form might not be significantly better than that which relays with definite time delays could provide. In such conditions, the latter relays could be used on sections remote from the source while relays with inverse time/current characteristics could be used in the remainder of the network.

It will be appreciated from the above that there is not a single time/current characteristic which would be ideal for all applications. Certainly the most commonly used IDMT relays have the characteristic which the earliest induction relays provided, namely the one specified in section 4.2.3 (Table 4.1). This is now designated as the standard normal inverse characteristic in BS142. Manufacturers have had to continue to produce relays with this characteristic to replace existing relays or for use on extensions to networks where they may be essential to provide correct discrimination with the relays already in use.

Relays with other characteristics are, however, now available, manufacturers in the USA having introduced two, one designated 'very inverse' and the other 'extremely inverse', the latter providing approximately the constant

I^2t performance, referred to earlier. These relays, like those with the standard inverse characteristic, do not operate below certain minimum current levels and they also have fixed or definite minimum operating times above certain current levels. The relative shapes of the above characteristics can be seen from Fig. 4.19, in which each of the curves is for a time multiplier setting of unity. In practice, performance may be improved in some applications by having some relays with one characteristic and the remainder with another.

It will be appreciated from the above that it is more difficult to choose the necessary settings for relays with inverse time/current characteristics than it is when relays with definite-time delays are to be used. To ensure that correct

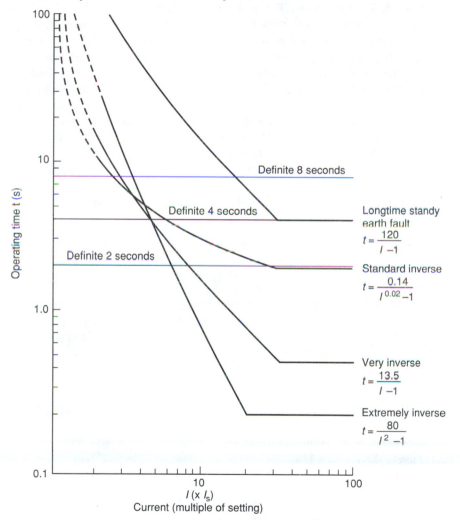

Fig. 4.19 Time/current characteristic (I_s = current setting). (Reproduced from *MCGG 22*, GEC Measurements with the permission of GEC Alsthom Protection and Control Ltd).

discrimination will be obtained under all conditions on a particular network a study of its behaviour should be done first to determine the currents which will flow in its various sections, in the event of faults at several different points on it. Because the relay operating times are relatively long, steady state calculations using well established techniques such as symmetrical components may be used.

As a guide, a possible procedure for determining the settings of overcurrent relays, with standard inverse time/current characteristics, needed to protect the network shown in Fig. 4.20 is outlined below.

Because the system is radial and currents cannot be fed back towards the source, the relays protecting the loads would operate most rapidly. The characteristics of the individual loads must be taken into account, however, when determining the performances required of the relays associated with them. Should loads contain large motors, for example, then relays with time-delayed operation may be needed to ensure that maloperation will not be caused by the surge currents which may flow when motors are started. In these circumstances thermal relays might be used. For other loads, instantaneous relays may be satisfactory and their use would be advantageous because the operating times of relays at points nearer the source would be lowered. In this example, it is assumed that instantaneous relays could be used to protect all the loads and that each relay would be set to operate at a current somewhat above the rated value of its load.

When determining the settings of the other relays on the network, the following factors must be taken into account:

1. That upstream relays may overrun after a fault is cleared by a downstream circuit-breaker.
2. That tolerances are allowed on the time/current characteristics of IDMT relays in the various standard specifications and in consequence an upstream relay may operate slightly faster than would be expected from its basic characteristic whilst a downstream relay may operate more slowly.
3. That a finite time is needed for a circuit-breaker clearing a fault to open and effect interruption.
4. That the current transformers feeding the relays may have errors within the limits allowed in national and international standards.

To allow for these factors it is usual to have a discriminating time margin between adjacent relays of about 0.5 s. This time interval may be reduced if an assessment of the minimum time grading interval t' is made. This may be done using the formula [14]

$$t' = \left[\frac{2E_R + E_{CT}}{100} \right] t + t_{CB} + t_0 + t_s, \text{ seconds}$$

where E_R is the percentage relay timing error,

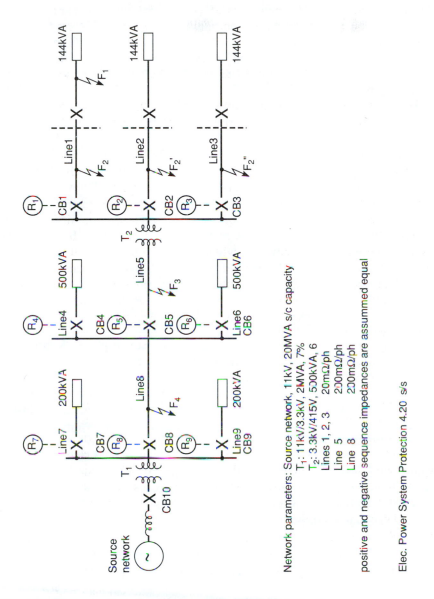

Fig. 4.20 The protection of a radially-connected network.

E_{CT} is the current transformer ratio error,

t is the nominal operating time of the relay nearer to the fault,

t_{CB} is the circuit-breaker interrupting time,

t_0 is the relay overshoot time

and t_s is a suitable safety margin.

For simplicity, in the example that follows a discriminating margin of 0.5 s has been used, i.e. $t' = 0.5$ s. In this example, therefore, relays R_1, R_2 and R_3 must not operate in less than 0.5 s for faults beyond the circuit-breakers in the load circuits, i.e. for a fault at a point such as F_1.

To ensure that this requirement will be met the maximum current flowing in line 1 for a fault at point F_1 must be determined. It can be seen that this condition would occur if the circuit-breakers CB2, CB3, CB4, CB6, CB7 and CB9 were open and if the source impedance Z_s had its lowest possible value. With the parameters shown, the current flowing in line 1 in the event of a three-phase fault at point F_1 would be 3.9 kA. Line 1 is rated at 200 A and relay R_1 is fed from current transformers of ratio 200/1. Relay R_1 could thus be given a 100% setting as it would not operate at line currents of less than about 260 A (i.e. 1.3×200 A). With this setting, the above fault current of 3.9 kA would represent a multiple of 19.5 (i.e. 3900/200) for which relay R_1 would operate in 2.29 s with a time multiplier setting of 1.0. To provide operation in 0.5 s a time multiplier setting of 0.22 would be needed.

In the event of a phase to phase fault at point F_1 the maximum possible current in line 1 would be 3.38 kA which would represent a multiple of 16.9 (i.e. 3380/200) to relay 1 giving an operating time of 2.4 s for a time multiplier setting of 1.0. For this condition a time multiplier setting of 0.21 could be used to obtain operation in 0.5 s. Similar calculations should be done for other types of faults at point F_1, e.g. two phase to earth and single phase to earth and clearly the highest of the time multiplier settings determined would have to be used to ensure correct discrimination.

It is improbable that the network would ever be operated with so many of its circuits disconnected or unloaded and hence the currents flowing in line 1 for faults at point F_1 would never reach quite such high values as those determined above.

Although, in most cases, the presence of loads is likely to lead to only small reductions in the currents flowing in lines during fault conditions, the above calculations could nevertheless, if desired, be repeated to determine whether significantly lower time multiplier settings could be used, as this would speed up the operation of relays near the source. For the network, represented in Fig. 4.20, the current in line 1 would reduce from the above value of 3.9 kA to 3.5 kA, if all the loads were connected, each of them being assumed to have the impedance corresponding to its rated load and a lagging power factor of 0.6. In this case, relay R_1 could have a time multiplier setting of 0.21 instead of the value of 0.22 determined earlier. This is an insignificant reduction but in other situations larger reductions might be possible.

The settings of relays R_2 and R_3 could clearly be determined in the above manner.

Relay R_5 protecting line 5, rated at 75 A and fed by current transformers of ratio 100/1 could be given a current setting of 75% or 0.75 A, thus allowing line overcurrents up to about 100 A to flow. This relay must have operating times greater than those of relays R_1, R_2 and R_3 by 0.5 s or more for faults on lines 1, 2 and 3. To ensure that this will be so, the operating times of relays R_1, R_2, R_3 for faults at positions F_2, F_3 and F_4 must be determined for a range of conditions. In this case the maximum fault current possible at point F_2 on line 1 should be determined, i.e. using the lowest value of source impedance (Z_s) likely to be present in service, a fault impedance of zero and with circuit-breakers CB4, CB6, CB7 and CB9 open.

Using the parameters shown in Fig. 4.20, a fault current of 5.8 kA would flow under these conditions in the event of a three-phase fault. This would cause the greatest possible current multiple of 29 in relay R_1 and this would cause it to operate in 0.44 s. The corresponding current for a phase to phase fault would be 5.0 kA, i.e. a current multiple in relay R_1 of 25. Operation in this case would occur after 0.46 s. For these two conditions the current multiples in relay R_5 would be 9.7 (i.e. 727/75) and 8.4 (i.e. 629/75) which would lead to operation in 3.0 and 3.2 s respectively using a time multiplier setting of 1.0. To allow a discriminating time margin of 0.5 between relays R_1 and R_5 the latter relay would require multiplier settings of 0.31 and 0.30 for the above two faults. Clearly other types of faults should be studied similarly and the highest value of time multiplier setting would have to be used.

This procedure could be repeated for conditions producing lower fault currents at point F_2. For example, the highest possible source impedance likely to be encountered could be used, all the circuit-breakers could be closed and then fault impedances could be introduced to obtain fault currents at point F_2 down to the minimum level needed to operate relay R_1, i.e. about 260 A. Unless lines 2 and 3 and their loads are very similar to line 1 and its load, the above procedure should be repeated for faults at points F_2' and F_2''. For each condition the time multiplier setting of relay R_5 needed to provide a 0.5 s discriminating time margin should be determined and the highest value calculated would then be the one required.

Similar studies could then be performed to determine the settings of all the other relays.

Repeating the studies assuming relays with other characteristics, e.g. very inverse, would enable the arrangement which would provide the optimum performance to be determined. Similar studies could also be used to determine the settings of earth fault IDMT relays.

Although the above procedure appears to be lengthy, it can be performed very quickly using modern fault computing programs.

Because the shortest relay operating times are obtained when large fault currents are flowing, the time multiplier settings needed to provide the necessary discriminating intervals are generally determined by the maximum fault-current conditions and the amount of computation may be reduced if this is assumed. Should it not be possible to undertake the above studies, then simpler approximate methods may be used.

As an example, for the network shown in Fig. 4.20, the current fed to line 1 in the event of a fault at point F_2 could be taken to be m times its rated or full load current, i.e. $I_f = mI_{f\ell}$. Should lines 2 and 3 each have the same rating as line 1 and assuming that they could be supplying full-load current in the presence of the above fault, and further assuming that the three line currents were of the same phase, then the current in line 5 would be equal to $(m + 2)I_{f\ell}$. In these circumstances, the rating of line 5 would be three times that of each of the lines it supplies, i.e. $3I_{f\ell}$. As a result, relay R_5 would be operating at a current multiple of $(m + 2)/3$ when relay R_1 was operating at a multiple of m.

The value of m could vary from 1.9, the minimum level which would cause relay R_5 to operate up to the maximum value possible on the protected circuit (20 in the earlier example).

A curve of the required operating times of relay 5 against either its own current multiple or the value of m could thus be derived, such a curve being shown in Fig. 4.21. From it the required time multiplier setting of relay R_5 could be obtained. The value needed for the relationship in Fig. 4.21 would be 0.29. Again it is the maximum fault current level which governs the time multiplier setting which must be used.

It should be noted that the approximations made in this latter method tend to increase the discriminating margins slightly and they would not therefore introduce the risk of incorrect tripping of circuit-breakers.

In some applications, high-set, instantaneous elements are used in conjunction with IDMT overcurrent elements to provide rapid operation in the event of very high current faults. Such relays are usually used near sources where current grading can be achieved, only IDMT relays being employed on the other downstream circuits.

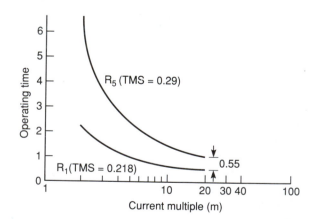

Fig. 4.21 Time/current characteristic of relays R_1 and R_5 (Fig. 4.20).

4.5 THE APPLICATION OF DIRECTIONAL AND CURRENT-OPERATED RELAYS

Satisfactory discrimination cannot be achieved for all the fault conditions which can occur on highly-interconnected networks when using only overcurrent and earth fault relays. This can be seen from a study of the very simple network containing a pair of parallel-connected lines, which is shown in Fig. 4.22.

Because of the symmetry of the network, relays R_1 and R_3 would both have the same time and current settings. Relays R_2 and R_4 would also have the same settings. In the event of a short circuit on line A near circuit-breaker CB2, all four relays (R_1–R_4) would carry similar currents. Because relays R_2 and R_4 would have lower time settings than relays R_1 and R_3 which are nearer to the source, relays R_2 and R_4 would operate together opening their respective circuit-breakers CB2 and CB4. Relay R_1 would then operate later opening circuit-breaker CB1 to clear the fault. As a result both lines would have been disconnected, causing a complete loss of supply to the load and the advantage of having duplicate lines would have been lost.

This incorrect behaviour could clearly be avoided by providing directional elements in addition to the other elements at the ends of lines A and B remote from the source. The directional relays would have to be set to operate only when currents were being fed back towards the source, an abnormal condition during which the current I_2 would lead the voltage (V_2) at the position of relay R_2 by an angle somewhat greater than $\pi/2$ radians. Normal currents flowing to the load would appear to the relay to be displaced by an angle over $\pi/2$ radians from that during the fault. A suitable operating characteristic for the directional relays would thus be of the form shown in Fig. 4.23. In practice, a directional element would have to be provided with each of the overcurrent and earth fault elements fitted near circuit-breakers CB2 and CB4.

It is usual to arrange that directional elements only allow their associated overcurrent or earth fault elements to operate when faults occur on the appro-

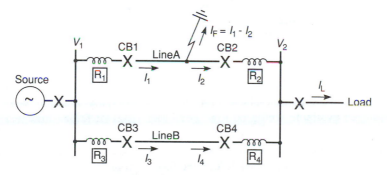

Fig. 4.22 The protection of a network containing two similar lines connected in parallel. For $I_L \ll I_F$ it follows that $\bar{I}_1 \simeq \bar{I}_3 \simeq \bar{I}_4 \simeq -\bar{I}_2$.

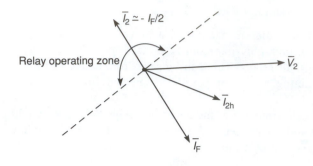

Fig. 4.23 Conditions associated with relay R_2 (Fig. 4.22) when a fault is present on line A and during healthy conditions.
$\overline{I}_{2h}$ = typical current during healthy conditions
$\overline{I}_2$ = current when line A is faulty
$\overline{I}_f$ = fault current
$\overline{V}_2$ = voltage applied to relay.

priate side of the relaying position, i.e. on line A or B in the above example. In applications in which IDMT induction-type elements are being used, the directional element contacts are connected in the circuits of the windings on the second electromagnets or in the shading-winding circuits. In this way the disc of an induction element will not rotate unless the current is in the direction for which operation should occur.

If the alternative of energizing both a directional element and its associated IDMT element at the same time were adopted and the contacts of the two elements were connected in series in the trip-coil circuit, incorrect tripping could result. This might happen, for example, if an overcurrent element completed its operation while its associated directional element was restraining during a fault in the reverse direction. If after clearance of the fault current flowed in the opposite direction before the overcurrent element began to reset, maloperation would occur.

Directional elements should clearly be suitable for operation over the same current ranges as their associated elements and, as stated earlier in section 4.2.4, they should be capable of operating correctly even when the voltages applied to them are very depressed. This feature would be essential in the unlikely event of a three-phase fault near the relaying position when all the voltages available would be low. For other types of faults, however, the situation may be imporoved by obtaining the polarizing voltages for elements from a phase or phases different to that from which their currents are supplied. Clearly there are several possible connections available and these were considered in detail by Sonnemann in a paper entitled 'A study of directional element connections for phase relays' [15].

As an example, one connection is considered below. As stated earlier in section 4.2.4, an induction-type directional element may be set to provide maximum operating torque at any desired phase angle between the current and voltage supplied to it, by introducing the necessary phase shift in the circuit of the voltage winding. The term maximum torque angle (MTA) was introduced, this angle being defined as the angle by which the current must lag the voltage fed to an element to produce maximum torque, say 30° or $\pi/6$ radians.

A connection which is often used because it tends to ensure an acceptable voltage level during faults is to connect the current windings in one phase (say phase a) and the voltage windings between the other two phases (say b to c). This is termed a 90° or $\pi/2$ connection because the voltage V_{bc} is normally in quadrature with the 'a' phase voltage. In these circumstances when using an element with an MTA of $-\pi/6$ radians, the maximum torque would occur when the element current (I_a) lagged its phase voltage (V_a) by $\pi/3$, a condition likely to be present under fault conditions. This arrangement, which is illustrated in Fig. 4.24, is described as a (90°–30°) or ($\pi/2$–$\pi/6$) characteristic.

Although static or electronic relays clearly do not have maximum torque angles, they nevertheless do have operating zones corresponding to those of electro-mechanical relays and they may be specified in the above manner.

The necessary time and current settings of the overcurrent and earth fault elements to be used on a given network could be determined using the methods described earlier in section 4.4.3. In addition the optimum settings of directional relays could readily be found.

Directional overcurrent and earth fault relays are widely used to protect ring mains and therefore this application is considered below to show how satisfactory discrimination can be achieved.

A ring-main circuit with a single source and five substations is shown in Fig. 4.25.

The double-headed arrows shown adjacent to the relays R_1, R_2 and R_5

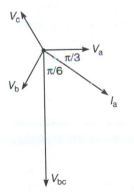

Fig. 4.24 Condition which provides maximum torque in a directional element with an MTA of $-\pi/6$ rad.

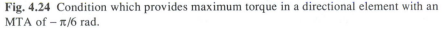

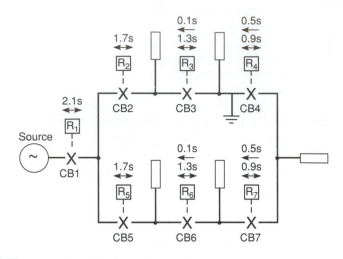

Fig. 4.25 The protection of a ring-main circuit.

indicate that these relays would operate for current flows in either direction, i.e. these relays do not incorporate directional elements. This is acceptable because these relays have higher minimum operating times than the others, the values being 1.7 s and 2.1 s, i.e. the times shown adjacent to the arrows. In this connection it will be appreciated that current would not be fed back through circuit-breakers CB2 and CB5 in the event of a fault on or near the source.

The other relays (R_4–R_7) each have directional elements so that operation is obtained in short times for current flows in the directions indicated by the single-headed arrows and longer times for current flows in either direction. As an example, relay R_3 would operate in a minimum time of 0.1 s for currents flowing towards the source (right to left) and a minimum time of 1.3 s for current flowing in either direction. To achieve this, two current-operated relays and one directional relay are needed.

It will be seen that a discriminating margin of 0.4 s has been allowed between the relays at adjacent substations.

For the short circuit shown between the circuit-breakers CB3 and CB4, correct discrimination would be obtained. Circuit breaker CB4 would open after a minimum time of 0.5 s followed by circuit-breaker CB3 opening after at least a time of 1.3 s to clear the fault. The remaining circuits would remain in service and all loads would continue to be supplied. Similar conditions would obtain for faults in other positions.

It must be appreciated, however, that completely satisfactory behaviour cannot be obtained in highly-interconnected networks fed from several sources because the directions of current flow and the current levels at the individual relaying positions do not enable the positions of faults to be correctly determined.

4.6 CURRENT AND VOLTAGE TRANSFORMERS

It is clearly essential that the accuracies of transformers supplying relays should be high enough to ensure that satisfactory performance will always be obtained. The requirements, which depend on the types of relays being used, are considered below.

4.6.1 Electro-mechanical relays with a fixed current setting

These relays may operate either instantaneously or after definite time delays. To ensure correct operation it is necessary that the current transformers supplying them have small errors at the operating current levels. At higher currents, the errors are unimportant because they will not cause the secondary currents to fall below the operating levels. In addition, transient errors caused by saturation of cores during the first few cycles of fault current will not cause maloperation and therefore only steady state conditions need be considered.

As an example, a relay with a 2 A setting (200%) fed from a current transformer of ratio 200/1 would operate satisfactorily if the transformer current error for a secondary current of 2 A was 2% or less. Assuming a secondary circuit of impedance 5Ω, i.e. 5 VA at rated current, a current error of not more than 2% would have to be achieved at a secondary winding e.m.f. of 10 V.

This would certainly be achieved by a current transformer rated at 15 VA with an accuracy classification of 10P5 as defined in BS 3938 : 1973. This classification corresponds to a current error of ± 3% at rated current (1 A) for a 15 VA burden, i.e. 15 V and a composite error (exciting current) of 10% or 0.1 A at five times rated current (5 A).

4.6.2 Electro-mechanical IDMT overcurrent relays

To ensure that satisfactory grading will be obtained over the whole operating characteristic of an IDMT relay it is necessary that the current transformer supplying it will maintain high accuracy over the range from the minimum current level needed for relay operation up to the current at which minimum operating time is achieved, i.e. 1.3–20 times the relay current setting.

These relays impose a burden of about 3 VA at rated current, i.e. the voltage drop across a 1 A relay operating with a plug setting of 100% is 3 V. In addition, allowance must be made for the voltage drops in the current transformer secondary winding and the interconnecting leads. Assuming, as an example, that such a relay is supplied by a current transformer of ratio 200/1 and that the total voltage drop at rated current is 10 V, then core saturation should not occur in steady state at a secondary winding e.m.f. of 200 V. This performance would be provided by a transformer rated at 15 VA with an accuracy classification of 10P20 as defined in BS 3938 : 1973.

4.6.3 Electro-mechanical directional relays

The operation of directional relays is not dependent on the magnitudes of the voltages and currents supplied to them but only on the phase differences between these quantities. It is nevertheless necessary that the waveforms of these inputs should not be so distorted during fault conditions by saturation effects in the cores of the transformers which feed them, that the fundamental components become significantly displaced in phase. While such displacements are not likely to be produced in voltage transformers because the voltages applied to them tend to be depressed when faults are present, they could occur in current transformers when large system currents are flowing. To avoid such situations current transformers capable of supplying their connected burdens at the highest possible fault current levels without core saturation occurring in steady state should be used.

The total VA burdens on current transformers including those of the secondary windings, the connecting leads and both the current-operated and directional relays should be determined at the rated-current level (e.g. 15 VA). Because high transformation accuracy is not needed Class 10P transformers are adequate and indeed this is the recommendation in BS 3938 : 1973. In most cases, the transformers needed to ensure correct operation of IDMT relays should prove satisfactory when directional elements are also to be used, i.e. say 15 VA Class 10P20.

4.6.4 Electronic relays

Transducers supplying modern electronic relays must provide outputs which will enable correct operation to be obtained under all conditions, for example a current transducer supplying a relay with an IDMT characteristic must respond correctly at currents up to 20 times the relay setting current.

Because electronic relays are provided with power supplies, however, they present negligible burdens to their transducers and therefore the burdens are limited to those presented by the output windings of the transducers and the connections between these windings and the relays. This factor should generally reduce the difficulty of obtaining transducers with the required performance.

The situation can be further eased if current transducers are used which produce voltage outputs proportional to the input currents as they would then provide voltage signals and negligible currents to the relays.

4.7 STANDARD SPECIFICATIONS

BS 142 : 1982 [16] is a very comprehensive document covering electrical protection relays. It is based upon, and is in technical agreement with, the appropriate parts of IEC Publication 255 Electrical Relays.

BS 142 provides definitions of the various terms used in connection with electrical protection relays as well as specifying characteristics, constructions and testing procedures.

The basic requirements for IDMT and directional relays are summarized below.

4.7.1 IDMT relays

As stated earlier, the time/current characteristics of the early induction-type relays were adopted as standards. They resulted from the construction of the relays rather than from considerations of the withstand abilities of the equipment being protected. They were not originally specified in mathematical form. In recent times, however, the general time/current relationship over the normal operating range has been expressed in the form:

$$t = \frac{k}{\left(\dfrac{G}{G_b}\right)^{\alpha} - 1} \, s$$

in which

t is the theoretical operating time,

G is the relay current,

G_b is the basic or setting value of the current and

k and α are constants characterizing particular relays.

For the standard normal-inverse, the very-inverse and extremely-inverse relay characteristics, which are shown in Fig. 4.19, the values of k and α are as given in Table 4.2.

Manufacturers are required to specify the minimum or threshold operating current level which must not be greater than 1.3 times the basic or setting value. They must also specify the tolerances or error bands within which their relays will perform. This must be done by declaring an assigned error as either:

(a) a percentage of the theoretical time; or
(b) a percentage of the theoretical time or a fixed maximum time error (where this may exceed the percentage value), whichever is the greater, eg 5% or 20 ms; or
(c) a fixed maximum time error.

Table 4.2

	k	α
Normal Inverse	0.14	0.02
Very Inverse	13.5	1
Extremely Inverse	80	2

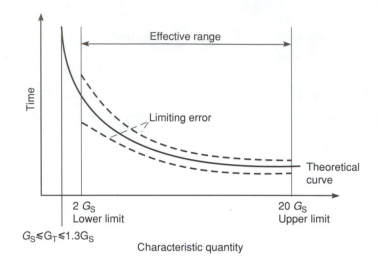

Fig. 4.26 Declaration of limiting error by means of graphical presentation.

The assigned error can be declared either by:

(i)   a theoretical time/current curve bounded by maximum and minimum curves as shown in Fig. 4.26; or
(ii) an assigned error at the maximum current level multiplied by the factors shown in Table 4.3 at lower values of current in the operating range.

As an example, if the manufacturer of a normal-inverse relay declared an assigned error of 5% or 0.15 s, whichever is the greater, then operation would have tó occur within the ranges shown in Table 4.4

4.7.2 Directional relays

BS 142, Section 4.2, 1984 specifies conditions with which directional relays must comply. The main requirements are listed below.

1.   The operating phase angle range of a relay must be within $\pm 2°$ of the range declared by the manufacturer.
2.   Relays must be capable of operating at voltages between 3% and 110% of the rated value except for earth fault elements to be used on systems which are not solidly earthed, for which the range must be extended to 3%–190%.
3.   A relay must not operate
     (a)   when rated voltage is applied and currents in the non-operating zone of up to 20 times rated current are applied or switched off, and

Table 4.3

Assigned error as a multiple of the declared error	Value of the current as a multiple of the setting value
2.5	2
1.5	5
1	10
1	20

Table 4.4

Multiple of basic value	Normal operating time (s)	Permissible error	Permissible operating time (s)
2	10	$2.5 \times 5\% = 1.25$ s	8.75–11.25
5	4.3	$1.5 \times 5\% = 0.32$ s	3.98–4.62
10	3.0	$5\% = 0.15$ s or 0.15 s	2.85–3.15
20	2.2	$5\% = 0.11$ s or 0.15 s	2.05–2.35

(b) when with zero voltage applied, currents up to 15 times the rated value are applied or switched off.

4.8 THE FUTURE

It seems probable that electro-mechanical overcurrent and directional relays will continue to be used for many more years because of their relative simplicity and proven reliability.

Electronic relays are now being installed in large numbers in many systems around the world. They have certain advantages over their electro-mechanical equivalents. For example, they can reset immediately when faults are cleared, a feature which enables reduced time-grading intervals to be used. Future developments such as the closer integration of protection and control functions will also favour the use of microprocessor-based relays.

REFERENCES

1. Andrews, L. (1898) The prevention of interruption to electricity supply, *J. Inst. Elec. Eng.*, **27**, 487–523.
2. Andrews, L. (1904) Automatic protective devices for electrical circuits, *Electrical Review*, **54** 933–935 and 972–975.

3. Frey, H. E. (1924) Relays for the protection of distribution systems, *The Brown Boveri Review*, **11**, 235–242.
4. Schuchardt, R. F. (1909) Protective features of high-tension systems, *Electrical World*, **53**, 1539–1543.
5. Wedmore, E. B. (1915) Automatic protective switchgear for alternating current systems, *J. Inst. Elec. Eng.*, **53**, 157–183.
6. Edgcumbe, K. (1920) The protection of alternating-current systems without the use of special conductors, *J. Inst. Elec. Eng.*, **58**, 391–416.
7. Wagner, C. F. and Evans, R. D. (1933) *Symmetrical Components*, McGraw-Hill.
8. Edgeley, R. K. and Hamilton, F. L. (1952) The application of transductors as relays in protective gear, *Proc. IEE*, **99**, 297.
9. Overcurrent relay for phase and earth faults, Type MCGG, GEC Alsthom, *Publ R-6054E*.
10. Directional relay, Type METI, GEC Alsthom, *Publ R-6003H*.
11. Single-phase directional overcurrent relay, Type SPAS 1K1J3, *Asea Brown Boveri, Publ 34 SPAS 3EN1*.
12. Sachdev, M. S. (ed.) Microprocessor relays and protection systems, *IEEE Tutorial Course, 88EH0269-1-PWR*.
13. Phadke, A. G. and Thorp, J. (1988) Computer relaying for power systems, RSP and John Wiley.
14. *Protective Relays – application guide* (1987) (3rd edn), GEC Measurements, England.
15. Sonneman, W. K. (1950) A study of directional element connections for phase relays, *Trans. AIEE*, **69**, (II).
16. British Standard Specification on *Electrical Protection Relays*, BS142 : 1982.

FURTHER READING

British Standard BS142 Electrical Protection Relays

Part 0 : 1982   General Introduction and list of parts.
Part 1 : 1982   Information and requirements for all protection relays.
Part 2 : 1982   Requirements for the principal families of protection relays.
Part 3 : 1990   Requirements for single input energizing quantity relays.
Part 4 : 1984   Requirements for multi-input energizing quantity relays.
Part 5 :   Supplementary requirements.

5

Current-differential protective schemes

It will be clear from the previous chapter that overcurrent and earth fault protective equipment employing time grading and directional detection cannot provide correct discrimination on all power networks and in many instances the clearance times for some faults would not be acceptable.

An alternative protective scheme which has been in use for many years to protect individual sections of networks or pieces of equipment, such as alternators, detects the difference between the current entering a section and that leaving it. Such schemes are designated as 'current-differential' or 'Merz–Price', the latter being a combination of the names of the two persons who proposed the basic principle of the scheme, which is expressed in Kirchhoff's first law, namely that the sum of the currents flowing to a node must equal the sum of the currents leaving it.

One phase of a basic current-differential protective scheme is illustrated in Fig. 5.1. Under normal conditions, the current entering the protected unit would be equal to that leaving it at every instant as shown in Fig. 5.1(a). The two current transformers, if ideal, would have equal secondary currents which would circulate around the interconnecting conductors, which are often called 'pilot wires'.

In the event of a fault on the protected unit, the input current would no longer be equal to the output current. For such a condition the input current would clearly be equal to the sum of the output and fault currents as shown in Fig. 5.1(b), i.e. $i_{pA} = i_{pB} + i_f$. The ideal current transformers would drive secondary currents ($i_{pA} N_p/N_s$ and $i_{pB} N_p/N_s$), proportional to their primary currents, and these could not circulate through only the interconnecting conductors. An interconnection must therefore be provided as shown, and this would carry the difference current, i.e. $i_f N_p/N_s$. A relay included in the interconnection would therefore carry a current proportional to the fault current at every instant and circuit-breaker operation could be initiated instantly for any fault current above a certain minimum level.

Should a fault occur at a point beyond the protected unit, that is at a point which is not between the current transformer primary windings, then each current transformer would carry the same primary current ($i_{pA} = i_{pB}$). This situation is shown in Fig. 5.1(c). As a result, no current would flow in the

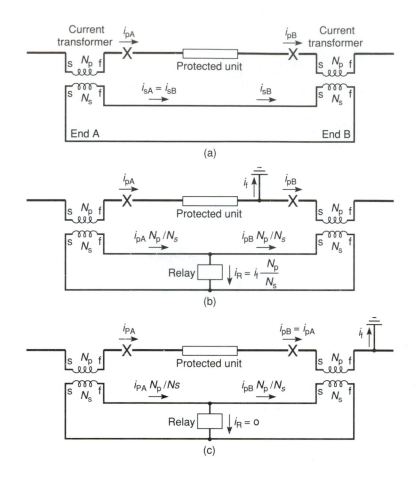

Fig. 5.1 Basic current-differential protective scheme. (a) healthy circuit; (b) internal fault; (c) external fault.

relay under ideal conditions, relay operation would not occur and correct fault discrimination would be achieved.

It will be clear that differential protective schemes provide the possibilities of rapid fault clearance coupled with correct discrimination. It must be recognized, however, that they only protect the circuit or zone between their current transformers and do not provide a measure of back-up protection to other parts of the network. They are therefore categorized as unit protective schemes.

Whilst the above principle is simple, several factors have to be taken into account when applying current-differential schemes to actual sections of networks or particular items of equipment. These will be considered both in this chapter and later chapters. Initially, however, some historical information is provided in the following section.

5.1 HISTORICAL BACKGROUND

In 1904, Charles H. Merz and Bernard Price obtained a British patent (No. 3896) entitled 'Improvements in the method and means for protecting apparatus on alternating current systems'. In the introduction of the patent specification it was stated that the 'cut-outs', based on detecting overcurrents, which were then in use had sometimes failed to act when they ought to have done and acted when they ought not to have done so. It was proposed that one or more electromagnetic devices or contact makers (relays) be connected by one or more conductors, referred to as a pilot wire or wires. Current transformers were to be used and so connected that no current would flow in the pilot wire or wires under normal conditions, but would do so in the event of a fault on the protected equipment.

It was stated that 'protective devices to act in the manner described can be variously constructed and arranged and be used not only in connection with high pressure (voltage) feeders, transformers or other apparatus, but also with any part of a distributing system'.

Because of its great historical significance, the circuit proposed in the patent specification, to protect a single-phase feeder, is shown in Fig. 5.2.

The proposed scheme required that the current transformers produce equal secondary winding e.m.f.s during normal or external fault conditions and that the connections be such that e.m.f.s opposed each other thus causing no current to flow in the interconnecting conductors (pilot wires) and relay windings. Because of this, the term 'voltage balance' was introduced to differentiate such schemes from those introduced later, in which the current transformers produce equal secondary currents during normal conditions. These currents flow in the interconnecting conductors and zero current flows in a relay connected between the interconnecting conductors as explained in the introduction to this chapter. The term 'circulating-current' is used in connection with these schemes.

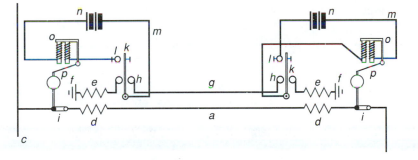

Fig. 5.2 Protective scheme proposed by C. H. Merz and B. Price. *a*, feeder; *b*, generator; *c*, substation; *d*, primary winding of CT; *e*, secondary winding of CT; *f*, earth or return conductor; *g*, pilot wire; *h*, relay windings; *i*, circuit breakers; *k, l*, movable and fixed relay contacts; *m*, circuit; *n*, battery; *o*, electromagnetic device with armature *p*.

The advantages of the scheme proposed by Merz and Price were soon recognized and it was applied to many circuits. This is clear from an article published in the *Electrical Review* of 28 August 1908 [1] under the title 'The Merz–Price system of automatic protection for high-tension circuits', in which it was stated that about 200 miles of high-tension network in the North of England were then protected by the scheme. It was further stated that Merz–Price protective gear had been in use for several years and that numerous faults had been isolated without visible shock on the system and, better still, without opening healthy sections even though the switches were closed repeatedly on to already matured faults. It was also stated that the scheme was in use to protect transformers on the same scale.

In spite of these early comments, which indicated that correct operation would be obtained under all conditions, difficulties did arise in later years as line lengths and current levels increased. Maloperations evidently occurred on occasions when protected units were healthy, due to several factors. Among these was the mismatching of current transformer outputs caused by them operating at different core permeabilities during healthy conditions. In an endeavour to overcome this particular problem current transformers having multi air-gapped cores, capable of performing linearly at primary current levels up to 10 000 A, were developed and introduced by 1925.

The introduction and use of special cables containing compensated pilot wires by 1925 indicates that incorrect operation must also have been caused by the currents which can flow in the capacitance between the interconnecting conductors of balanced-voltage schemes. During external faults these conductors may have quite high voltages between them and their capacity current is fed through the operating windings of the relays. This behaviour is considered in more detail in section 5.2.4.

Over the years various methods have been developed to ensure that correct operation will be obtained under all conditions, examples being the inclusion of biasing proportional to the current flowing through a protected unit and to the harmonics in power-transformer exciting currents.

5.2 FACTORS AFFECTING CURRENT-DIFFERENTIAL SCHEMES

The behaviour of the circulating-current differential scheme, described in the introduction to this chapter, during three different conditions, was based on ideal current transformers. In practice, however, there are several factors, including the errors of practical transformers, which could cause incorrect operation and these are examined separately, on a single-phase basis, in the following sections.

5.2.1 Current transformer errors

Current transformers must be used in differential schemes to enable the input and output currents of protected units to be compared and also to isolate the relays and interconnecting conductors from the high voltages associated with power systems.

In practice, all current transformers have transformation errors and therefore it is desirable that those at the input and output ends of a protected zone should be identical to try to ensure that their errors during healthy and external fault conditions will always be the same. In these circumstances, the secondary currents of balancing pairs of transformers would be equal at all times and zero current would flow in the relays except when internal faults were present. It must be recognized, however, that this behaviour might not be obtained even when using transformers which are physically identical, because mismatching can occur as a result of earlier internal faults. During such faults the currents (I_{pA} and I_{pB}) at the two ends of a protected zone could both have large magnitudes and be almost in antiphase with each other, as shown in Fig. 5.3. In these circumstances, the current transformers in which they flow would produce large e.m.f.s (E_{sA} and E_{sB}), also in antiphase with each other, and these would drive large currents (I_{sA} and I_{sB}) in the interconnecting cables. The difference of these currents, which would be very large, would flow through the relay, causing it to operate correctly to initiate the opening of the appropriate circuit-breakers. As a result, large residual fluxes could be left in the cores of the current transformers at the two ends of the protected unit after clearance of the fault, as explained in section 2.1 of Chapter 2.

These residual fluxes, which would be of opposite polarities, could persist and cause the two current transformers to operate at different points on their excitation characteristics during a later external fault and, indeed, they could cause the core of one of the transformers to saturate whilst the other did not.

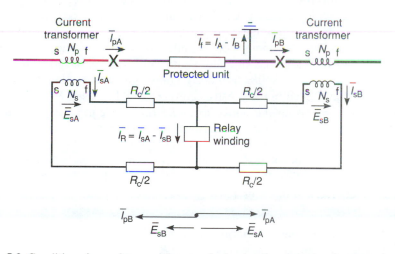

Fig. 5.3 Conditions in a scheme with a matched pair of current transformers when an internal fault is present. R_s, secondary winding resistance; R_c resistance of each conductor.

As a result, sufficient unbalance current could flow in the relay to cause it to operate incorrectly.

To minimize these effects, it is desirable that current transformers with relatively low errors and cores with low flux retentivities should be used.

5.2.2 Current transformer secondary ratings

During external fault conditions, the current transformers have to provide e.m.f.s to drive their secondary currents not only through their own windings but also through the cables that interconnect them. In applications where the two ends of a protected unit may be a considerable distance apart, the resistances of the interconnections may be quite high and the voltage drops along them may be correspondingly large when large fault currents are flowing. In such circumstances, it is desirable that current transformers with relatively low secondary ratings, e.g. 1 A rather than 5 A, be used. Although the lower rated transformers have higher secondary winding resistances they can produce greater e.m.f.s for given flux changes and the effects of the resistances of the connecting cables are reduced, as is illustrated in the example below.

Example

In an application where the maximum current which may be fed to a fault external to the protected zone is 15 times the rated current of the circuit, the connecting cables have a total resistance of 6 Ω. When using current transformers with 5 A secondary windings, each of 1 Ω resistance, a balancing pair would have to provide a total e.m.f. of 600 V, i.e. $15 \times 5(2 + 6)$ V. Current transformers with secondary windings rated at 1 A and each of resistance 25 Ω would have to provide a total e.m.f. of 840 V, i.e. $15 \times 1(50 + 6)$ V. Because of the greater number of secondary turns on the lower rated transformers, the flux variations in them would only be 0.28 of those in the 5 A transformers, i.e. $1/5 \times 840/600$. Indeed, cores of a cross-sectional area equal to 0.28 of that needed in the 5 A transformers could be used.

It will be seen from the above example that the e.m.f.s which must be produced by the transformers do tend to rise as their secondary winding ratings are reduced and the windings costs also rise significantly. The possible reductions in core area do, however, become very small as ratings are lowered further and, in practice, the optimum situation is likely to be achieved using transformers rated at 0.5 A or 1 A.

5.2.3 Interconnecting cables (pilot wires)

It will be apparent from the previous section that the resistances of the interconnecting cables affect the magnitudes of the e.m.f.s which must be provided by current transformers during fault conditions, the effect being very considerable when secondary windings of high rating (e.g. 5 A) are used. Although the effect is smaller when lower ratings (e.g. 1 A) are used, it is nevertheless desirable that the cable resistances should not be high compared with the secondary winding resistances of the current transformers. Clearly the cost of long cables with a large cross-sectional area might be unacceptably high and a compromise must be effected in such applications.

5.2.4 Symmetry of protective circuits

Figure 5.4(a) shows the conditions which would obtain in a protective circuit when an external fault is present.

If the circuit was symmetrical, i.e. $R_{cA} = R_{cB}$, and the current transformers were perfectly matched, the secondary currents i_{sA} and i_{sB} would be equal at all times. The current transformers would provide equal secondary winding e.m.f.s, i.e. $e_{sA} = e_{sB} = i_{sA}(R_s + R_{cA})$. The voltage ($v_R$) across the relay would be zero and no current would flow in it, i.e. $i_R = 0$.

The above ideal behaviour cannot be obtained, however, when protective circuits are asymmetric. For such a circuit, in which the conductor resistances

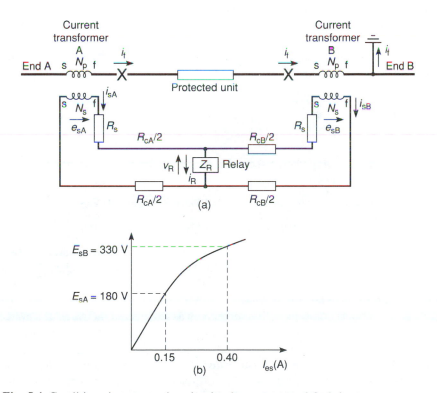

Fig. 5.4 Conditions in a protective circuit when an external fault is present.

R_{cA} and R_{cB} are unequal the following relationships would hold if the relay had zero impedance, i.e. if the voltage across it were zero at all times:

$$e_{sA} = i_{sA}(R_s + R_{cA})$$

$$e_{sB} = i_{sB}(R_s + R_{cB}) \tag{5.1}$$

and $$i_R = i_{sA} - i_{sB}$$

The relay current (i_R) is equal to the difference between the transformer secondary winding currents, i.e. ($i_{sA} - i_{sB}$), this being equal to the difference in the two exciting currents needed to supply the e.m.f.s e_{sA} and e_{sB}. In the limiting and impracticable case of cores with infinite permeability, the exciting currents would be zero and no relay current would flow. In practice, with cores having a finite permeability, a relay current would flow, its magnitude clearly increasing with the decrease of the permeability. In addition, the relay current would be greater, the greater the difference in the resistances of the two sections of the circuit.

It should also be noted that relay current would flow even when the resistances of the two sections of the circuit were equal, if the cores of the transformers had different permeabilities. Clearly, therefore, any lack of balance in the circuit will cause relay current to flow when the protected unit is healthy, the greatest values occurring in the event of large external faults.

Numerical examples based on the circuit shown in Fig. 5.4(a) are given below to illustrate the above behaviour.

1. When an external fault current of 3000 A flowed in the primary windings of current transformers with a ratio of 200/1, each of them would have secondary currents (I_{sA}, I_{sB}) of approximately 15 A. If the circuit resistances had the following values, $R_s = 2\,\Omega$, $R_{cA} = 10\,\Omega$, $R_{cB} = 20\,\Omega$, and the relay had zero impedance then the current transformers would have to provide the following e.m.f.s:

    Current transformer A: $E_{sA} = 15(5 + 5 + 2) = 180\,\text{V}$

    Current transformer B: $E_{sB} = 15(10 + 10 + 2) = 330\,\text{V}$

    Assuming that the current transformers had the excitation characteristic shown in Fig. 5.4(b), their exciting currents, referred to secondary levels, would be 0.4 A and 0.15 A. In these circumstances the relay current I_R would be approximately 0.25 A.

2. For the above conditions, if the relay was connected between the mid points of the conductors so that $R_{cA} = R_{cB} = 15\,\Omega$, then the current transformers would each have to produce an e.m.f. of $15(7.5 + 7.5 + 2) = 255\,\text{V}$. If, however, the current transformers had the excitation characteristics shown in Fig. 5.5 then the difference in the exciting currents and thus the relay current would be approximately 0.25 A.

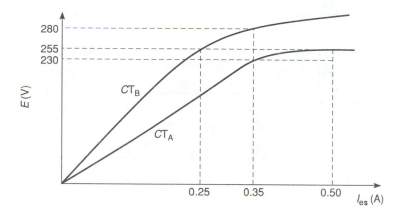

Fig. 5.5 Current transformer excitation characteristics.

Electro-mechanical relays require input energy to cause them to operate and therefore their windings possess finite impedance. As a result, the effects caused by asymmetry in their associated circuits must differ from that described above, but, as would be expected, the differences are small when so-called low-impedance relays, i.e. current-operated relays with low operating VAs, are used.

As is shown later in section 5.6, improved performance can be obtained in some applications by using high-impedance relays, i.e. voltage-operated relays. In practice, these relays have very low operating currents, but in the limiting case, for which the impedance would be infinite, zero current would be required and indeed the current flow in them would be zero at all times. Should such a relay be used in a current-differential scheme the current-transformer secondary currents and exciting currents would be constrained to be equal under all healthy and external fault conditions. In these circumstances, and for the first unbalanced condition considered earlier in connection with Fig. 5.4, each current transformer would produce an e.m.f. of 255 V to drive a secondary current of 15 A around the loop of resistance 34 Ω. As a result, a voltage of 75 V, i.e. $255 - 15(5 + 5 + 2)$, would be present across the relay.

Should the resistance be balanced by connecting the relay between the mid-points of the conductors (17 Ω on each side of the relay) but the transformers again have the excitation characteristics shown in Fig. 5.5, then, with equal exciting currents, the sum of their e.m.f.s would have to be approximately 510 V. In this instance, current transformers A and B would provide secondary e.m.f.s (E_{sA} and E_{sB}) of 280 and 230 V respectively and the relay at the mid-point would have a voltage of 25 V across it, i.e. $280 - 15 \times 17$ V.

It will be clear from the above that quite large voltages may be present between interconnecting conductors, particularly when large currents are

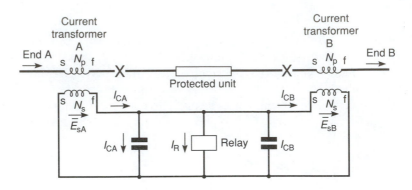

Fig. 5.6 Effect of capacitance between interconnecting conductors. $i_R = 0$ when $i_{cA} = i_{cB}$.

flowing to faults outside a protected zone. As a result, significant currents (i_{cA} and i_{cB}) will flow through the capacitance present between the conductors as shown in Fig. 5.6. In a completely symmetrical circuit this will not cause imbalance and zero current will be present in the relay. In an asymmetric circuit, however, in which the relay is not connected between the mid-points of the interconnecting cables, the capacitive currents flowing on the two sides of the relay (i_{cA} and i_{cB}) will not be equal and their difference will flow in the relay.

From the above, it is clear that the circuits of circulating current protective schemes should be symmetrical whatever types of relays are used. In particular, identical current transformers are desirable at both ends of a protected unit and the relay should be connected between the mid-points of the interconnecting cables. In practice, it is not always possible to comply with these requirements because different manufacturers may be supplying the equipment at the two ends of a protected zone and, in consequence, space or other considerations may preclude the use of similar transformers. Similarly it may be impractical to connect relays at a point midway along the run of the interconnecting cables. In the latter case, ballast resistors may be added in one section of a scheme to equalize the resistances in the two sections, and while this reduces the possible unbalance currents it does increase the e.m.f.s which the current transformers must be able to provide.

5.2.5 The setting of low-impedance relays

Current-differential schemes can in principle provide very low fault settings because they do not have to grade with any other schemes. There should therefore be no difficulty in obtaining operation for all interphase faults within a protected zone and indeed even quite highly-resistive faults to earth should be detectable.

In addition, very rapid or so-called instantaneous operation should be attainable.

It must be realized, however, that the minimum fault current at which operation will occur is not determined solely by the relay settings. This can be illustrated by considering the single-phase arrangement shown in Fig. 5.7 in which the fault current I_f is fed from end A of the circuit.

If an electro-mechanical relay which required a 2 VA input to operate at its setting of 0.2 A were used, the voltage drop across it at this level would be 10 V. This voltage, minus the probably insignificant voltage drop in the cable resistance, would appear across the secondary winding of current transformer B requiring an exciting current I_{esB} to induce the necessary e.m.f. (E_{sB}) (approximately 10 V) in the winding. The sum of the relay current (I_R) and the exciting current I_{esB} would flow through the resistances of the secondary winding of current transformer A (R_s) and the interconnecting cables (R_c). Current transformer A would thus have to provide an e.m.f. E_{sA} of approximately 12.4 V, i.e. $V_R + 0.2(5 + 5 + 2)$. The minimum fault current for which operation would occur would thus be $200(0.2 + I_{esA} + I_{esB})$, i.e. c.t. ratio $\times (I_R + I_{esA} + I_{esB})$ and this could be significantly greater than 0.2 pu or 20% unless the transformers had cores of low reluctance, i.e. cores with a large cross-sectional area and high permeability.

It is quite probable that full-load current (1 pu) could be flowing in the protected circuit when a highly-resistive earth fault is present. In such circumstances the fault and load conditions would be superimposed and higher voltage drops, transformer e.m.f.s and core fluxes would be present than those considered above. As a result, because of the non-linearity of the core-excitation characteristics, the increases in the exciting currents caused by the fault might exceed those above and require a higher minimum fault current for operation. Clearly the above effects should be taken into account when selecting

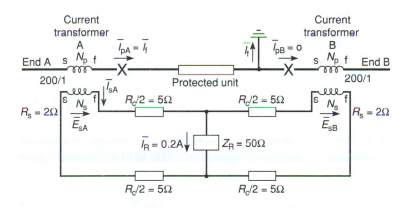

Fig. 5.7 Conditions in a symmetrical protective circuit when an internal fault is present.

relay settings. In some cases, the minimum relay-operating current would have to be significantly lower than that given by the product of the current transformer secondary rating and the minimum per-unit fault current level at which operation is required; for example, if transformers with rated secondary currents of 1 A are to be used and faults of 0.2 pu are to be detected, then the relay must be set to operate at a current of less than 0.2 A.

5.3 THE USE OF BIASING FEATURES

It will be appreciated from the preceding section that although current-differential schemes are basically very simple, great care must be taken in applying them, to ensure that they will operate correctly under all the conditions to which they may be subjected.

It is very desirable that they should be capable of detecting internal faults which cause low currents to flow, say down to 20% of the rated current of the protected unit. They must not operate, however, during external faults of the maximum current level which may occur, say 20 times the rated current, i.e. 2000%. To achieve such performance from the basic scheme would be difficult, because it would require the total imbalance resulting from mismatching of the current transformers and unequal circuit resistances to be less than 1%. Biasing features, which effectively increase the currents needed to operate relays when high currents are flowing, were therefore introduced many years ago to reduce the degree of matching needed. Biasing may be effected on an electro-mechanical relay by including one or more restraining windings in addition to the operating winding. Current flow in the restraining windings produces force or torque opposing relay operation whilst current flow in the operating winding produces force or torque in the direction needed to cause operation.

Because of the desirability of having symmetry in differential schemes, it is common practice to use relays with two restraining windings, these being connected as shown in Fig. 5.8. During healthy conditions or when external faults are present, restraint is required and therefore the forces or torques produced by the two restraining windings should be in the same direction when the currents in them (I_{sA} and I_{sB}) are in phase with each other.

It is clearly desirable that relay performance should not be greatly impaired when internal faults occur and the limiting case tends to be that of a highly-resistive earth fault on a unit which nevertheless continues to supply its rated current. This situation is illustrated in Fig. 5.9. Assuming, for simplicity, that the fault current (I_f) of 0.2 pu (40 A) is in phase with the normal current (I_{pB}) of 1 pu (200 A), then one restraining winding would carry 1.2 A and the other 1 A whilst the operating winding would carry 0.2 A. For operation to be obtained under this condition, the restraining force or torque would have to be less than the operating force or torque and indeed it would normally be very much less.

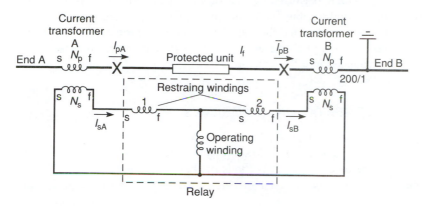

Fig. 5.8 A biased protective scheme.

In practice the amount of bias or restraint applied is normally expressed in percentage terms. As an example, with 2% bias applied, the minimum current needed for operation would increase by 0.02 A if both restraining windings were carrying a current of 1 A, hence a relay which operated at 0.178 A with zero restraint would operate at 0.20 A if rated current was flowing through the protected circuit on which a 0.2 pu fault was present. This performance is illustrated in Fig. 5.9.

Small amounts of bias are very effective when high currents are flowing to external faults. As an example, in the case considered above, 2% bias would raise the relay operating current from 0.178 A to 0.58 A if a current of 20 pu were flowing to an external fault, thus considerably increasing the allowable amounts of imbalance.

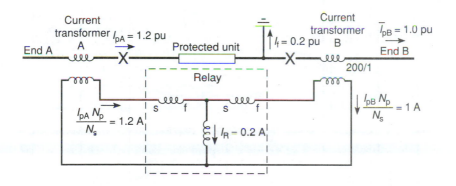

Fig. 5.9 Conditions in a biased protective scheme when an internal fault is present (minimum operating current of the relay without bias 0.178 A, bias 2%).

Clearly similar performance can be achieved using a relay with only one restraining winding, but to obtain the same percentage bias its restraining winding must have twice the number of turns of each of the two restraining windings considered above and there is little advantage to offset the asymmetry introduced into the scheme.

5.4 IMPLEMENTATION OF SCHEMES

The behaviour of current-differential protective schemes has been outlined above on a single-phase basis to simplify the treatment. Normally, however, such schemes are used to protect sections of three-phase networks and the arrangements adopted in such applications are considered in the following sections.

5.4.1 Units of short physical length

When units such as rotating machines are to be protected, the current transformers at the two ends of a protected zone are only a relatively short distance apart and each of the phases could be covered by independent schemes. It is normal, however, to have four interconnecting conductors, one of them forming a common neutral return, as shown in Fig. 5.10. Three relays, each of them having the same current setting, are needed and they are virtually inde-

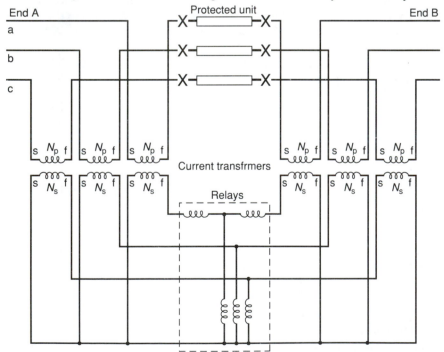

Fig. 5.10 Differential protection of a three-phase unit of short physical length.

pendent of each other. In the event of an internal single-phase to earth fault only one of the relays will operate, whereas for interphase faults two or all three relays will operate.

An advantage of this arrangement over one with three independent phases is that balanced components of the phase currents, i.e. positive- and negative-sequence components, do not flow in the fourth, or neutral, interconnecting conductor and thus the e.m.f. to be provided by each of the current transformers is reduced, there being a voltage drop on only one conductor instead of two.

5.4.2 Units of considerable physical length

Implementation of the basic current-differential scheme to protect units, such as distribution lines or cables, would often be too complex or expensive. As examples, the cost of providing four relatively long, interconnecting conductors to enable comparisons to be effected on each of the phases could be unacceptably high and the installation of relays at the mid-point of a unit could be impractical. These factors and modifications to the basic scheme which have been introduced to enable it to be used for these applications are considered below.

Reduction of the number of interconnecting conductors

It is clearly desirable that schemes should only require two rather than four interconnecting conductors when the ends of a zone are a considerable distance apart. Independence of the phase comparisons cannot, however, be retained using two conductors, because only one quantity can then be conveyed between the ends of a zone. This single quantity must be produced from the three phase currents at one end of a protected zone for comparison with a corresponding quantity produced at the other end. Such a quantity could be obtained by feeding the output currents of the three transformers to a network which would provide an output related to one or more of the sequence components of the input currents. Because positive-sequence components are present during all faults, signals derived from only this component could be used. During single-phase to earth faults, however, which occur much more frequently in practice than other types of faults, the positive-sequence component is only one third of the magnitude of the actual fault current and therefore the relays would have to be given very sensitive settings. As an example, to detect a fault current of 20% of the rated current of a protected circuit, the relays would have to be set to operate at 67 mA if current transformers with secondary windings rated at 1 A were used. This would increase the difficulty of meeting the requirement that relay operation should not occur when external faults, including large interphase faults, are present. A similar situation would arise if a signal related to the negative-sequence component of the currents were used and clearly a signal related to the zero-sequence component alone could not be used as it would not enable interphase faults clear of earth to be detected. In practice, therefore, a signal produced from two or three of the components must be used.

While such techniques have been used in some protective schemes, sequence networks tend to be somewhat complex because phase shifting is necessary to determine the 'a' phase positive- and negative-sequence current components ($\overline{I}_{1a}$ and $\overline{I}_{2a}$) as can be seen from the following equations:

$$\overline{I}_{1a} = \tfrac{1}{3}\,(\overline{I}_a + a\overline{I}_b + a^2\overline{I}_c)$$

and

$$\overline{I}_{2a} = \tfrac{1}{3}\,(\overline{I}_c + a^2\,\overline{I}_b + a\overline{I}_c)$$

in which a is the operator $1\,\lfloor 2\pi/3$ and $\overline{I}_a$, $\overline{I}_b$ and $\overline{I}_c$ are the phase currents.

In addition, the above equations are only valid when the currents are sinusoidal and therefore not applicable during the periods when transient components are present.

Simpler alternatives, which are suitable for all current waveshapes, have been widely used in current-differential schemes.

The simplest arrangement, which has been employed since the earliest years of the century, is to sum the outputs of transformers having different ratios in the three phases, say 300/1.25, 300/1 and 300/0.8 in phases 'a', 'b' and 'c' respectively, i.e. 240/1, 300/1 and 375/1.

As shown in Fig. 5.11, this would give the following output signals for primary currents of 300 A of each of the sequences.

Positive-sequence   = $(0.35 - j0.173)$A or $0.39\,\lfloor -0.46\,\text{rad}$ A
   output   relative to the 'a' phase primary current

Negative-sequence = $(0.35 + j0.173)$A or $0.39\,\lfloor 0.46\,\text{rad}$ A
   output   relative to the 'a' phase primary current

Zero-sequence   = 3.05 A in phase with the primary currents
   output

It is evident that this arrangement is much more sensitive to zero-sequence currents than to those of the other sequences and as a result it gives larger output signals for earth faults than for interphase faults, thus making it easier to obtain the necessary stability during external faults. With transformers of the ratios considered above, a signal of 0.39 A would be produced for a balanced three-phase fault of 300 A, whereas a signal of 1.25 A would be produced for an earth fault of 300 A on phase 'a'. It should be realized, however, that the behaviour obtained is phase dependent, the outputs for earth faults of 300 A on phases 'b' and 'c' in this instance being 1.0 A and 0.8 A respectively.

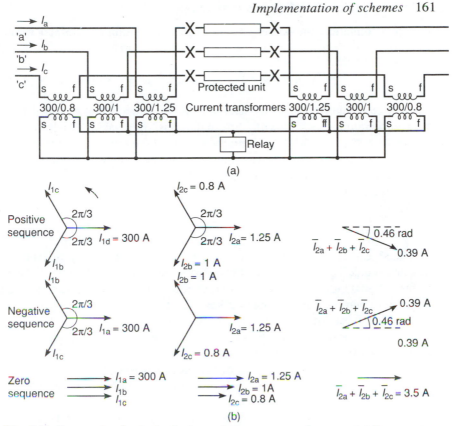

Fig. 5.11 Output signals obtained when using current transformers of different ratios.

Another alternative, which has been used widely for many years, is to feed the outputs of current transformers, of the same ratio in each of the three phases, to a summation transformer arranged as shown in Fig. 5.12. Assuming

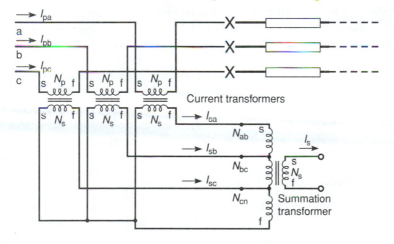

Fig. 5.12 Current transformers feeding a summation transformer.

ideal transformation, the output current of the summation transformer would be given by:

$$i_s = \frac{1}{N_s} \{ (N_{ab} + N_{bc} + N_{cn})i_{sa} + (N_{bc} + N_{cn})i_{sb} + N_{cn}\, i_{2c} \}$$

It will be recognized that this behaviour is the same as that obtained from three current transformers with different ratios. By making the turns N_{cn} greater than N_{ab} and N_{bc} the sensitivity to earth faults may be made greater than that to interphase faults. Tappings are usually provided on the primary windings of summation transformers to enable the sensitivities to be varied to suit particular applications.

The positions and connections of relays

As stated earlier, it is necessary that the circuits of current-differential schemes should be as symmetrical as possible to assist in achieving balance during external fault conditions. Ideally, therefore, relays should be connected at the mid-points of units. In many cases this is impractical because the ends of a unit will be at different substations and the provision of housing for relays at some other position could be expensive. In any event, if relays were so positioned extra cables would have to be installed to enable the relays to trip the circuit-breakers at the two ends of the protected zone.

To overcome these difficulties, relays could be installed at each end of a unit, using either of the arrangements shown in Fig. 5.13. With the connection represented on a single-phase basis in Fig. 5.13(a), i.e. with the relays effectively in parallel, each relay would carry current during external fault or healthy conditions and clearly these currents would increase with either decrease in the resistance of the relays or increase in the length or resistance of the interconnecting conductors. Numerical values are shown on Fig. 5.13(a) to illustrate this situation. The voltage across the relays of 150 V produced by an external fault current of 15 pu is very high and unlikely to be acceptable.

The arrangement illustrated in Fig. 5.13(b), with the two relays connected in series, is completely balanced and no current would tend to flow in the relays during healthy conditions or when external faults were present. This arrangement, which requires an extra interconnecting conductor per phase, increases the burden imposed on the current transformers, because they have to provide extra e.m.f.s to drive current through two relays in series. It has, nevertheless, been used quite widely. Again, biasing features are necessary to prevent maloperation occurring because of asymmetries in the circuit.

5.4.3 Balanced-voltage schemes

The above schemes are described as 'circulating-current' schemes, because currents flow around the loop formed by the interconnecting conductors dur-

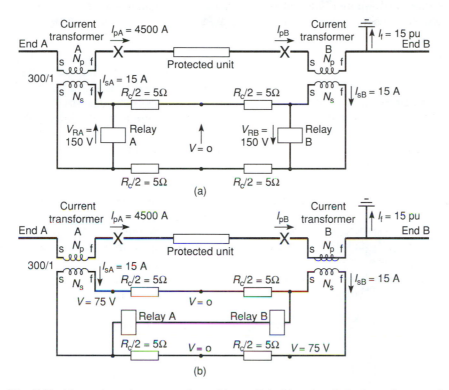

Fig. 5.13 Alternative relay connections. (a) parallel; (b) series. R_c is the resistance of each conductor.

ing healthy and external fault conditions. Such behaviour tends to require that the e.m.f.s which must be produced by the current transformers become very high for applications where the ends of protected zones are quite long distances apart, i.e. several miles, and therefore alternative schemes based on the original proposal by Merz and Price and designated as 'balanced voltage' or 'opposed voltage' are preferred for such applications.

A basic scheme of this class is shown in single-phase form in Fig 5.14(a), from which it will be seen that the secondary windings of the current transformers at the two ends of a unit are connected, via interconnecting conductors, in opposition to each other. In the event of a fault within the protected zone, both the primary currents and the e.m.f.s induced in the secondary windings of the transformers will be unequal and current will circulate through the interconnecting conductors. To enable the circuit-breakers at the two ends of the protected zone to be opened when this condition occurs, two relays, with their operating windings connected in series with the interconnecting conductors, are included. When the protected unit is healthy, however, and current is flowing through it, the current transformers will each have the same primary current and will produce e.m.f.s which oppose each other. Provided that the transformers are identical and performing at the same points on their excitation characteristics, they will produce equal instantaneous e.m.f.s, as shown in Fig. 5.14(b), and current will not circulate through the secondary windings and interconnecting conductors. It will be clear that the voltage between the two interconnecting conductors will then be equal to the e.m.f. provided by each current transformer and this will cause a capacitive current to flow between the conductors, its magnitude increasing with the length of the conductors. Half this current will be fed from each end of the circuit through the operating windings of the relays and unlike the situation which occurs with circulating-current schemes, the relay currents cannot be reduced to zero by making the circuit symmetrical.

As in the schemes described earlier, the relays may be set to operate for internal faults with current levels below the rated value of the circuit. Again, however, relay operation must not occur during external faults up to the highest level possible on the protected network. To satisfy this requirement

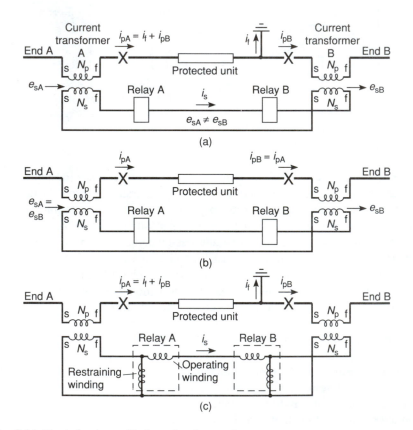

Fig. 5.14 The behaviour of balanced-voltage schemes.

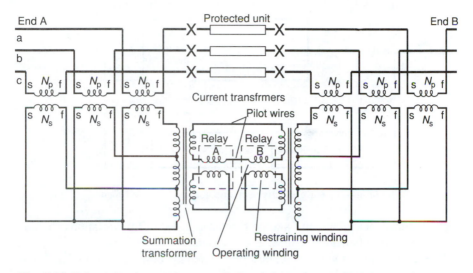

Fig. 5.15 Balanced-voltage scheme applied to a three-phase unit.

when low relay settings are used it is essential that the current transformers be well matched. In addition, biasing must be employed, this being achieved by connecting the restraining windings of each of the relays between the interconnecting conductors at their respective ends of the protected zone. This arrangement is shown in Fig. 5.14(c) and it will be realized that the voltages across the restraining windings will be high when large currents are flowing through protected units to faults outside their zones. The percentage of bias applied must be sufficient to ensure that maloperation will not be caused by imbalances in the protective circuit and current transformers and by the capacitive currents between the interconnecting conductors. In practice the percentage of bias may be varied using tappings on the restraining windings or by varying the impedances of the restraining-winding circuits.

Balanced-voltage schemes may be implemented for three-phase applications in similar ways to those used with the circulating-current schemes described earlier. Summation-transformers are widely used, a typical arrangement which only requires two interconnecting conductors being shown in Fig. 5.15.

5.4.4 Schemes to protect zones with more than two ends

In the previous sections, schemes suitable for protecting zones with only two ends have been considered. In such applications, pairs of current transformers which carry the same primary current under healthy or external fault conditions are required to balance with each other, e.g. the two transformers associated with phase a should ideally produce identical secondary currents. In

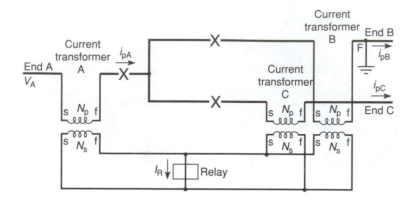

Fig. 5.16 Circulating-current protective scheme applied to a teed feeder.

such circumstances, the e.m.f.s to be produced and the flux variations should be the same in a balancing pair of transformers.

These conditions do not apply, however, when the outputs of more than two transformers are required to balance. Such situations arise when teed feeders or busbars are to be protected. Considering the teed feeder shown in Fig. 5.16, it can be seen that the input current (i_{pA}) must be equal to the sum of the output currents (i_{pB} and i_{pC}) at every instant when the protected unit is healthy. Under these conditions, the three current transformers would be producing different e.m.f.s at any instant and therefore operating over different flux ranges. As a result, zero current would only flow in the relay if the transformer cores were infinitely permeable or if their excitation characteristics were linear and the permeabilities of the cores were the same. Such conditions do not obtain in practice and a degree of unbalance must always exist. Clearly the resulting relay currents will be greater, the higher the currents flowing in the feeder. The behaviour is, however, dependent on the ratios of the currents flowing in the sections of such a feeder and in practice the current levels for different fault and operating conditions should be determined when such schemes are to be applied. Two examples are considered below to indicate the situations which can arise.

A teed feeder fed from a single source

Should an external fault occur just beyond one of the output ends of a teed feeder, say at point F at end B in Fig. 5.16, and there was only a source at end A, then the voltages at the output ends (v_b and v_c) would tend to be depressed and the load currents at those ends (I_{pB} and I_{pC}) would probably be well below the rated values. Currents well above rated levels would flow, however, in the primary windings of two of the current transformers (A and B) and these currents would have similar magnitudes and phases. The primary current of the third transformer would be low and in such circumstances it is probable that the degree of unbalance would be small.

A teed feeder fed from two or more sources

In a highly-interconnected network, it is possible that all the ends of a teed feeder would be connected to sources. In such circumstances, large currents could be fed into two ends of a feeder in the event of a fault just beyond the third end. The current leaving the third end which would be the sum of the other two currents could require an e.m.f. in the secondary winding of the current transformer at its end, necessitating very high flux density levels in its core whilst the other transformers would be operating at much lower flux density levels. Under such a condition significant non-linearity in the core excitation characteristics could lead to a significant unbalance current flowing in the relay.

5.4.5 The protection of busbars

Similar but even more extreme conditions than those described in the previous section may arise in schemes used to protect busbars. In these applications, several sources may feed a fault, external to the protected zone, one such situation being illustrated in Fig. 5.17. To avoid maloperation during this fault condition, balance would have to be achieved between the outputs of five current transformers, one of which would have a primary current much greater than those flowing in the others.

In these circumstances and also in teed-feeder applications, biasing must be employed to ensure that stability will be maintained when large currents are

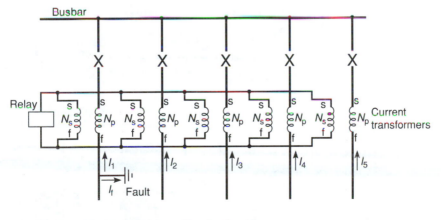

$$I_1 = -(I_2 + I_3 + I_4 + I_5)$$

Fig. 5.17 Scheme applied to a busbar.

being fed to faults external to the protected zone. Many different combinations of currents may be present during such faults, two being shown in Fig. 5.18. For the condition shown in Fig. 5.18(a) circuits 1 to 4 carry equal input currents (5 pu) to feed a fault fed by circuit 5. If bias were obtained by summing the input currents in circuits 1 to 4 and the current fed out of circuit 5, i.e. bias was proportional to $I_1 + I_2 + I_3 + I_4 - I_5$, then a bias proportional to 40 pu would be obtained.

If the same biasing arrangement was used for the fault condition shown in Fig. 5.18(b), in which a fault was fed by circuit 1, circuit 1 would provide a bias of -20 pu. Because of the direction of its current flow, circuits 2 to 4 would each provide a bias proportional to 5 pu and circuit 5 would provide -5 pu. As a result the magnitude of the bias would only be proportional to 10 pu, i.e. $(-20 + 5 + 5 + 5 - 5)$.

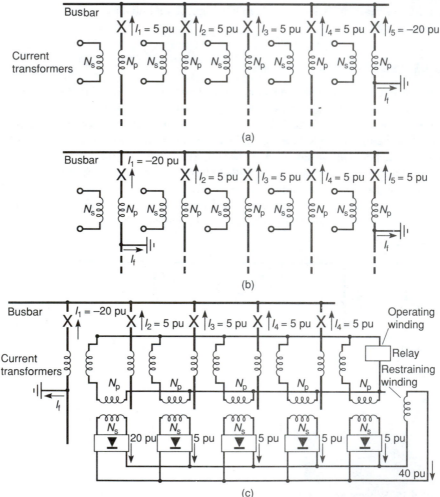

(a)

(b)

(c)

Fig. 5.18 Current distribution during faults and a biasing arrangement.

Two similar fault conditions would therefore produce very different levels of bias. This situation is clearly undesirable and in practice the larger value of 40 pu could be produced for a fault on any of the circuits by rectifying each of the circuit currents before summing them to obtain the biassing current. This arrangement is shown in Fig. 5.18(c).

5.5 EARTH FAULT PROTECTIVE SCHEMES

It was shown in section 4.4 of the previous chapter that earth faults may be detected by summing the currents flowing in the three phases of a circuit. In many applications, however, it is necessary that tripping of appropriate circuit-breakers should only occur for earth faults within a certain zone. As an example, the star-connected windings of power transformers are usually protected with sensitive earth fault protection. It is common to earth the star points solidly or through resistors and in these cases four current transformers connected as shown in Fig. 5.19 are employed. Such schemes, which have been described as 'balanced earth fault' or 'restricted earth fault' protection, should only operate for earth faults on the star-connected windings or the connections between the current transformers. Their behaviour differs from that of the schemes described earlier, in that balance must be maintained between current transformers in the three phases and neutral when faults occur outside the protected zone, rather than between two or more current transformers in the same phase.

This feature of summing the phase currents in earth fault schemes is significant. Under balanced conditions the phase currents have equal magnitudes and are displaced in phase from each other by $2\pi/3$ rad. These currents, which

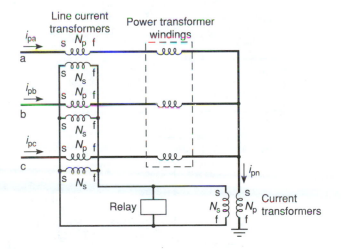

Fig. 5.19 Earth-fault protective scheme applied to the star-connected windings of a power transformer.

are fed to the primary windings of the current transformers, sum to zero. The excitation currents of the transformers will have fundamental components similarly displaced from each other, but in addition there will be harmonic components if the excitation characteristics are non-linear. Any third harmonic components which are present will be in phase with each other and will thus sum directly and flow in the relay. While this effect will normally be very small at rated currents, it could be significant in the unlikely event of an external three-phase short circuit.

As with other protective schemes, it is a requirement that relay operation should not occur during healthy conditions or when faults external to the protected zone are present and in addition it is desirable that the relay settings are low enough to detect small fault currents. The latter feature is particularly important when star-connected windings are to be protected as it determines the percentage of the windings that can be protected. This can be seen from Fig. 5.20, which shows a transformer winding with an earth fault on one phase.

The smaller the number of turns (N_f) between the star point and the fault position the lower is the voltage available to drive the fault current (I_f). If the impedance of the winding is neglected, the fault current is given by:

$$I_f = \frac{N_f V_{ph}}{N_p R_e} \tag{5.2}$$

in which R_e is the resistance in the fault path and N_p is the number of turns in the phase winding.

The fault current (I_f) is made up of I_{f1} flowing through the faulted part of the winding (N_f turns) and I_{f2} flowing in the remainder of the winding ($N_p - N_f$ turns), the latter current, which is needed to counterbalance the ampere-turns $I_{f1} N_f$, returning via a neutral point associated with the source, i.e. $I_{f1} N_f = I_{f2} (N_p - N_f)$ and $I_f = I_{f1} + I_{f2}$.

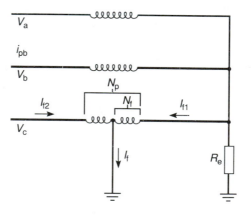

Fig. 5.20 Earth fault on a transformer winding.

The scheme would therefore sense the total fault current (I_f), I_{f2} flowing in one of the current transformers connected in the phases and I_{f1} flowing in the transformer in the neutral connection.

The simple expression for the fault current given in equation (5.2) clearly shows that small currents will flow in the event of winding faults to earth near the star point and therefore it is desirable that low fault settings be used. Nevertheless it must be accepted that the windings cannot be completely protected.

As an example, if a relay set to operate at 0.2 A were used and this protected 80% of the winding, then reducing the setting to 0.1 A would provide protection for faults on 90% of the turns. Such an increase in sensitivity would, however, increase the difficulty of maintaining stability during external faults.

A further factor, to which reference has already been made, is that the balancing current transformers may be carrying different currents during external faults and could thus be operating at different points on their excitation characteristics. In such circumstances, the core of one of the current transformers might be saturated for much of the time whilst the other transformer cores might not saturate. As a result, large unbalance currents could flow in the relay. Whilst such an extreme situation would probably not arise in balanced earth fault protective schemes, it could occur in schemes associated with busbars because several circuits could then be feeding relatively low currents to a circuit on which there is a fault.

In many cases, where the above conditions arise, the required stability during external faults may not be obtainable in schemes employing low-impedance relays which are set to provide high sensitivity to internal faults. Much improved performance may be obtained in such situations by using relays having high impedances, a practice introduced and proved over 40 years ago.

5.6 SCHEMES EMPLOYING HIGH-IMPEDANCE RELAYS

The behaviour of these schemes is quite complex. The basic principles on which they are based are simple, however, and these are illustrated in this section, using the balanced-earth fault scheme shown in Fig. 5.21(a) as an example.

To simplify the treatment, the following assumptions are made:

(a) that all four current transformers are identical;
(b) that the resistances of the connections to the secondary windings of the transformers are the same;
(c) that the relay circuit has an infinite impedance.

The last assumption requires that the relay must operate when carrying zero current and thus must be voltage operated, i.e. it must have a voltage setting rather than a current setting.

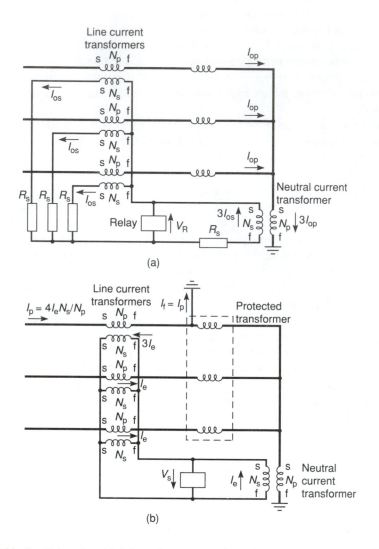

Fig. 5.21 Conditions in a high-impedance protective scheme applied to a transformer winding. (a) external fault; (b) internal fault.

With these protective schemes it is necessary that the voltage setting of the relay be high enough to ensure that maloperation will not occur when faults occur at points external to the protected zone. Clearly, the relay setting must be determined by considering the most onerous condition that may be encountered in service. Although such a situation could not arise in practice, the behaviour which would be obtained if high zero-sequence currents flowed as a result of an external fault is considered below because this represents a particularly onerous condition which may be readily examined and it is a condition which is often used in proving tests in laboratories, because of its simplicity.

For such a fault, the primary windings of the three current transformers in the phases would each carry currents of the same magnitude and phase whilst the primary winding of the neutral transformer would carry the sum of these three currents as shown in Fig. 5.21(a). As a result, the neutral current transformer should, under ideal conditions, produce an e.m.f. three times that produced in each of the other transformers. If it is further assumed that this situation would in practice cause the core of the neutral current transformer to be saturated throughout each half cycle, whilst the other transformers did not saturate, the most unbalanced condition possible would exist. Under this condition, the neutral current transformer would produce zero e.m.f. and each of the transformers in the phases would produce the e.m.f. needed to drive their secondary currents through their own individual circuits, i.e. $I_{0s} R_s = I_{0p} N_p R_s/N_s$, and in addition to drive the sum of their currents through the secondary circuit of the neutral transformer, i.e. $3I_{0s} R_s = 3I_{0p} N_p R_s/N_s$.

It is clear from Fig. 5.21(a) that this latter voltage would also appear across the relay circuit and that operation would not occur if the setting voltage exceeded the above value. It will be appreciated that the limiting conditions assumed above will not obtain in practice. For example, the neutral transformer will produce some e.m.f. and not only zero-sequence currents will flow, and therefore a factor of safety would exist if the above relay settings were used.

Having determined the minimum relay-setting voltage needed to maintain stability, the current needed to cause operation during internal faults must be determined to ensure that it is sufficiently sensitive. When high-impedance relays are used the fault setting is basically dependent on the excitation characteristics of the current transformers, as can be seen from Fig. 5.21(b).

When the setting voltage (V_s) is present across the relay it is also present across the secondary windings of each of the current transformers, each of which must then have an exciting current, referred to the secondary winding, of I_e. In the event of an 'a' phase to earth fault, the current transformers in phases 'b' and 'c' and in the neutral must each have secondary currents of I_e, supplied by the 'a' phase current transformer, i.e. its secondary winding must carry a current of $3I_e$. Because it also needs an m.m.f. of $I_e N_s$, it will need a primary current of $4I_e N_s/N_p$ to cause relay operation.

It is clear that satisfactory operation would thus be obtained if the proposed transformers had an excitation characteristic where for $I_e = I_{f\,min} N_p/(4N_s)$ the induced e.m.f. is equal to V_s, and the minimum fault current at which operation is required is given by:

$$I_{f\,min} = 4I_e N_s/N_p \qquad (5.3)$$

If the above setting is lower than that required, then the proposed current transformers could be used but the relay setting voltage would have to be increased so that higher exciting currents would flow. If, however, the setting determined ($I_{f\,min}$) is higher than that required then transformers with the lower exciting currents needed would have to be used.

In practice electro-mechanical relays require a finite operating current (I_R) and therefore they cannot be of infinite impedance. As a result, the expression for the minimum fault setting given in equation (5.1) must be modified to include the current supplied to relay, as follows:

$$I_{f\,min} = \frac{|4\bar{I}_e + \bar{I}_R| N_s}{N_p} A \qquad (5.4)$$

Whilst satisfactory performance can be achieved when the number of current transformers is small, as in balanced earth fault schemes, it can be difficult to obtain low fault settings when a relatively large number of transformers form a group, as in busbar-zone schemes, because the fault setting is dependent on the number of transformers (n), as can be seen from equation (5.5):

$$I_{f\,min} = \frac{|n\bar{I}_e + \bar{I}_R| N_s}{N_p} A \qquad (5.5)$$

5.7 RELAYS USED IN CURRENT-DIFFERENTIAL SCHEMES

Various forms of electro-mechanical relays have been used over the years since current-differential schemes were introduced.

In schemes of the basic form shown in Fig. 5.22, current circulates around the interconnecting conductors during healthy conditions and when external faults are present. Internal faults are detected by current imbalance and there-

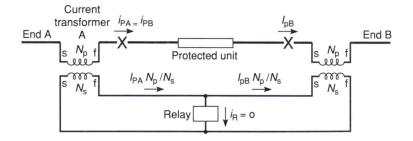

Fig. 5.22 Conditions in a current-differential scheme incorporating a low-impedance relay, during healthy conditions or when an external fault is present.

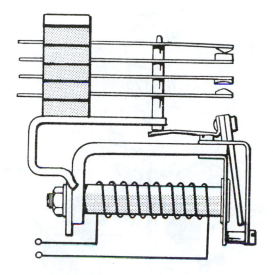

Fig. 5.23 Hinged-armature relay.

fore ideally the relay should have zero impedance so that the current through it tends to be proportional to the fault current.

Simple relays, with hinged armatures, of the form shown in Fig. 5.23 typically require 2 VA at the operating current and therefore a relay with a 0.2 A setting would have an impedance of 50Ω, i.e. the operating voltage would be 10 V.

Whilst relays of this type are suitable for some applications, they cannot present near zero or infinite impedances, i.e. they are not truly current or voltage operated, and therefore over the years more complex relays capable of operating with lower volt-ampere inputs have been produced. One such relay, which is shown in Fig. 5.24, was designated as being 'rotary sensitive' by its manufacturers, A. Reyrolle and Co Ltd. It is described in some detail below to illustrate the constructional features which it incorporated to enable the desired performance to be obtained. The armature, which was a vane mounted on a vertical spindle, was able to rotate in its own plane. The spindle had hardened pivots working in jewelled bearings, the lower bearing being spring loaded to reduce the stress on the pivot. The field-system consisted of two electromagnets in opposition so that when the operating windings were energized, the resulting flux threaded the armature, causing it to rotate, against the restraint of the control spring, into slots in the magnets. The ends of the armature were so shaped that, as it moved, the driving torque increased more rapidly than the restraining torque so ensuring positive operation at the minimum operating current.

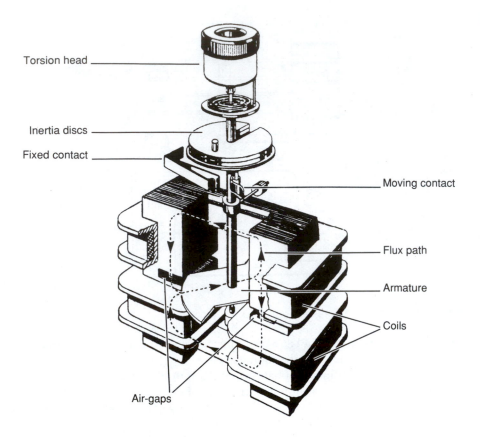

Torsion head

Inertia discs

Fixed contact

Moving contact

Flux path

Armature

Coils

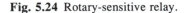
Air-gaps

Fig. 5.24 Rotary-sensitive relay.

Rotation of the armature and of the spindle closed sensitive leaf contacts, the contact travel being limited by two adjustable stops which, acting on both sides of the armature, balanced the operating shock and reduced the stress on the pivots when large currents relative to the operating value flowed in the windings during internal fault conditions. To relieve the leaf contacts of the full tripping-current duty, an auxiliary d.c. operated contactor was included.

Because of its construction, this relay required an input of only 30 m VA at its minimum setting.

Biasing features were possible by including an extra electromagnet which produced flux opposing rotation of the armature in the operating direction. This electromagnet had either a single tapped winding to allow the percentage of bias needed to be selected or a number of windings could be provided to allow biasing from several circuits to be effected. The burden of the restraining windings was lower than 0.5 VA.

In addition weights could be placed on a pan fitted to the spindle to increase the mass of the movement and thereby slow the relay operation slightly to ensure that it would be less liable to maloperate when high surges of current were experienced.

Relays of this type have been widely used in both balanced-voltage and circulating-current schemes.

Simple hinged-armature relays are used in high-impedance schemes, such as balanced earth fault protection. As stated earlier, such relays must operate at low current levels and therefore have windings of many turns. They therefore have quite a large impedance. An impedor (usually resistive) is connected in series with the winding to set the operating voltage level to that needed to maintain stability during external faults.

Modern electronically-based relays capable of operating at either particular current or voltage levels may also be used in current-differential schemes and because such relays have external power sources they may be arranged to have imput impedances of any desired values including zero and infinity.

5.8 APPLICATION OF CURRENT-DIFFERENTIAL SCHEMES

Because of their ability to discriminate between faults in a particular zone of a power system and those external to it, without relying on time delays, these schemes are widely applied.

The conditions they encounter in practice are dependent on the equipment or unit they protect. As an example, high exciting current surges may flow in power transformers for considerable periods after they are energized. In consequence, schemes of varying forms are used, as indicated earlier in this chapter, which has been confined to considerations of features and behaviour.

More detailed information on the schemes used for particular applications is provided in the following chapters.

REFERENCE

1. *Electrical Review*, 28 August 1908, The Merz–Price system of automatic protection for high tension circuits.

FURTHER READING

GEC Measurements, *Protective Relays Application Guide* (3rd edn), 1987, Chapter 10. *Power System Protection*, Vol 3, edited by The Electricity Council, 1981, (2nd edn), Peter Peregrinus Ltd.

Royle, J. B. (1990) Differential relay concepts and applications, *Canadian Electrical Association – Transaction of Electrical and Operating Division*, Vol 29, Paper No. 90-SP-157.

Further references are included in chapters dealing with applications.

6

The protection of transformers

It was stated in Chapter 2 that power transformers were first included in power supply systems in the last decade of the nineteenth century to enable electricity to be distributed from central stations at relatively high voltages to consumers spread over quite wide areas. This practice, which proved to be economically sound, has been continued around the world.

As the demand for electricity has grown, large generating stations have been built, each containing a few large three-phase alternators of high volt-ampere ratings, a typical value being 600 MVA. Because the output windings of these machines are fitted in slots in the steel stators, it is not practical to insulate their conductors to withstand the voltage levels which are used to obtain high efficiency transmission over long distances. Consequently, alternators are now wound to generate voltages of the order of 25 kV (line) and these are stepped up to transmission levels, such as 400 kV (line), by a transformer directly connected to each alternator, the combination forming a generator–transformer unit. Clearly the volt-ampere rating of a transformer must be the same as that of its associated alternator.

A so-called unit transformer is also connected to each alternator to step down the machine voltage and provide supplies to the auxiliary plant associated with the driving turbine and boiler or hydro plant. The electrical layout of a typical generator–transformer unit is thus of the form shown in Fig. 6.1.

In recent years relatively small gas-turbine driven alternators and machines in combined heat and power plants providing relatively low voltage outputs have been commissioned and again they are connected to large high voltage networks or loads via transformers.

Transformers of various sizes and ratings are provided to step down voltages from transmission to distribution levels, e.g. 400 kV/132 kV and further transformers are used to provide the voltage levels used by consumers (e.g. 415 V (line) for domestic supplies).

Many other transformers, both three-phase and single-phase, are used within power systems.

The behaviour of transformers and methods of protecting them are examined in this chapter after the following section in which historical information is provided.

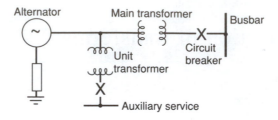

Fig. 6.1 Alternator–transformer unit.

6.1 HISTORICAL BACKGROUND

As stated earlier, power transformers were being quite widely used in electrical power systems in the last decade of the nineteenth century. Initially, fuses were the only form of protection available. Satisfactory circuit-breakers were produced shortly afterwards, enabling protective relays to be applied to transformers. Induction-type relays with inverse time/current characteristics were being manufactured by Brown Boveri, as stated in Chapter 4 and C. H. Merz and B. Price patented the current-differential scheme named after them, in 1904. They stated that the scheme could be applied to both transmission and distribution lines and also transformers. IDMT relays and differential schemes have been applied to relatively large transformers since that time and a later very significant development was the introduction of harmonic biasing in 1938 to ensure that differential schemes would not maloperate during the exciting-current surges which may flow when transformers are energized.

Earlier, L. M. Buchholz had developed relays which detect the production of gases and oil vapour and also oil surges produced by large arcs resulting from major breakdowns of insulation. These relays, which were named after their inventor, have been fitted to most large oil-filled transformers since that time and they are still being applied.

6.2 THE CONSTRUCTION AND BEHAVIOUR OF TRANSFORMERS

As stated earlier, many transformers, ranging from small single-phase units to extremely large, three-phase units associated with alternators and transmission lines, are in use in power-supply networks around the world. They all need to be protected, the complexity of the equipment used for this purpose being related to the size and importance of the individual transformers. In each application, the protective equipment should ensure that a transformer is not allowed to remain in service when abnormal conditions exist, either within it or external to it, which may cause it to suffer consequential damage, but it should not be taken out of service unnecessarily. As an example, a fault on one of the circuits supplied by a transformer should be cleared by opening the appropriate load circuit-breaker, rather than by disconnecting the transformer. To ensure, when selecting protective equipment for particular applications, that the above aims will be achieved, it is necessary that the construction of the transformers and their behaviour during both normal and abnormal conditions, be studied. In the following sections these factors are examined in a general manner.

6.2.1 Construction

All transformers whether they be required to step-up or step-down voltage levels, operate on the principle enunciated by Michael Faraday, namely that an e.m.f. is induced in a winding when the magnetic flux linking with it changes, i.e. the well-known relationship:

$$e = -N \, d\varphi/dt \tag{6.1}$$

By having two windings with different numbers of turns (N_p and N_s), both linking with the same changing flux (φ), two different e.m.f.s may be produced, i.e.

$$e_p = -N_p \, d\varphi/dt \quad \text{and} \quad e_s = -N_s \, d\varphi/dt.$$

Physical dimensions

In practice the windings are usually mounted concentrically around a limb of a core made up of laminations of a magnetic material. Such materials may not be operated satisfactorily at flux densities above about 1.7–1.8 T. As a result, cores of large cross-sectional areas and windings with large numbers of turns are needed in transformers which are to be connected to high voltage transmission networks. This is illustrated below:

Equation (6.1) may be rewritten as $e = -NA_c \, dB/dt$ in which A_c is the cross-sectional area of a core and B is the flux density in it. For sinusoidal conditions the flux density must be of the form:

$$B = B_{pk} \sin(2\pi f t)$$

allowing an e.m.f. $e = -2\pi f N A_c \, B_{pk} \cos(2\pi f t)$ to be produced.

The r.m.s. value of this e.m.f. is:

$$E = \sqrt{2} \pi f N A_c \, B_{pk} \, V$$

A transformer to operate on a three phase, 400 kV, 50 Hz network must therefore have a NA_c product for each of its star-connected high voltage windings of

$$NA_c = \frac{400 \times 10^3}{\sqrt{3} \times \sqrt{2} \times 50\pi \times B_{pk}}$$

i.e. $NA_c = 611.5$ if B_{pk} is to be limited to 1.7 T.

To convey the significance of this value, a transformer with a core of 0.5 m square cross-section, i.e. $A_c = 0.25\,\mathrm{m}^2$, would require a high voltage winding of 2446 turns. Should the transformer be rated at 600 MVA, the high voltage winding would need to have a conductor capable of carrying a current of 866 A. Clearly such a winding together with its insulation would be very large, as also would be the lower voltage winding associated with it. As a result, the magnetic core would not only be of large area but the flux path would be quite long.

Winding connections and core configurations

Small single-phase transformers usually have two or more windings, i.e. primary, secondary, tertiary, etc., mounted concentrically around one of the limbs of their magnetic cores, these being of either the core type or shell type which are shown in Fig. 6.2(a) and (b).

Three-phase transformers usually have three-limb cores of the form shown in Fig. 6.2(c), windings being mounted over each of the limbs. This is a very economical arrangement as it reduces the volume of magnetic material to the lowest possible level. It will be appreciated, however, that some magnetic flux must exist in non-iron paths if the sum of the three-limb fluxes in such a core is not zero. This condition will not arise when the three-phase voltages are balanced but it can be present during faults which cause zero-sequence components to be produced.

As pointed out above, transformers of high ratings are very large and in some cases three-phase units of the very highest ratings would be impracticably large. Their weights and bulk would be too great to allow them to be

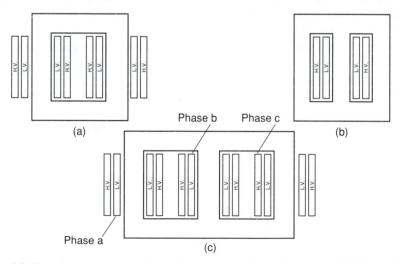

Fig. 6.2 Transformer constructions. (a) core-type, (b) shell-type and (c) three-phase transformer with a three-limb core.

transported from the manufacturer's premises to the sites where they are to be installed. In such cases, three separate single-phase transformers are produced and installed.

The windings of three-phase transformers may be connected either in star or delta, the arrangements adopted in particular applications being governed by the conditions which will exist in the network in which they are to be used.

It is normal practice for windings which are to operate at very high voltages to be connected in star. The star points are usually solidly earthed and, as a result, the voltages to earth at points on the windings increase linearly with their distances from the star point. This enables the insulation to be graded along the windings and minimizes the total amount of insulating material needed.

In many applications, lower voltage windings are connected in delta. This practice is, for example, almost invariably adopted on the transformers directly connected to the alternators in power stations. The output voltages of the machines are typically of the order of 25 kV (line) and the machine star points are earthed. It will be appreciated that no points are at earth potential on delta-connected windings, the two ends of each phase being at full phase voltage to earth and the centre points at half that value. This increases the amount of insulation needed above that required on a star-connected winding but at relatively low voltage levels the resulting extra cost is not great.

An advantage obtained by having delta-connected low voltage windings on the step-up transformers is that any third or higher-order triplen harmonic voltages produced by the alternators are not transmitted to the high voltage network. A further advantage is that the zero-sequence current components which will flow in the star-connected, high voltage windings in the event of an earth fault on the transmission network will be counterbalanced by a zero-sequence current circulating around the delta-connected, low voltage windings and no current will flow in the neutral connection of the alternator. These advantages are considered to be great enough to justify the extra insulation costs referred to above.

Whilst the majority of single-phase and three-phase power transformers are of the two-winding type, there are applications where the use of other types may be advantageous. As examples, three-phase auto transformers are used to interconnect high voltage transmission networks because they are cheaper than equivalent two-winding transformers and in some cases three-phase, three-winding transformers connected in star–delta–star are used.

Insulation and cooling

Many transformers and certainly all those of high ratings are mounted in tanks or casings and the cores and windings are immersed in oil. The ingress of moisture is prevented to ensure that the insulants will not be adversely affected. The oil not only forms part of the overall insulation but also conducts heat away from the cores and windings.

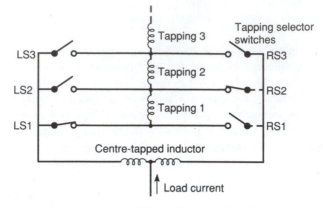

Fig. 6.3 Tap-changing arrangement. Note: The situation shown occurs during the change from tapping 1 to tapping 2.

Whilst modern transformers operate at very high efficiencies, the actual power losses which occur in the winding conductors and the magnetic core materials may nevertheless be very large. As an example, the power loss in a transformer rated at 600 MVA with an efficiency of 99.6% is about 2 MW. To dissipate such amounts of power without temperatures being raised to unacceptable levels, radiators are fitted and equipment is provided to circulate the oil.

Tap-changing equipment

Many transformers used in power networks are provided with tap-changing equipment to enable their ratios to be changed over a range, typically ± 15% of the nominal ratio, so that the voltages supplied to consumers may be maintained within the statutory limits. To enable reasonably fine control to be achieved sufficient tappings are provided on the windings to provide changes of ratio in steps of 1.5%.

Supplies to loads must be maintained during the switching operations needed to effect the ratio changes. One method of achieving this is to include a centre-tapped inductor which is brought into circuit, as shown in Fig. 6.3, during transition periods when connections are made to two adjacent tappings.

The tappings may be provided on either the low voltage or high voltage windings, it being common practice to use the latter alternative because of the lower currents to be carried by the switching equipment.

6.2.2 Operation during normal and external fault conditions

The voltage and current relationships which would be provided at every instant by an ideal, single-phase, two-winding transformer are:

$$v_s = -\frac{N_s}{N_p} \cdot v_p \quad \text{and} \quad i_s = -\frac{N_p}{N_s} \cdot i_p$$

Because the windings of practical transformers possess resistance and inductive reactance, the latter, as a result of fluxes which do not link both windings, and because there are power losses in cores and also that m.m.f.s are needed to set up the necessary core fluxes, the above ideal transformations are not achieved. The actual behaviour that will be obtained during any particular condition can be determined by using the well-known equivalent circuit shown in Fig. 6.4(a). The corresponding phasor diagram for steady state sinusoidal conditions is shown in Fig. 6.4(b).

In steady state, the exciting currents (I_e) are small relative to the rated currents, this being particularly the case with very large transformers; for example, values of 1%, i.e. $I_e = 0.01$ pu, are typical. The equivalent circuit components R_ℓ and X_m therefore have high values.

The series resistors (R_p and R_s) have very low values in highly efficient transformers but the series inductive reactances (X_p and X_s) are usually significant, particularly in large transformers, being made sufficiently large to limit the currents which will flow during external faults to acceptable levels. It is

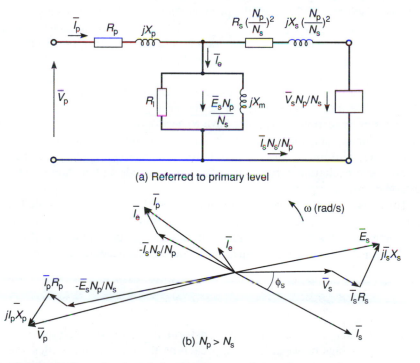

(a) Referred to primary level

(b) $N_p > N_s$

Fig. 6.4 Equivalent circuit and the steady-state phasor diagram of a two-winding transformer.

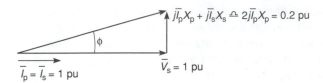

Fig. 6.5 Conditions when the output is at unity power factor. Note: The per-unit values of the phasors are shown.

common for each of the reactances to have a value up to 0.1 pu (i.e. $X_p = X_s = 0.1$ pu) in which event the maximum possible current fed to an external fault would not exceed 5 pu.

When the above conditions apply, the current transformation normally approaches the ideal value, i.e. $i_s \triangleq -i_p N_p/N_s$ but the voltage transformation is usually more removed from the ideal value at any instant. For example, during full-load conditions, a 0.2 pu voltage drop will be present across the inductive reactances. When operation is at unity power factor this drop primarily causes a phase displacement between the primary voltage and the reversed secondary winding voltage as shown in Fig. 6.5, i.e. $V_p/V_s \triangleq N_p/N_s$, the phase displacement being approximately $2|\bar{I}_p|X_p$ radians, I_p and X_p being in pu values. A phase displacement of about 0.2 rad would clearly be present with the reactances values quoted above.

When operation is at relatively low power factors the voltage transformation ratio may be significantly different to the turns ratio.

As stated in section 6.2.1, cores have to be worked over a wide flux-density range in steady state, e.g. between ± 1.7 T, to minimize their sizes. As there is considerable non-linearity in excitation characteristics over wide ranges, steady state excitation currents are not pure sinusoids and contain quite large odd-harmonic components.

The above comments have been concerned with normal steady state behaviour. There are a number of other conditions, however, which can occur and which must be taken into account when considering the suitability of protective schemes for particular applications. They are considered separately in the following sections.

Excitation-current surges

It has been stated above that many transformers are so designed that symmetrical flux swings between levels approaching the saturation flux densities of the core materials are present during steady state operation, to enable their costs and sizes to be minimized. It must therefore be accepted that saturation densities will be reached in the cores of such transformers during some normal and abnormal operating conditions. Particularly severe saturation may occur in periods after a transformer is connected to a source of supply, the degree of saturation being dependent on the instant in the supply voltage cycle at which the transformer is energized.

The flux-density variation (B) which would be required in the core of an ideal loss-free transformer after energization is given by:

$$B = \frac{1}{N_p A_c} \int_0^t v_p \, dt = \frac{V_{ppk}}{\omega N_p A_c} \left[\cos \alpha - \cos(\omega t + \alpha) \right] + B_{res} \qquad (6.2)$$

in which the supply voltage is $v_p = V_{ppk} \sin(\omega t + \alpha)$

N_p = primary winding turns

A_c = cross-sectional area of the core

B_{res} = Residual flux density at time $t = 0$

If the initial flux density (B_{res}) was zero and switching occurred at a voltage zero, the peak flux density would be twice that of the peak value of the sinusoidal component and indeed a multiple of three times could be approached if significant residual flux was present at the instant of energization.

The associated exciting current (i_e), which would be of the form shown in Fig. 6.6(a), would reach peak values of several times the rated current of the transformer and it would continue to do so indefinitely. This current contains a fundamental component, a quite large unidirectional component and also harmonic components, the latter decreasing the higher their order. The passage of the unidirectional current component through the resistance which must be present in the primary windings of practical transformers causes a unidirectional voltage drop and a corresponding component of opposite polarity in the e.m.f. which must be induced in the winding. This causes the unidirectional component in the flux-density variation to reduce from an initial value of

$$\left(\frac{V_{ppk}}{\omega N_p A_c} + B_{res} \right)$$

and eventually become zero. It must be recognized, however, that the rate of reduction of this component is very low in large transformers with very high efficiencies and consequently the durations of their exciting-current surges are long. Figure 6.6(b) shows a typical exciting-current waveform.

The above treatment, which was based for simplicity on a single-phase transformer, shows that the greatest current surges are experienced when transformers are energized at an instant when the voltage is zero. The same situation arises with three-phase transformers but with these there are six voltage zeros per cycle, i.e. two on each phase, and therefore it is highly likely that saturation associated with one of the phases will occur. A further factor which must be recognized is that whilst the fluxes of the three phases will

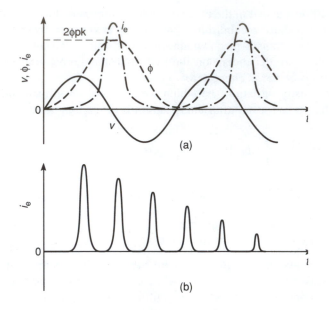

Fig. 6.6 Conditions when a transformer is energized.

always sum to zero and may therefore circulate in the magnetic material of the cores of three-limb transformers, the permeabilities of the limbs and yokes will differ considerably at any given instant. The m.m.f.s provided by the phase windings must clearly allow the required fluxes to be set up, but it must be appreciated that the phases are not independent of each other, and as a result the windings mounted on the limbs with higher permeabilities will contribute m.m.f.s to supplement that provided by the winding on a deeply-saturated limb.

Current relationships during healthy conditions

The steady state current relationships associated with healthy single-phase transformers are very simple, the magnitudes being given by:

$$|I_s| \; \Omega \; \frac{N_p}{N_s} |I_p|$$

The phase relationships are dependent on the winding connections and conventions, the primary and secondary winding currents being almost in phase or antiphase with each other, depending on the polarities adopted.

The same relationships apply to the currents in the windings mounted on the individual limbs of three-phase transformers, i.e. the primary and secondary winding currents have magnitudes almost proportional to the ratios of the turns in the windings in which they flow and they are almost in phase or antiphase depending on the conventions adopted. The currents in the windings are the 'phase' and 'line' currents when windings are star-connected whereas in delta-connected windings they are the 'phase' currents, the line currents being the differences of pairs of phase currents.

In the case of a transformer with delta-connected low voltage windings and star-connected high voltage windings, the ratio of the magnitudes of the line currents (I_{Lp} and I_{Ls}) under steady state balanced conditions is given by:

$$|I_{Lp}| \; \Omega \; \sqrt{3}\,\frac{N_s}{N_p}\,|I_{Ls}|$$

in which N_p and N_s are the numbers of turns in the low voltage and high voltage windings respectively. In addition, pairs of low voltage and high voltage line currents are displaced in phase from each other by $5\pi/6$ or $-5\pi/6$ rad depending on the connections of the windings. This situation is illustrated in Fig. 6.7(a,b). Clearly the general relationships which apply under all conditions, except when core saturation occurs, for the connections and conventions shown in Fig. 6.7, are:

$$i_{pa} \triangleq -\frac{N_s}{N_p}(i_{sa} - i_{sb})$$

$$i_{pb} \triangleq -\frac{N_s}{N_p}(i_{sb} - i_{sc})$$

$$i_{pc} \triangleq -\frac{N_s}{N_p}(i_{sc} - i_{sa}) \tag{6.3}$$

The comparisons made in differential protective schemes utilizing line currents should take the above factors into account and should be based on equations (6.3). It is evident that balance must be maintained, during healthy conditions, between three groups of currents, each group containing three currents; for example, balance must be obtained between i_{pa}, i_{sa}, i_{sb}.

It is clearly vital that balance be maintained when faults occur on the networks connected to transformers.

A common condition which may arise is that of a single-phase to earth fault on the circuit fed by a star-connected winding of a transformer. If the other winding is delta-connected, the resulting current distribution would be as shown in Fig. 6.8(a). It can be seen that equal and opposite currents would flow in two of the line conductors connected to the delta-connected windings, a condition normally associated with an interphase fault. During this fault, balance would have to be maintained in a differential protective scheme separately between the fault current (i_{sc}) and two of the input line currents (i_{pb} and i_{pc}), i.e.

$$i_{pb} \triangleq \frac{N_s}{N_p} i_{sc} \quad \text{and} \quad i_{pc} \triangleq -\frac{N_s}{N_p} i_{sc}$$

In the event of an interphase fault on the circuit connected to the star-connected winding of a transformer with a delta-connected primary winding, the input line currents would be in the ratio $2 : -1 : -1$, as shown in Fig. 6.8(b). In these circumstances, balance would have to be maintained in a differential scheme separately between each input line current and one or two secondary currents, i.e.

$$i_{pa} \triangleq -\frac{N_s}{N_p}(i_{sa} - i_{sb}) = \frac{2N_s}{N_p} i_f ; \quad i_{pb} \triangleq -\frac{N_s}{N_p} i_{sb} = -\frac{N_s}{N_p} i_f$$

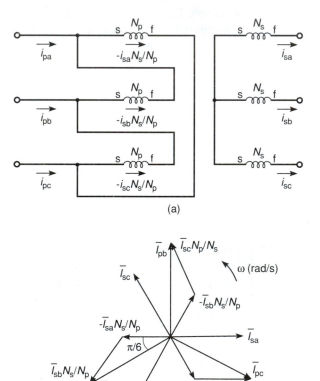

(a)

(b)

Fig. 6.7 Conditions in a delta-star connected transformer and the steady-state phasor diagram.

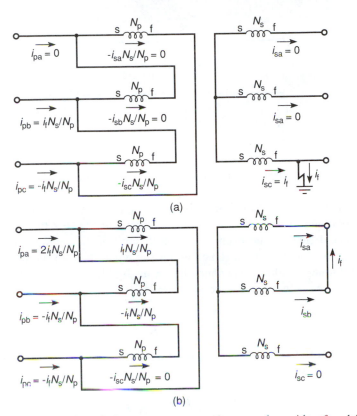

(a)

(b)

Fig. 6.8 Conditions when faults are present on the secondary side of a delta-star connected transformer.

and

$$i_{pc} \triangleq \frac{N_s}{N_p} i_{sa} = -\frac{N_s}{N_p} i_f$$

6.2.3 Behaviour during internal fault conditions

The majority of internal faults are associated with the windings. These may result from a number of causes such as overloading for long periods or overheating of cores. The fault currents which may flow during winding faults are dependent on the winding connections and earthing arrangements. These factors are considered separately below.

Earth faults on windings

In all cases, the current which will flow when an earth fault occurs will be dependent on the voltage to earth at the fault position and the impedance in the fault path.

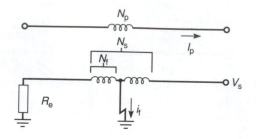

Fig. 6.9 Conditions when an earth fault is present on a winding.

When a single-phase transformer is earthed through a relatively high resistance at one end of one of its windings as shown in Fig. 6.9 then the current which would flow in the event of an earth fault on that winding would be almost proportional to the distance along the winding from the earthed end, the transformer winding impedance being negligible relative to the earthing resistance. The actual current would be given by:

$$i_f \simeq \frac{V_s N_f}{N_s R_e}$$

The current in the other winding to cancel the m.m.f. produced by the fault current would be:

$$i_p = \frac{N_f}{N_p} \cdot I_f \simeq \frac{V_s N_f^2}{N_p N_s R_e}$$

Clearly the current in the healthy winding is more affected by the position of a fault than that in the faulted winding. As an example, if an earthing resistor was fitted to permit full-load current (1 pu) to flow for an earth fault at the end of the winding remote from earth, then a fault current of 0.2 pu would flow for a fault 20% along the winding ($N_f/N_s = 0.2$) but a current of only 0.04 pu would flow in the healthy winding.

Obviously much higher currents flow in the event of earth faults on windings which are solidly earthed. Whilst the resistance of a section of a winding through which a fault current flows is proportional to the turns affected (N_f), the leakage reactance does not vary in such a simple manner, tending to vary with the square of the number of turns affected and also the path through which the leakage flux passes. The leakage reactance normally exceeds the resistance and thus has the greater effect on the fault current. When necessary, information on the variations of fault current and current in the healthy winding with fault position should be obtained from the transformer manufacturer so that the suitability of protective schemes may be determined.

Very similar conditions obtain when earth faults occur on earthed star-connected windings of three-phase transformers. As stated in section 6.2.2

(page 189), equal and opposite currents flow in two of the line connections if the other winding is connected in delta.

Delta-connected windings which are fed from earthed sources or which feed earthed loads have no points on them which are at earth potential and therefore currents will flow in any earth faults which occur on them. Faults at the ends of the phase windings will normally cause large currents to flow to earth, the actual levels being dependent on the impedance in the earthing connections and that of the source or load. The voltage to earth at the mid-point of a phase winding is normally half the phase voltage and again quite large currents should flow in the event of a fault in such a position. This fault level, which is the lowest that may be experienced, during an earth fault on a delta-connected winding may be calculated in any given case, using the appropriate system parameters. Little difficulty should be experienced in setting protective equipment to operate at levels which will detect these faults.

Interphase faults on windings

Interphase faults not involving earth clearly cannot occur on large three-phase transformers formed of three separate units mounted in their own housings. They may, of course, occur in three-phase transformers in which the windings are mounted on a single core, but in practice they rarely do so.

Because the voltages which exist between phases are high, the currents which will flow in the event of an interphase fault will usually tend to be large. If, however, a fault did occur between two points on different phases of a star-connected winding, each of them being near the star point, the voltage difference normally present between the points could be low but nevertheless quite a high fault current would flow because the leakage reactances of the faulted parts of the windings would be low. In these circumstances, the m.m.f.s produced by the fault current (i_f) would be cancelled by currents flowing in the faulted phase windings (i_{pa}, i_{pb}), as shown in Fig. 6.10. These currents

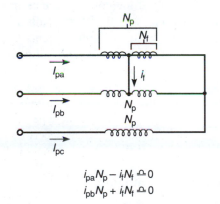

$$i_{pa}N_p - i_fN_f \triangleq 0$$
$$i_{pb}N_p + i_fN_f \triangleq 0$$

Fig. 6.10 Conditions when an interphase fault is present on a star-connected winding.

would approach zero as the fault positions approached the star point.

Similar conditions would clearly apply if interphase faults occurred on delta-connected windings and again the currents in the connections to the winding would be small if a short circuit occurred between two phases at positions near the point of connection of the phases affected.

Interturn faults

Steep-fronted voltage surges propagated along transmission lines to the terminals of transformers may, as explained later in section 6.2.4 (page 197), over-stress the insulation of the end turns of windings and cause interturn faults. The low leakage reactances of such groups of faulted turns may permit large currents to be driven through them by the power frequency e.m.f.s induced in them. As in the cases considered earlier, the m.m.f. produced in the faulted turns of a winding must be cancelled by one resulting from a current flowing through the whole winding. Once again, this current will usually be small, its magnitude decreasing as the number of faulted turns reduces.

Although the input currents to windings may be small when interturn faults are present, the current flowing in the fault path will nevertheless be high, as stated above, and as a result local heating will occur. If the condition were allowed to persist, further damage to the winding insulation would occur and a fault to earth would be established. This in turn could cause the core to be damaged. Clearly, therefore, it is desirable that the presence of interturn faults should be detected soon after their inception to keep the resultant damage and the associated repair costs to a minimum.

Interwinding faults

The windings of small single-phase transformers are usually concentrically positioned or they may be mounted in several sections adjacent to each other along the axial length of a limb of the core. Adequate insulation is provided between the individual windings and therefore faults rarely occur between them. When they do so, however, quite large currents may flow, because of the large voltages which may exist between the fault points on the windings.

Interwinding faults may also occur between the windings on one limb of a three-phase transformer. Very high voltages may be present between points on the affected windings and therefore large currents could flow in such circumstances. The likelihood of faults of this type is very low, however, because there is always much insulating material between the windings, this including spaces through which oil flows to cool the windings.

6.2.4 Causes of internal electrical faults

There are a number of operating conditions which can be permitted to continue for limited periods but which will cause damage that will lead to breakdown and electrical faults if they persist for long periods. These conditions, which should ideally be detected before their effects become serious, are considered below. There are also conditions which will cause faults to develop very quickly.

Overloading

Transformers may be operated at currents above their rated values for limited periods. The power losses in the winding conductors are proportional to the square of the currents in them and therefore overcurrents cause the temperatures of the conductors to rise to cause extra energy to be dissipated from them to their surroundings. The rate of rise of temperature tends to be very low, however, this being particularly so in large transformers and, as a result, low levels of overloading may be permitted to continue for fairly long periods without unacceptably high temperatures being reached. The actual periods allowable are dependent on the designs of the transformers and the operating conditions prior to overloading. Clearly if a transformer has been lightly loaded for a long period before it is overloaded then the temperatures of the parts within it will all be well below the limiting values when overloading commences and it may then be allowed to continue for a long period before the temperatures which can be produced under full-load conditions are reached. If, however, a transformer has been operating at its rated output level for several hours before it is overloaded then this new condition may only be allowed to continue for a limited period.

In practice, the permissible operating temperatures and their durations are dependent on the insulating materials which are used, because their lives are shortened if they are heated to temperatures above certain levels and indeed the deterioration is dependent on the periods of operation at high temperatures.

It will be appreciated, therefore, that regular overloading for long periods will eventually lead to breakdowns within insulation, even at normal voltage levels and that faults between windings, between turns or to earth may occur as a result.

Guidance on permissible overloading should be obtained from transformer manufacturers if it is not included in the original specification documents.

Overvoltage operation

When a transformer is operated in steady state at its rated frequency the flux density in its core is proportional to the e.m.f.s which must be induced in its windings and therefore to the voltages at its terminals.

Overvoltage operation of a transformer, because of the greater flux density swings associated with it, causes increased eddy-current and hysteresis power losses in the core material and therefore in turn it causes the temperatures of the core and other parts to reach above normal levels. In addition, the winding currents may also be increased by the overvoltages and certainly in large transformers, which operate over large flux ranges during normal conditions, the exciting currents will be relatively high and distorted because of the measure of saturation which will be caused. In addition, flux may be diverted away from the saturated sections of cores into iron paths in which the flux densities are normally very low. For example, core clamping-bolts, around which insulation is provided to ensure that the core laminations are isolated from each other, have low flux densities in them under normal conditions because of the high reluctance presented by the insulation. When core saturation is present, however, the flux densities in the bolts may be much higher leading to large power losses and heating in the bolts. If this condition was allowed to persist, the bolt insulation could be damaged to such an extent that currents might flow between the core laminations during subsequent normal operation causing further overheating which could damage the windings.

Operation at frequencies above and below the nominal value

At a given operating voltage, the peak flux densities in steady state are inversely proportional to the supply frequency and therefore, at below-normal frequencies, some saturation of the cores will occur in transformers which operate over large flux density ranges under normal conditions.

Generally the eddy-current losses in cores are not greatly affected by frequency variations because these losses are proportional to the product of the squares of the peak flux density and the frequency, i.e. $(B_{pk}^2 f^2)$, and they thus tend to be constant for a given voltage as the frequency varies. Hysteresis losses are proportional to the product of the area of the hysteresis loop and the frequency. In practice, these losses tend to reduce as the supply frequency falls because the areas of the loops do not tend to increase significantly at densities above the saturation level. At higher frequencies, the increase due to the increased frequency tends to be offset by the reduced areas of the loops at the lower-peak flux densities which are reached at the rated voltage level.

Nevertheless, the saturation caused by low-frequency operation can cause overheating of components such as core bolts, as described in the previous section.

In general the ratio of the per-unit values of the operating voltage to frequency should be maintained near unity to ensure that damage due to overheating will not occur.

Faults on systems connected to transformers

Short circuits on a network fed by a transformer can cause large currents to flow in it, resulting in relatively rapid heating of its windings. To ensure that the magnitudes of such currents are limited, transformers are designed to have particular leakage reactances ranging up to about 0.2 pu. Should this be the only impedance in the fault, then clearly the maximum current during a short circuit would be at least 5 pu. In general the higher the rating of a transformer, the greater is its per-unit leakage reactance. In addition, transformers are designed to carry their maximum fault currents for specified periods without being damaged by overheating. Relationships between these withstand periods and the fault current levels and leakage reactances are included in documents such as BS 171 : 1936 [1]. As an example, a transformer with a leakage reactance of 0.05 pu would limit the level of external fault currents to 20 pu and these could be withstood for 3 seconds. If necessary, such data should be obtained from manufacturers.

High fault currents, in addition to causing overheating, can cause large mechanical forces between winding conductors and stresses in other components. These forces and stresses reach their highest levels when the currents have their highest instantaneous values. Because these conditions occur during the first cycle of fault current when maximum asymmetry may be present, and cannot be prevented, transformers must be so designed that they will not be damaged by them.

A further factor which must be recognized is that steep-fronted voltage waves may be propagated along overhead lines as a result of lightning strikes near them, or to them or because of flashovers of insulators. Should these waves reach a transformer and be applied across one of its windings, they will distribute non-linearly across the winding in a manner dependent on the interturn and turn to earth capacitances it possesses. In practice, a very high percentage of the voltage would appear across a few of the turns near the high voltage end of the winding. Because of this, it is the usual practice to apply extra insulation to these turns, but even so the arrival of steep-fronted waves could cause interturn breakdowns. Whilst normal protective schemes would detect such breakdowns they could not prevent them. As a further measure, devices such as arcing horns and voltage limiters, which are outside the scope of this book, are fitted to overhead lines, and although these will operate when steep-fronted waves reach them, they nevertheless take a finite time to do so and it must be accepted that the initial sections of such waves may be applied across the windings of transformers.

Mechanical failures

Many transformers are mounted in oil-filled tanks, the oil being needed to maintain the insulating materials in good condition and also to conduct heat from the windings. In large transformers the oil usually circulates through air-cooled radiator pipes and in many cases the circulation rate is raised to the necessary level by pumps.

Clearly a reduction of the effectiveness of the cooling caused by the blockage of pipes by sludge or by a pump failure would not affect the transformer performance immediately but it would lead eventually to unacceptable heating and damage to insulation.

Similar unacceptable conditions would arise if oil leaked, for example, from a faulty drain valve.

6.3 THE APPLICATION OF PROTECTIVE SCHEMES AND DEVICES TO POWER TRANSFORMERS WITH TWO OR MORE WINDINGS PER PHASE

The protective equipment used with transformers is dependent on their importance and ratings. Single-phase and three-phase transformers used at voltage levels up to 33 kV and with ratings up to 5 MVA are often protected by fuselinks which must in all cases be capable of carrying the maximum exciting-current surges which may occur without operating. Guidance on such applications was provided earlier in section 1.9.8 (page 00).

The other forms of protective equipment which are used with transformers are examined in the following sections.

6.3.1 Inverse time/current relays

IDMT overcurrent and earth fault relays are used as the main form of protection on both the primary and secondary sides of relatively small power transformers. As with fuses, the current and time settings of these relays must be such that they will not operate when the maximum exciting current surges flow. They must also be such that they will enable correct discrimination with other protective equipment on associated networks to be achieved.

Relays of this type are also used to provide back-up protection for large transformers.

6.3.2 Current-differential schemes

As stated in earlier sections, current-differential schemes have been used to protect transformers since the early years of this century. They are now applied to all large transformers.

The basic principle of current-differential protection is that the current entering a circuit is equal to that leaving it under healthy conditions. To implement this principle when protecting a single-phase two-winding transformer, it would be necessary to compare the currents entering and leaving both the primary and secondary windings separately using four current transformers and two relays.

To protect a three-phase, two-winding transformer in this way would require the use of twelve current transformers and six relays, and should one of the windings be connected in delta, its current transformers would have to be positioned as shown in Fig. 6.11. With this arrangement, faults on the connections between the phases would not be detected and nor would faults on the connections between the windings and the external circuit.

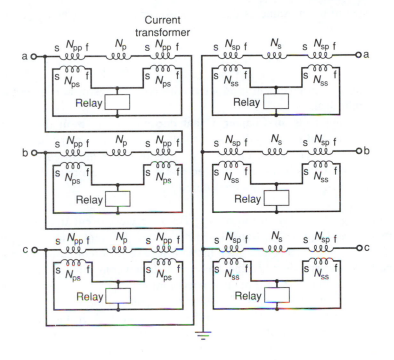

Fig. 6.11 The protection of individual windings.

To reduce the number of current transformers and relays required and also provide more complete protection, a modified form of differential protection is used in transformer applications, the principle employed being of m.m.f. or ampere-turn balance. The basic scheme and factors which must be taken into account when applying it to a single-phase transformer are examined in detail below and then its application to three-phase transformers is considered in later sections.

Basic scheme for use with single-phase two-winding transformers

Two current transformers must be provided, their primary windings being connected in series with the connections to the high and low voltage windings of the transformer being protected, as shown in Fig. 6.12. Because the currents in the power transformer windings will be different at all times, unless its ratio is unity, the primary ratings of the current transformers must also be different, as must their transformation ratios to enable balance to be maintained between their secondary currents during healthy conditions.

If the power transformer were ideal, it would require no exciting m.m.f. and therefore at every instant the sum of the winding ampere turns would be zero, i.e. $i_p N_{pp} + i_s N_{ps} = 0$. In this case the ratio of the current transformer ratios would have to be the same as that of the power transformer being protected, i.e.

$$\frac{N_{pp}}{N_{ps}} : \frac{N_{sp}}{N_{ss}} = N_p : N_s \quad \text{or} \quad \frac{N_{pp}}{N_{ps}} \times \frac{N_{ss}}{N_{sp}} = \frac{N_p}{N_s}$$

In practice, this arrangement is adopted with power transformers even though they are not ideal.

It was pointed out in Chapter 5 that current will flow in the relays of current-differential schemes during normal conditions or when external faults are present if the protective circuit is asymmetric. The modified scheme described above will be similarly affected and it must be recognized that it will also be affected by the factors considered below.

Mismatching of current transformers Because the current transformers on the two sides of a protected transformer must have different ratios, they cannot be of identical designs. Quite often they must have different numbers of turns in their secondary windings and their core dimensions cannot be the same because of the numbers of turns to be accommodated on them and because of the difference in the amounts of insulation needed on the primary windings. As a consequence, the exciting m.m.f.s needed at each instant by the two current transformers will usually be different, thus causing their transformation errors to be unequal. The resulting differences in the secondary currents will cause imbalance current to flow in the relay.

Mismatching caused by the power transformer steady state exciting current Power transformers are not ideal and do require exciting m.m.f.s to set up the necessary fluxes in their cores. The basis of the modified differential scheme, namely that the primary and secondary winding m.m.f.s sum to zero, i.e. $i_p N_p + i_s N_s = 0$, does not therefore apply in practice. As a result, an imbalance current (i_R), proportional to the power transformer exciting current, will flow in the relay at all times, i.e.

$$i_R = i_e N_{pp}/N_{ps}$$

In steady state, the exciting-currents of power transformers are quite low, this being particularly so for transformers with high ratings, values of 0.02 pu or less being usual. Using this value as an example, it can be seen that a relay current of 20 mA could flow in a relay employed in a scheme in which the current transformers had secondary current ratings of 1 A.

This imbalance, which will be present even when a protected transformer is lightly loaded, cannot be counteracted by percentage-bias windings on the relays.

Mismatching caused by power transformer exciting current surges Although the magnitudes of the exciting currents of power transformers are usually very small during healthy steady state conditions, they may rise to values many times the rated currents for significant periods after transformers are energized, as explained in section 6.2.2 (page 186). These currents, which are present only in one winding of a transformer, cause large currents to flow in the operating windings of the relays used in current-differential schemes and would cause operation unless effective biasing arrangements were provided. Their effect cannot be overcome by applying a relatively small percentage bias and therefore a large amount of bias obtained from one or more quantities which are only present in exciting currents is necessary.

Mismatching caused by tap-changing   Tap-changing is effected by operating switches to select a particular number of turns in either the primary or secondary winding of a transformer so that the desired output voltage is obtained and, as stated in section 6.2.1 (page 184), the turns ratio of a transformer may thus be varied in steps over a range.

To retain balance in current-differential schemes which compare the winding m.m.f.s, the current transformers associated with the tapped windings should also have tapped windings and their ratios should be changed appropriately when tap-changes are effected. If this practice were to be adopted it would be necessary to ensure that the secondary windings of the current transformers would never become open circuited and also that the sections of their windings being switched would not become short circuited. In addition, the changes would have to be effected at the same instants as those on the transformers being protected.

Because of the complexity and cost of such switching and the general feeling that switches should not be included in current transformer circuits, it is the practice to use current transformers with fixed ratios which will balance when the transformers being protected are operating at the mid-points of their tapping ranges. As a result, the flow of imbalance currents in the operating windings of the relays must be accepted at other times. These may clearly reach a percentage value equal to half the total tapping range of a protected transformer.

Implementation of the scheme

The basic scheme, as indicated above, employs two current transformers with their primary windings in series with the primary and secondary windings of the protected transformer as shown in Fig. 6.12. The current transformer ratios must be related as follows:

$$\frac{N_{pp}}{N_{ps}} = \frac{N_p}{N_s} \cdot \frac{N_{sp}}{N_{ss}}$$

They must be so connected that the currents in their secondary windings circulate around the loop they form with the interconnecting conductors when the protected transformer is healthy. The operating winding of the relay should be connected between the mid-points of the interconnecting conductors to form a symmetrical circuit.

Relays   Rough balance protective schemes are applied to relatively small transformers, where rapid operation in the event of faults is not essential. These schemes, which balance m.m.f.s, incorporate standard IDMT relays having current settings of 20–80%. Their time settings should be such that they will not be operated by the exciting current surges which may flow into the transformers protected by them and they should also grade satisfactorily with other IDMT relays connected to their associated networks.

Schemes applied to large transformers must operate rapidly when faults occur within their protected zones and therefore they must detect relatively low fault currents and must not operate when their protected transformers are healthy. The settings and biasing arrangements needed to provide the necessary performance are considered below.

Settings   Relays must have a minimum operating current which will ensure that faults on most of each of the windings will be detected. Clearly the percentages of the windings protected in the event of earth faults is dependent on the method of earthing. When solid earthing is employed very high percentages are protected by a relay with a setting of 20%, i.e. 0.2 A when current transformers with 1 A secondary ratings are used. It was shown, however, in section 6.2.3 (page 191), that should an earthing resistor be used, which would allow rated current to flow for an earth fault at the high voltage end of a winding, then a fault at a point 20% along the winding from its earthed end would cause a current of only 0.04 pu to flow in the other (healthy) winding. As a result a relay setting of 4% (0.04 pu) would be required to ensure that faults on 80% of the earthed winding would be detected.

Percentage-bias windings   As stated in the previous chapter (section 5.3, page 156), small percentages of bias are very effective in preventing incorrect relay operations occurring as a result of mismatching when faults are present at points outside a protected zone. Such biasing will certainly ensure that maloperations do not occur because of mismatching resulting from current transformer errors, protective-circuit asymmetry and the normal exciting current of a protected transformer.

It is also advantageous to have two bias windings, as shown in Fig. 6.12, to maintain symmetry.

Higher bias percentages are necessary when a transformer with tap-changing facilities is to be protected because significant imbalances will occur when operation is taking place at the upper or lower ends of the tapping range. As an example, if a transformer is operating at full load on a + 15% tapping then

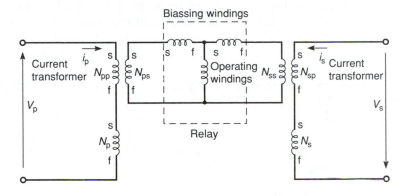

Fig. 6.12 Protective scheme based on m.m.f. or ampere-turn balance.

a current of 0.15 pu will flow in the relay, i.e. 0.15 A when current transformers with secondary ratings of 1 A are used. In such circumstances, a 20% relay setting (0.2 A) would not provide a significant margin to allow for other possible mismatching.

Whilst higher relay settings could be used, this would have the serious disadvantage of increasing the amount of the windings on which earth faults would not be detected. The alternative of using higher amounts of percentage bias is preferable.

Harmonic biasing   As explained in section 6.2.2, an exciting current with peak values of many times the rated current of a transformer may be present during significant periods after it has been energized and, as a result, large currents will flow during such periods in the operating winding of a relay used in a scheme of the form shown in Fig. 6.12.

It was realized many years ago that this effect could not be counteracted by using percentage-bias windings and methods of dealing with the situation were sought.

One possible solution was to open the relay tripping circuit for a short period each time a protected transformer was energized. During these periods reliance would have had to be placed on other protective equipment operating should a transformer be faulty. This was not considered to be an acceptable practice and therefore novel biasing methods were examined.

In 1938, Kennedy and Hayward published a paper [2] in which they quoted values, obtained experimentally, of the unidirectional, fundamental and harmonic components present in typical internal fault currents and exciting current surges. Because the values of the unidirectional, second, third and fifth harmonic components in the exciting current surges were much higher than those present during internal fault conditions, they developed a relay which contained two filter circuits. One of these allowed only the fundamental component of any imbalance current to flow in the operating winding of the relay whilst the second filter was tuned to block the fundametnal component of the current but permit all the other components to be fed, after rectification, to a bias winding. The relay was tested and it was shown to perform satisfactorily.

Subsequently the effectiveness of the individual components in biasing relays was studied further.

It was recognized that large slowly decaying unidirectional components are present in exciting current surges but it was found that these components could not be reproduced faithfully by current transformers because of the large unidirectional changes of core flux required by them. It was realized that the secondary winding currents of current transformers would fall to zero and become of the opposite polarity after relatively short times when their primary windings carried decaying unidirectional currents and that bias derived from unidirectional components could thus reach very low values and be ineffective, whilst quite large currents resulting from exciting current surges could still be flowing in the operating windings of relays. It was apparent, therefore, that acceptable performance would not be obtained from schemes employing only bias derived from the unidirectional components of currents.

It was also realized that only a limited amount of bias obtained from these components could be applied because of the quite large unidirectional components which can be present initially in the currents fed to internal faults on power transformers.

It was shown in other studies that the large second harmonic components present in exciting current surges could be reproduced faithfully by current transformers and that these current components could provide the necessary bias to ensure that relay operation would not occur as a result of such surges. It was found, however, that significant second harmonic components can be present in the secondary winding currents of current transformers if their cores saturate when high currents are flowing as a result of internal faults on power transformers. It was then clear that either the degree of second harmonic biasing applied in a scheme should be limited to ensure that relay operation would not be prevented under such conditions or that separate unbiased high-set relays should be provided to clear these faults.

The possibility of obtaining bias from the relatively large third and fifth harmonic components present in exciting current surges was also considered. Whilst such components are not present in the currents which flow when internal faults are present on power transformers, they can nevertheless be present in the secondary winding currents of current transformers when their cores are driven into saturation for parts of half cycles. It was recognized therefore that any bias obtained from these components would have to be limited.

As a result of these studies, several schemes employing relatively simple analogue filters were developed to provide biasing for current-differential schemes which were produced about 1950. In one arrangement filters were used to extract the second harmonic components from the imbalance currents flowing in the protective circuit, the extracted currents being fed through bias winding on the relays which had the number of turns needed to ensure satisfactory performance. Because the filters did not have narrow enough pass bands, small proportions of the unidirectional and third and fifth harmonic components of the exciting currents also flowed through the bias windings of the relays. The extraction or diversion of some of the imbalance current to the bias windings reduced the current flowing in the operating windings of the relays, thus further helping to reduce the possibility of incorrect operation being caused by exciting-current surges.

In an alternative arrangement [3], simple LC series-tuned circuits were used to permit only the fundamental component of the imbalance currents to flow in the operating windings of the relays. The other components were fed to the bias winding. In this scheme the relays were set so that they would not operate when their bias winding current exceeded 15% of the current in the operating windings. The minimum operating current of the relays, i.e. when there was zero bias, was 15% of the secondary rating of the current transformers and the minimum operating time was about 40 ms.

When high-current internal faults occur, the cores of the current transformers may saturate, causing their secondary winding currents to contain second and third harmonic components, as stated earlier. These components together with the unidirectional components present in the fault currents may provide sufficient bias in schemes of the above type to delay relay operation until the transients fall to low levels. Unbiased relays were therefore often connected in series with the operating windings of the main relays, the extra relays having current settings above the levels produced by exciting current surges. These relays then provided rapid clearance of high current internal faults.

Recent developments   Current-differential protective schemes of the type described above, with both percentage and harmonic bias, have been widely applied to power transformers since 1950. Developments in electronic equipment in recent years have enabled relays with both non-linear percentage biasing and an improved method of biasing associated with exciting current surges, to be introduced.

In older schemes, percentage biasing was obtained by having bias windings, with fixed numbers, of turns connected so that the current circulating in a current-differential scheme flowed through them. As a result, the bias was directly proportional to the current and a fixed percentage bias was thereby provided. There are, however, advantages in having non-linear or variable percentage biasing. At currents up to rated level a relatively low percentage bias, say 20%, is desirable to maintain high sensitivity for internal faults and it is adequate to counter the small mismatches which occur at low current levels and also any imbalance caused by tap changing. During external faults when currents well above the rated level are flowing, saturation may occur in one or more of the current transformer cores and large imbalance currents may flow as a result. To counter such situations a high percentage bias slope, of say 80%, is desirable. Relays [4,5] with such characteristics, e.g. 20% bias up to rated current and a bias slope of 80% above rated current, are now produced.

Because of the possibility of the clearance of high-current internal faults being delayed by harmonic biasing, which was referred to earlier, an alternative method of inhibiting relay operation during exciting current surges has been introduced recently. It is based on detecting the significant periods in each cycle during which an exciting surge current is almost zero. This feature, which can be seen in Fig. 6.6(b), is very pronounced and as it is not present in the currents associated with internal faults it can be used to inhibit relay operation. In practice, the imbalance currents in current-differential schemes are monitored because they tend to be proportional to the exciting currents when switching surges are present.

The current transformers feeding schemes using this technique should be so designed that they will not saturate when internal faults occur with current magnitudes up to the levels of exciting current surges. This is necessary to ensure that their secondary currents, when such internal faults are present, will not be near zero for significant periods as a result of core saturation. In addition, internal faults which cause very high currents to flow should be detected separately by employing techniques which will provide rapid operation and do not involve bias.

A relay which incorporates the above features is described in more detail in reference [4].

Application to three-phase transformers with two windings per phase

When current-differential schemes are applied to three-phase transformers account must be taken of the connections of their windings. Six current transformers must be used and their secondary windings must be so connected that significant currents do not flow in the three relays unless internal faults are present.

As an example, if a transformer has delta- and star-connected windings arranged as shown earlier in Fig. 6.7, and for which the winding-current relationships under healthy conditions are as expressed in equation (6.3), then the secondary windings of the current transformers on the low voltage side must be connected in star and those on the high voltage side must be connected in delta.

The ratio of the sets of currents transformers on the two sides must be different satisfying the condition given below:

$$\frac{N_{pp}}{N_{ps}} = \frac{N_s}{N_p} \cdot \frac{N_{sp}}{N_{ss}}$$

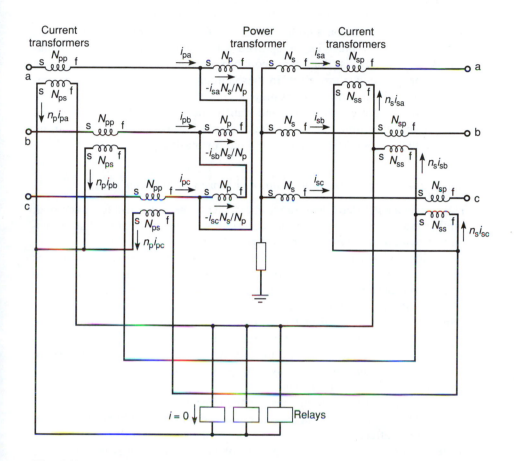

Fig. 6.13 Protective scheme applied to a delta-star connected power transformer.

To balance $n_p i_{pa} = n_s(i_{sb} - i_{sa})$, but $i_{pa} = \dfrac{N_s}{N_p}(i_{sb} - i_{sa})$ $\therefore$ $n_s/n_p = N_s/N_p$

i.e. $\dfrac{N_{sp}}{N_{ss}} = \dfrac{N_s}{N_p}\dfrac{N_{pp}}{N_{ps}}$

Note: The above assumes that the power transformer and the current transformers are ideal.

and the interconnections between the two groups of current transformers must be as shown in Fig. 6.13.

The relays should be set as sensitively as possible whilst ensuring that they will not operate during healthy conditions including exciting-current surges or when faults are present at points external to the protected zone. To enable this objective to be achieved, both percentage and harmonic or other biasing must be employed.

Operation of one or more of the relays should be obtained for all types of faults on the delta-connected winding, for interwinding faults, and for inter-phase and earth faults on a high percentage of the star-connected windings.

A factor which must be recognized when a transformer has a delta-connected winding is that zero-sequence currents and triplen harmonic currents may circulate around it. In these circumstances third harmonic components in exciting current surges may be confined to a delta-connected winding and not be present in the current transformers included in the input and output line connections to a star-delta connected transformer. As a result, no third harmonic components can be derived for biasing purposes.

Application to transformers with three or more windings per phase

The arrangements described above may clearly be extended to cover both single-phase transformers with three or more windings and three-phase transformers with three or more windings per phase.

Considering, as an example, a three-phase transformer with three windings per phase, nine current transformers are necessary, and to enable balance to be achieved during healthy conditions there will be groups of current transformers with three different ratios.

In such applications, mismatching similar to that considered earlier will be present, indeed it may be greater, and again imbalance currents will flow during exciting current surges. Satisfactory behaviour may nevertheless be obtained by adopting suitable relay settings and employing adequate percentage and harmonic or other biasing.

It has only been possible to consider the most commonly used connections of three-phase transformers in the preceding sections but it will be appreciated that current-differential schemes can be applied successfully to all transformers. Basically, the currents in each of the windings must be monitored using current transformers and their ratios and secondary winding interconnections must be such that relay operation will not occur when the protected transformer is healthy.

6.3.3 Restricted earth fault protection

It will be appreciated from the previous section that extremely low relay settings may not be used in overall current-differential schemes of the type described because of the mismatching which may occur and because of the need to ensure that operation will not be caused by exciting current surges.

It will also be realized that the percentage of a star-connected winding on which earth faults will not be detected by such schemes may be quite high when relays with relatively high current settings are employed. This is particularly so when significant impedance is present in the neutral connection to limit the levels of earth fault currents.

The above behaviour is not acceptable in many cases and therefore restricted earth fault protection is applied to star-connected windings. This form of protection, which was described in section 5.5, also belongs to the current-differential group. As stated earlier, it employs four current transformers, each of the same ratio, one having its primary winding in series with the neutral connection, the others being connected in the connections to the three phases. When the protected winding is healthy, or in the event of interphase faults not involving earth, the secondary currents should circulate and balance at all times, the current in the operating winding of the shunt-connected relay being zero. When an earth fault occurs, however, balance is not achieved and a current proportional to the fault current would flow in the relay winding if it had zero impedance.

In practice, some mismatching occurs, possibly due to residual fluxes being present in the cores of the current transformers at the instant when a fault occurs. To prevent relays with low-impedance windings operating incorrectly in these circumstances, percentage bias is usually necessary.

The alternative arrangement, shown in Fig. 6.14, in which the relay circuit has a high impedance enables improved performance to be obtained. The basic behaviour of this form of protection was explained in the previous chapter (section 5.6, page 171). In principle the relay is set to operate at a voltage higher than those which can be produced across it during external faults as a result of one or more of the current transformers saturating whilst the others remain unsaturated. A detailed theoretical and experimental study of this protective scheme, made by Wright [6], confirmed that satisfactory performance can be obtained under all conditions, provided that appropriate relay settings are used and that the current transformers have the required characteristics.

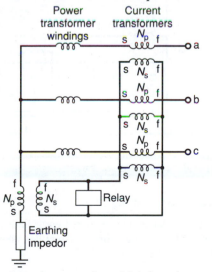

Fig. 6.14 Protective scheme incorporating a high-impedance relay.

This particular scheme has been widely applied since 1950 and it is still used successfully today. It is usual to arrange for operation of the relay to occur when fault currents of 20% or more of the rating of the protected winding are flowing from it to earth. Clearly this scheme, which effectively compares the current flowing into the phase windings with that flowing to the neutral, is not affected by the exciting currents in the windings and the conditions needed to ensure that relay operation will not occur at current levels up to the maximum which can be reached during faults can readily be met.

6.3.4 Combined differential and restricted earth fault protection

The application of separate current-differential and restricted earth fault protective schemes to a three-phase, two-winding, power transformer requires that ten current transformers be provided, six for the current-differential scheme and four for the earth fault protection. In addition, further current transformers will be necessary for other purposes such as metering, back-up protection and unit-type protective schemes covering adjacent zones.

In some cases, therefore, because of the difficulties of accommodating large numbers of current transformers within switchgear or other power equipment, it is considered to be advantageous to reduce the number of transformers needed by feeding current-differential schemes and restricted earth fault

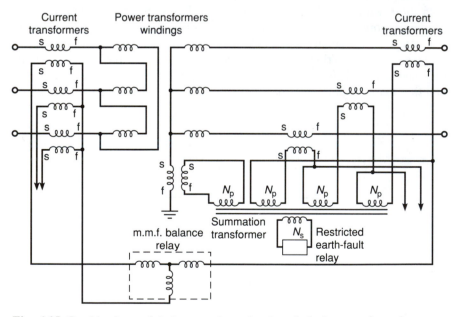

Fig. 6.15 Combined m.m.f. balance and restricted earth fault protective scheme applied to a delta-star connected power transformer. Note: For simplicity only one m.m.f. balance relay is shown.

schemes from the same line current transformers. As a result, the number of current transformers may be reduced from ten to seven. One arrangement which is commonly adopted to achieve this reduction when protecting a three-phase, delta/star connected power transformer is shown in Fig. 6.15, from which it will be seen that an auxiliary summation current transformer is included to supply the operating winding of the earth fault relay. Clearly the summation transformer and the other current transformers must be so designed that the necessary current balances will be achieved in both protective schemes unless faults are present within the protected zone.

6.3.5 Earth fault protection of delta-connected windings

Should earth faults occur on delta-connected windings which are connected to networks on which there are earthing transformers or on which there are earthed neutral points, then quite large currents will normally flow because all points on the winding will be at voltages of at least half the phase value, as stated earlier in section 6.2.3 (page 193). Such faults should normally be detected by current-differential protective schemes, but nevertheless separate earth fault protection is often applied to delta-connected windings. This protection is basically the same as the restricted earth fault protection described in section 6.3.3 (page 208) except that only three current transformers are used, their primary windings being connected in the line connections to the winding, as shown in Fig. 6.16. Either high-impedance or percentage-biased low-impedance relays may be used with fault setting of 20% or less. These will operate for earth faults on the protected winding because the three line currents will not sum to zero for these conditions. They will not operate for faults on the associated network, however, because delta-connected windings cannot provide zero-sequence currents and the other current components add to zero at all times, i.e. the line currents must always sum to zero unless there are faults on the windings.

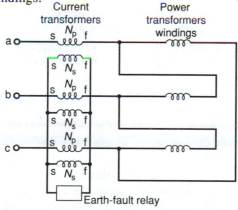

Fig. 6.16 Earth-fault protective scheme applied to a delta-connected winding.

6.3.6 Tank earth protection

When an earth fault occurs on a winding of a transformer protected by a differential protective scheme in which comparisons are made of the m.m.f.s associated with the windings, a current will flow in the operating winding of one of the relays. This current, as shown earlier in section 6.2.3, will not be solely dependent on the magnitude of the fault current, but also on the fault position. As a result, faults on a significant part of a winding may not be detected even when a relatively sensitive relay setting is used. This situation will certainly apply if a large earthing impedor is present in the circuit.

Restricted earth fault protection, as stated in section 6.3.2 (page 208), can detect faults on star-connected windings which cause currents above the setting value to flow to earth. Because of the need to ensure that such schemes will not maloperate when earth faults occur outside their zones, relay settings of the order of 20% are normally used, i.e. 0.2 A when the current transformers have secondary windings rated at 1 A. In those applications where impedance is included in transformer neutral connections to limit earth fault currents to the rated current (1 pu), operation of relays with 20% settings would not occur for faults on the 20% of the winding adjacent to the neutral point.

The amount of winding protected can be increased in such applications if the transformer tank can be lightly insulated from earth and a current transformer can be installed with its primary winding connected between the tank and earth. A current-operated relay connected to the secondary winding of the current transformer can be given a very low setting, say 5%, because current can only flow in it when winding faults to the tank occur. With such a setting, 95% of the windings would be protected if an earthing impedor limiting fault currents to 1 pu were present.

6.3.7 Overfluxing protection

As explained in section 6.2.4, it is desirable that the ratio of winding voltage to frequency should not exceed a particular value at which the required flux swings are approaching saturation levels. In some applications, therefore, a relay is provided to monitor this ratio and provide an alarm when the set value is exceeded.

One relay [7], which has been used for this purpose for some time, is supplied by the output (V) of a voltage transformer connected to a winding of the power transformer being monitored. The relay contains a resistor (R) of high ohmic value connected in series with a capacitor (C) of relatively low reactance. The input voltage (V) is applied across this combination and a current proportional to the voltage flows through the capacitor, i.e. $I \simeq V/R$. The resulting voltage V_C across the capacitor is proportional to V/f i.e. $V_C = I/\omega C \simeq V/2\pi f CR$. A voltage-operated device connected across the capacitor is set to operate when the ratio exceeds a critical level.

More recently, a relay [8] has been introduced which monitors the imbalance current in current-differential protective schemes. This current has a large component caused by the exciting current of the protected transformer, and when the V/f ratio rises core saturation occurs and the exciting current has high peaks and periods when it is very low. This behaviour, which is detected using modern processing techniques, is a direct indication of overfluxing.

6.3.8 Protection against overheating

The temperatures within a power transformer at any instant are not dependent solely on the currents then flowing in its windings. Certainly in some instances, because of inadequate oil circulation resulting from faulty pumps or blockages in ducts or pipes, local or general overheating can occur even when transformers are lightly loaded. It will therefore be clear that reliance cannot be placed on using overcurrent relays to detect unacceptable overheating, and that devices which monitor temperatures directly are preferable.

Over the years, temperature sensors have been positioned in the oil and within the windings of power transformers. In the past, thermostats or bulbs containing volatile liquids which operated remote pressure-operated devices to which they were connected by small diameter tubes were used. In recent years, however, it has become common practice to install a thermal-sensing element in a small compartment positioned near the top of a transformer tank where the oil tends to be hottest. A small heater, which is also installed in the compartment, is fed from a current transformer connected in series with one of the phase windings. A local oil temperature comparable with that present in the main windings is thereby produced in the compartment. Heat-sensitive resistors are now used as sensors and these form parts of resistance bridges which produce imbalance output signals that initiate either alarms or the opening of the appropriate circuit-breakers when unacceptable overheating occurs. A relay type TTT was produced for this purpose by GEC Measurements [9].

Because the life of insulation is dependent on both the temperatures at which it operates and the periods spent at each temperature, relays, designated temperature–time integrators, have been developed to measure the periods for which transformers operate within a number of temperature bands; for example, the number of hours for which a transformer operates at a temperature within the band 100–110°C may be recorded. From such information an estimate of the degree of ageing of the insulation in the transformer may be made. Further information on these relays is available in reference [10].

6.3.9 Buchholz relays

Protective schemes which monitor circuit currents or voltages can operate rapidly and open the appropriate circuit-breakers when certain conditions exist. As an example, a differential scheme will operate when the current imbalance is at or above a particular level, a situation which only arises when a significant fault current is flowing in the protected equipment. In many cases serious damage will have been caused before the faults are cleared in this way and expensive repairs will be needed.

It is clearly desirable therefore that, where possible, other methods be employed to detect potentially serious conditions when they begin so that action may be taken before extensive damage is caused. For many years this objective has been achieved by fitting Buchholz relays in the pipes between the main tanks and the conservators of oil-filled transformers. A typical relay and mounting arrangement is shown in Fig. 6.17.

Gases such as hydrogen and carbon monoxide are produced when decomposition of the oil in a transformer occurs as a result of localized heating or arcing associated with winding faults. Gases produced in this way rise from the transformer and pass up the sloping pipe, shown in Fig. 6.17(a), towards the conservator. They pass into the Buchholz relay, which is normally filled with oil, and rise and become trapped in the top of the casing so displacing oil. As a result, a pivoted float or bucket, depending on the relay design, falls and operates a mercury switch or closes contacts. This situation is shown in Fig. 6.17(b). An alarm signal can thereby be provided for a number of conditions of low urgency, including the following:

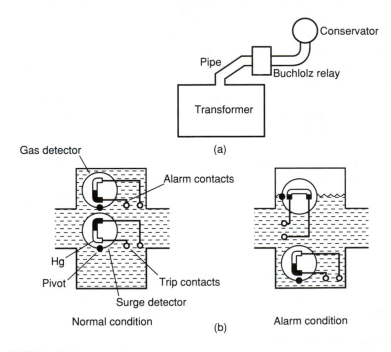

Fig. 6.17 Buchholz protection.

(a) Hot spots in a core caused by short circuiting of the laminations.
(b) Breakdown of the insulation on core clamping-bolts.
(c) Winding faults causing only low currents to flow through the current transformers.

In addition, an alarm would be given if the oil level fell below that of the Buchholz relay as a result of a small leakage from the main tank.

Buchholz relays are provided with valves or cocks to allow samples of gas collected in them to be analysed. This is a valuable feature because it allows the cause of gas generation to be determined and may assist in determining the remedial action which is needed. As examples, the presence of H_2 and C_2H_2 indicates arcing in oil between constructional parts and the presence of H_2, C_2H_4, CO_2 and C_3H_6 indicates a hot spot in a winding.

Major winding faults, either to earth or between phases or windings, involving severe arcing, cause the rapid production of large volumes of gas and oil vapour which cannot escape. They therefore produce a steep build up of pressure and displace oil, setting up a rapid flow towards the conservator. Such flows cause either a vane or a second bucket or float in a Buchholz relay to move and close contacts which initiate the tripping of the appropriate circuit-breakers. Whilst such relays do not operate as rapidly as those in current-differential schemes, they nevertheless operate much more quickly than IDMT relays, and they therefore provide a very satisfactory back-up feature.

The physical size and the settings of a Buchholz relay are clearly dependent on the transformer to which it is to be fitted. Firstly, it must be suitable for mounting in the pipe between the main tank and the conservator of the transformer. Secondly, it must be set to provide an alarm when a particular volume of gas has been collected and thirdly it must initiate circuit-breaker tripping when the oil velocity exceeds a particular value. These values tend to increase with the ratings of transformers, typical values being shown in Table 6.1.

A number of factors have to be taken into account when selecting the settings of Buchholz relays, in particular the oil–velocity setting must be high enough to ensure that relay operation will not occur as a result of pressure surges which may be produced when oil-circulating pumps are started. This and other factors are considered in more detail in reference [11].

It is now European practice to provide Buchholz protection on all transformers fitted with conservators and suitable relays are either recommended

Table 6.1

Transformer rating MVA	Pipe diameter (in)	Alarm Volume of gas (cm^3)	Trip min. oil velocity (cm/s)
up to 1	1	110	70–130
1–10	2	200	25–140
> 10	3	250	90–160

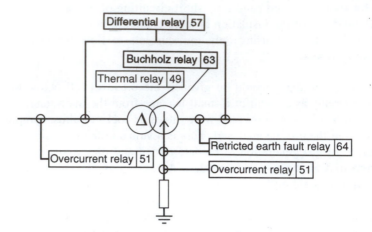

Fig. 6.18 The protective equipment applied to a large transformer.

by transformer manufacturers or installed by them during manufacture or erection.

6.3.10 The protection of large three-phase transformers

Several of the protective schemes and relays described in the preceding sections must be applied to each large transformer to ensure that all the unacceptable conditions which may arise in it will be correctly detected. A typical arrangement, which will provide adequate cover, is shown in Fig. 6.18.

6.4 THE PROTECTION OF EARTHING TRANSFORMERS

Step-down transformers used to feed distribution networks from higher voltage transmission circuits usually have star-connected primary windings and delta-connected secondary windings because the amount of insulation on the high voltage windings is thereby minimized. Clearly the delta-connected, low voltage windings cannot be directly earthed and therefore a separate earthing transformer is often provided to ensure that the distribution circuits are not floating free from earth.

Three-phase earthing transformers have six windings, each of the same number of turns, mounted in pairs on the limbs of the core and they are connected to the delta-connected windings of power transformers as shown in Fig. 6.19(a). During normal balanced conditions, the winding voltages are as shown in Fig. 6.19(b) and the fluxes in the earthing transformer limbs are each of the same magnitude but displaced from each other by $2\pi/3$ radians. Only the small currents needed to provide the m.m.f.s flow in the earthing-transformer windings.

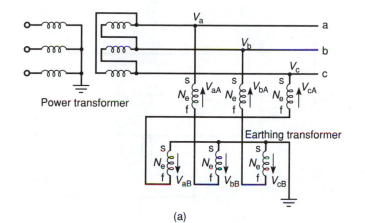

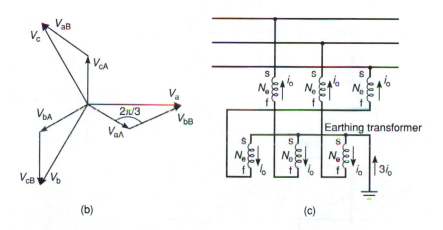

Fig. 6.19 Earthing transformer connections and behaviour.

When an earth fault occurs on the low voltage circuit, positive- and negative-sequence current components are fed by the power transformer and the fault current, which is equal to the sum of the zero-sequence currents, flows into the neutral of the earthing transformers as shown in Fig. 6.19(c). Because of the distributions of the currents in the windings, no resultant m.m.f.s are produced by the currents and therefore no zero-sequence fluxes or e.m.f.s are produced. As a result, the earthing transformer effectively presents zero reactance to the earth fault current.

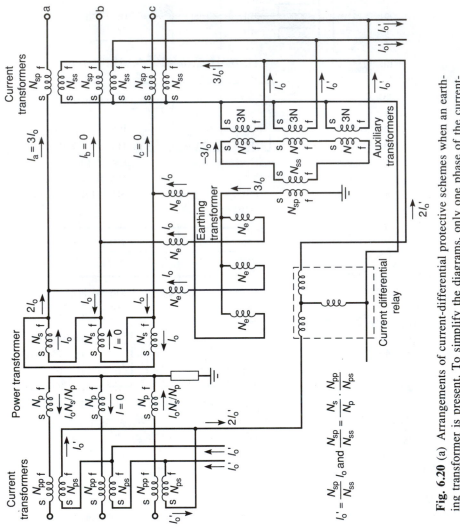

Fig. 6.20 (a) Arrangements of current-differential protective schemes when an earthing transformer is present. To simplify the diagrams, only one phase of the current-differential relays is shown.

An earthing transformer, when it is mounted near a power transformer, may be included in the zone of a current-differential protective scheme. When this is done, the sum of the three zero-sequence current components flowing to an earth fault on the distribution network through the line current transformers will return up the neutral connection of the earthing transformer. To ensure that these components will not cause imbalance an extra current transformer must be included in the neutral connection of the earthing transformer. This transformer may have the same ratio as the line current transformers on the delta-connected side of the power transformer, in which case its secondary winding must feed three auxiliary current transformers with a step-down ratio of 3 to 1 to provide the appropriate zero-sequence currents to balance with those in secondary windings of the line current transformers. This arrangement is shown in Fig. 6.20(a).

An alternative arrangement, which may be adopted when the current-differential protection applied to a transformer is not combined with restricted earth fault protection, filters out the zero-sequence components present in the secondary windings of the line current transformers on the delta-connected side of a power transformer by including a three-phase, star–delta–star-connected interposing transformer, as shown in Fig. 6.20(b).

Should an earthing transformer be mounted at some distance from a power transformer or if it is felt desirable to exclude it from the zone of the current-differential scheme protecting a power transformer, it must be provided with its own protection. Because of its relatively low rating, the use of a separate current-differential scheme is not justifiable. A relatively simple scheme which is used in these applications employs three current transformers whose primary windings are connected in the lines between the network and the earthing transformer, as shown in Fig. 6.21. The secondary windings, being connected in delta, allow the zero-sequence currents which are present when there are earth faults on the network to circulate around them and do not cause currents to flow in the relays during these conditions. The relays may therefore be given relatively low current settings, which enables operation to be obtained for most of the faults which may occur within the earthing transformer. IDMT overcurrent relays are usually employed in these applications because rapid operation is not necessary unless fault current levels are very high.

6.5 AUTO TRANSFORMERS AND THEIR PROTECTION

In many networks it is necessary to provide isolation between circuits operating at different voltage levels. As an example, transmission lines operating at voltages such as 400 kV are isolated from alternators, producing voltages of about 25 kV, by transformers with separate primary (lv) and secondary (hv) windings.

In some situations, however, where the ratio of the voltage levels of circuits is not high and where isolation is not necessary, it is economically advant-

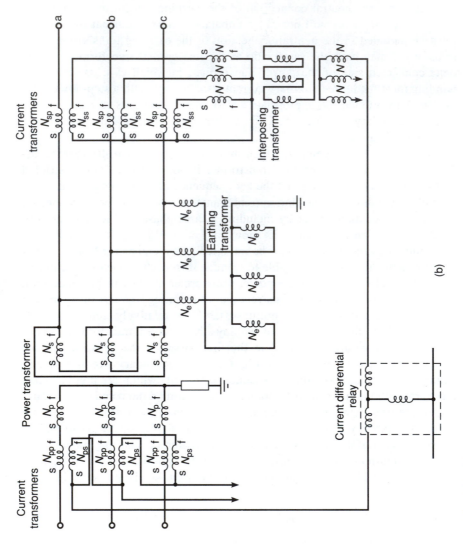

Fig. 6.20 (b)

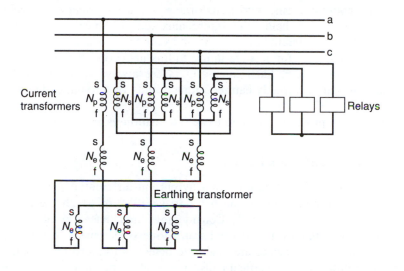

Fig. 6.21 The protection of an earthing transformer.

ageous to use auto transformers. This can be seen by comparing the ideal auto and two-winding transformers, shown in Fig. 6.22, which both provide a step-down ratio of $2:1$. The auto transformer has a total of N_p turns of conductor capable of carrying a current of I_p, whereas the two-winding transformer has a primary winding of N_p turns which must carry a current of I_p and also a secondary winding of $N_p/2$ turns which must carry a current of $2I_p$. The volume of conductor needed for the auto transformer in this case would be half that needed for the two-winding transformer and the power loss in the auto transformer would also be half that in the two-winding transformer. The advantages of the auto transformer clearly increase as the required transformer ratio reduces towards unity but decrease as it rises above two.

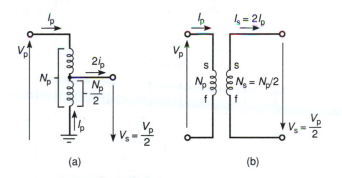

Fig. 6.22 Auto- and two-winding transformers.

For the above reasons, auto transformers are used whenever possible, a common application being the interconnection of transmission networks which operate at different voltages. As an example, a considerable number of three-phase auto transformers are in use to link networks operating at 275 kV and 132 kV in Britain. Such transformers are always star-connected. The neutral points are invariably earthed either solidly or through earthing impedors.

The forms of protection applied to auto transformers are considered below.

6.5.1 Current-differential schemes

Auto transformers could be protected by current-differential schemes of the type considered earlier in section 6.3.2 (page 198), in which the input and output m.m.f.s are compared. Considerable imbalances occur in these schemes, which require two current transformers per phase. In addition, as stated earlier, these schemes are unable to detect earth faults on a significant number of turns near the earthed ends of windings, this number representing a considerable percentage of the total number of turns in a winding when earthing resistors are present.

A superior alternative, which is normally used on auto transformers, is a current-differential scheme which directly sums the currents entering a winding. Three current transformers, each of the same ratio, must be provided per phase, the primary winding of one being in series with the neutral connection and the others in series with the high and low voltage connections. Their secondary windings and the operating winding of the relay must be interconnected as shown in Fig. 6.23.

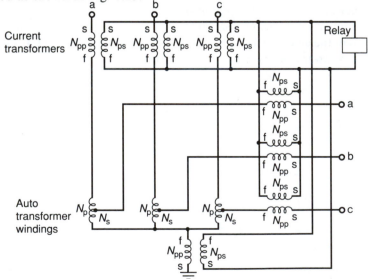

Fig. 6.23 The protection of an auto-transformer.

Unbalance is not produced in this scheme by exciting current surges, and because the interconnections tend to be short other imbalances are usually small. Either percentage-biased low-impedance or unbiased high-impedance relays may be employed and they may be set to operate at low per-unit values, 0.2 or less, i.e. 0.2 A when current transformers with secondary windings rated at 1 A are used. With such settings, earth faults on the windings which cause currents of 0.2 pu or more to flow will be detected as also will interphase faults of the same magnitudes. In the latter case, two or more relays will operate depending on the type of fault.

It must be recognized that this scheme will not operate in the event of interturn faults and preferably some other form of protection should be provided to detect them. If this is not possible, then it must be accepted that these faults will persist until they cause breakdowns to earth or between phases.

6.5.2 Tank earth protective equipment

As stated in section 6.3.6 (page 212), the presence of earth faults on windings can be detected by lightly insulating a transformer tank from earth and connecting the primary winding of a current transformer in the connection between the tank and earth. The secondary winding of the current transformer may then supply the operating winding of a relay. Because only currents flowing from the windings to the tank can flow in the primary winding of the current transformer and no current comparisons are involved, the relay can be set to operate at very low current levels without increasing the possibility of maloperation. The application of this form of protection to auto transformers allows the numbers of turns near the earthed ends of windings on which earth faults cannot be detected to be reduced below the levels obtainable with current-differential schemes. The reduction can be particularly great when earthing impedors are present in the neutral connections to earth and the use of tank earth protection in such applications should be considered.

6.5.3 Other protective equipment

Because the physical constructions of auto and multi-winding transformers are similar, the methods of detecting overfluxing and overheating which were described in sections 6.3.7 and 6.3.8 respectively may be applied to auto transformers. In addition, IDMT relays are applied to provide back-up protection and Buchholz relays, which were described in section 6.3.9, are applied to large auto transformers which are oil-filled and have conservators.

6.6 THE FUTURE

The basic principles of transformer protection are well established. Changes which are envisaged in the future relate to the use of digital techniques for the extraction of relaying parameters. In the case of transformers, it is likely that restraint (or bias) signals, which hitherto have been obtained using analogue filtering techniques, will in the future be obtained by digital filtering. To illustrate this development, if the third and fifth harmonic of the current is required to form a restraint quantity, these could be obtained by sampling the current at regular time intervals T, and processing these sample values to obtain the desired harmonic components. There are a number of algorithms suitable for this purpose, e.g. those based on the discrete Fourier Transform. The sampling frequency must exceed twice the highest frequency of interest as described in section 4.3.1 (page 118). Hence, if the 5th harmonic of a 50 Hz signal is required the sampling frequency must be at least 500 Hz, or 10 samples per period. Samples from all the current transformers or transducers in a differential protective scheme could be processed by a single unit using multiplexing techniques. An advantage of such methods over conventional techniques is that it is easier to combine restraint signals from a number of transducers or from different harmonic components to form a more secure restraint signal. A description of these techniques may be found in references [12, 13].

A further development, which may affect transformer protection, is the wider availability of relaying signals in an integrated protection and control environment. Current practice relies on current transducers for transformer protection. Voltage transducers are too expensive to be justified in terms of transformer protection alone. In an integrated environment, however, voltage signals may be available and shared by control, monitoring and protective systems. In such cases, relaying algorithms which provide restraint signals derived from voltage transducers may be more widely used [12, 14].

REFERENCES

1. *BS171 Power Transformers, Parts 1 to 5: 1978*, British Standards Institution (equivalent to IEC 76: 1976).
2. Kennedy, L. F. and Hayward, C. D. (1938) Harmonic restrained relays for differential protection, *Trans. AIEE*, **57**, 262–271.
3. Hayward, C. D. (1941) Prolonged inrush current with parallel transformers affect differential relays, *Trans. AIEE*, **60**, 1096–1101.
4. Type MBCH biased differential protection for transformers, GEC Measurements, *Publication R-6070B*.
5. Type DT92 three-phased biased differential relay, Asea Brown Boveri, *Publication CH-ES 65–60*, Vol. 1, *Buyers Guide 1989–90*.
6. Wright, A. (1954) The performance of current transformers and relay circuits employed in the earth fault protection of power-systems. MSc Thesis, University of Durham.
7. Type GTT overfluxing relay, GEC Measurements, *Publication R-5159C*.
8. Type RATUB 2 V/Hz overexcitation relay for transformers, Asea Brown Boveri, *Publication B03-5011*.

9. Winding temperature relay and indicator, Type TTT, GEC Measurements, *Publication R-5074G*.

10. *Protective Relays Application Guide*, GEC Measurements, (3rd edn), 1987, p. 282.

11. Ibid., pp. 94 and 296.

12. Phadke, A. G. and Thorp, J. S. (1988) *Computer Relaying for Power Systems*. RSP–J Wiley.

13. Rahman, M. A. and Jeyasurya, B. (April 1988) A state of the art review of transformer protection algorithms, *IEEE Trans. on Power Delivery*, **3**, (2), 534–544.

14. Microprocessor relays and protection systems, IEEE Tutorial Course, 88EH0269-1-PWR, 1987.

FURTHER READING

IEEE Trans. on Power Delivery, **PD-3**, 525–533.

GEC Measurements, Transformer protection, *Publication R-4025*.

Murty, Y. V. V. S. and Smolinski, W. J. (1988) Design and implementation of a digital differential relay for a three-phase power transformer based on Kalman filtering theory.

Reyrolle Protection (1991) *Digital transformer protection-Duobias M*, Data Sheet Duobias-M, 10/91.

7

The protection of rotating machines

A very large number of electrical machines of a wide range of types and ratings are used in power systems around the world. The vast majority of them have a rotating member, i.e. a rotor and a stator, and both members usually have windings associated with them.

The largest machines are the three-phase alternators used in generating stations. In the UK, many stations have machines which provide 50 Hz voltages of 22 kV (line) and power outputs of 500 MW. Their star-connected output windings, which are mounted in the stator, are thus capable of carrying currents of 15 kA. Their rotor windings, which provide the m.m.f.s needed to set up the magnetic fields, carry direct currents up to 4.24 kA, these being supplied by exciters rated at 2.5 MW. The field windings of these machines are supplied in turn by pilot exciters. The alternators and their exciters are driven by steam turbines supplied by boilers. Associated with such plants are motors which drive pumps, fans and other items including coal pulverizers.

Even larger alternators are now in use. As an example those installed in the Itaipu hydro-electric power station on the Parana River which flows between Brazil and Paraguay are rated at 700 MW. This station contains 18 of these machines, each of which weighs 2700 tonnes.

In an article [1] published in the IEE *Power Engineering Journal* in 1990, Creek described machines rated at 985 MW, 1158 MVA which were being produced at that time by GEC Turbine Generators for the Daya Bay nuclear power station in the People's Republic of China. These machines are designed to produce voltages of 26 kV (line) at a frequency of 50 Hz, the full-load current of the output windings being 25.7 kA. Their stator and rotor cores are to be hydrogen cooled and water flowing in a closed circuit will cool the stator windings.

It will be appreciated that power units, such as those referred to above, are very complex and it is necessary that their performances are completely monitored at all times to ensure that continuity of supply is maintained and that any faults or breakdowns which may occur do not cause unnecessary consequential damage.

Machines with ratings ranging down from the very high levels quoted above to levels of only a few watts or less are in service and many are produced annually for widely ranging applications. Both three-phase and single-phase a.c. synchronous and induction machines are produced as well as motors suitable for operation from sources of direct voltage or rectifiers. In addition, in recent years, reluctance motors [2] operating from controlled-rectifier banks have been introduced to provide efficient variable-speed drives.

Over the years several forms of protective equipment have been developed to enable the whole range of rotating machines to be protected at cost levels which are acceptable. Clearly cheap devices must be used with small machines, whereas quite costly equipment is justifiable when very large and important plant is to be protected.

In all applications, it is desirable that the most appropriate protective equipment is used and to enable this to be achieved it is necessary that adequate information about the machine to be protected is available when protective schemes or devices are to be selected. As an example, if fuses or relays with inverse time/current characteristics are being considered for use with a particular motor, then it is essential that the magnitude and duration of the current surges which will flow in it during starting periods are known.

A significant factor which should be recognized at this point is that although machines which are to be used as motors may be physically the same as others which are to be used as generators, their protection must take into account the use to which they are to be put.

Motors, clearly, are loads and therefore only their operating characteristics need to be considered when selecting the protective equipment to be used with them. Should time-graded schemes be used, then the protection on the circuits supplying the motors must be set so that the necessary co-ordination will be achieved under all conditions, i.e. the devices on the supply circuits should operate more slowly than those on the motors.

Generators are, of course, sources and therefore the settings of their protective equipment must be such that operation will not occur in the event of faults on the circuits being supplied by them and, in addition, the generators should not suffer consequential damage.

It will be appreciated that it is not possible in this work to provide detailed guidance on the protective arrangements which are most suitable for each of the many types and sizes of machines which may be encountered in practice. In this chapter, therefore, the protective schemes and devices currently available are examined and a number of applications of them are considered, after some details are given of the early relays which were introduced to protect machines.

7.1 HISTORICAL BACKGROUND

When motors, generators and alternators were being produced in the final decades of the last century for installation in power systems the only electrical protective devices available were fuses. Whilst they were suitable for use with some small motors, it was apparent that they could not provide adequate protection for generators and alternators. To have ensured that faults on an output winding connected to a busbar would be cleared, fuses would have had to have been included at each end of the winding and their operating times would have had to be high enough to ensure that they would not operate in the event of faults on the load circuits. They were particularly unsuitable for use with three-phase machines because the operation of a single fuse in the event of a phase to earth fault on a machine would have caused it to carry unacceptably high negative- and zero-sequence currents. In addition the operation of fuses in the three-phase circuits could not have initiated the opening of field-winding circuits.

H. W. Clothier stated in a paper entitled 'The construction of high-tension central station switchgears with a comparison of British and foreign methods', which was presented to the IEE in 1902, that for large a.c. generators it is feasible to do without fuses or other automatic devices.

This view was also expressed by W. B. Woodhouse during the discussion on a paper by H. L. Riseley entitled 'Some notes on continental power house equipment', which was presented to the IEE in 1903.

In spite of these views, there were those who felt that protective relays should be used and certainly reverse-current relays were referred to in a paper by C. H. Merz and W. McLellan entitled 'Power station design' which was presented to the IEE in 1904. These relays were directional relays and their purpose was to detect current fed from busbars into faulted machines. The authors stated that these relays would not detect short circuits within machines because the voltages would be too low for the relays to function correctly. The authors nevertheless felt that automatic protective devices were required. In the discussion on this paper, L. Andrews stated that he had demonstrated a reverse-current device in 1898 but recognized that it still was not entirely satisfactory. Reverse-current relays, developed by Brown Boveri, were referred to in a paper entitled 'Protection devices for H. T. electrical systems' which was presented to the British Association in 1903. Five years later C. C. Garrard referred to the use of reverse-current relays with 0.5 s time lags to allow for transient effects in a paper presented to the IEE entitled 'Apparatus and relays for a.c. circuits'. It is clear, therefore, that most machines were operating without effective protective equipment up to this time.

In 1910 K. Faye-Hansen and G. Harlow presented a paper entitled 'Merz–Price protective gear and other discriminative apparatus for alternating-current circuits' to the IEE. They advocated the use of circulating-current protective schemes on alternators and showed arrangements suitable for delta- and star-connected machines. Even at that time several speakers in the ensuing discussion expressed the view that such schemes were unreliable and not suitable for application to alternators. One speaker, A. E. McKenzie, stated that two 4000 kW, three-phase machines in Manchester had then been protected by Merz–Price schemes for two years and that similar schemes were being applied to new 6000 kW machines. He also expressed the view that devices should be provided to detect loss of excitation. Eventually it became the standard practice to apply current-differential protective schemes to the main windings of all large machines and other protective devices were applied to detect the other unacceptable conditions which are examined in the following sections of this chapter.

7.2 PROTECTIVE DEVICES AND SCHEMES

The performance required of the protective equipment to be used with a particular rotating machine is dependent on its type, size and characteristics, and therefore a variety of schemes and devices, including fuses and thermal relays, have been developed to cater for the wide range of machines produced for use in power systems.

The types of protective equipment suitable for machine applications are considered briefly in the following sections.

7.2.1 Fuses

The operation of fuselinks results from the heating of their elements to their melting temperatures and they therefore have operating time/current characteristics similar in form to the withstand time/current characteristics of machines. Cartridge fuselinks are therefore an inherently suitable form of protection for small and medium-sized machines. As a result, many are produced annually for this purpose and a vast number must now be in use around the world.

For some applications, standard cartridge fuselinks are suitable, indeed domestic fuses with current ratings of 13 A or less are used to protect motors fitted in the many appliances present in homes and other buildings.

For other applications, special motor-protection fuselinks are produced with voltage and current ratings up to 11 kV and 1000 A respectively. These fuselinks, which have special elements to enable them to cope with the current surges which will flow through them when the motors they are protecting are started, are described in section 1.9.8 (page 29).

7.2.2 Thermal relays

Thermal relays usually incorporate bimetal elements which deflect by amounts dependent on the currents flowing in heating elements mounted adjacent to them. The deflections tend to be proportional to the square of the current flowing and above a certain current level closure of associated contacts occurs. As a result, an inverse operating time/current characteristic is obtained above a minimum operating-current level. Such relays therefore behave similarly to fuses and they may be used in conjunction with contactors or circuit-breakers to protect rotating machines. A relay of this type is described in section 4.2.5 (page 116).

These relays, which are produced in single-phase and three-phase forms, may have their heating elements connected directly in series with the windings of the machines they protect or they may be supplied from current transformers. They may therefore be used to protect machines of a wide range of power outputs and operating voltages.

7.2.3 Thermal devices

The above items are mounted externally to the machines they protect and their settings must be co-ordinated. In some applications, however, there are advantages in mounting protective devices within machines during their manufacture; for example, a temperature-sensing device may be sited in the stator core of a machine or near to or within a winding. Such a device, which is not directly and solely affected by the machine currents, would operate and disconnect its machine if overheating occurred because the cooling was not adequate. It would also operate if the overheating was caused by overloading.

Similarly overheating of a short-time rated starting winding, caused by a slow build-up of motor speed, could be detected by a temperature sensor mounted near the winding.

Devices of the above types are clearly selected by the manufacturers of machines who are fully aware of the setting levels needed because they have all the necessary design details of their products.

Bimetallic devices have been used as temperature sensors but more recently thermistor probes have been introduced for this duty. The resistances of these probes vary with temperature and thus the limiting temperature may be detected by an electronic circuit which measures the probe resistance and an output relay can thus be operated to disconnect a machine from its supply. Clearly such a device must have its own in-built power supply.

7.2.4 Instantaneous electromagnetic relays

The devices considered above allow overcurrents or abnormal conditions to continue until almost the point where the protected machine will be damaged. This is desirable as it allows machines to be kept in service as long as possible. In the event of a fault on a machine, such as a breakdown of the insulation on a winding, it is desirable that the machine be disconnected immediately so that further damage is not caused and thus the cost of repair and the period out of service are both minimized. This can be achieved by using instantaneous electromagnetic relays with operating currents above the levels which may be encountered under healthy conditions including starting currents, but below those produced by short circuits.

7.2.5 Current-differential schemes

Current-differential schemes, as explained in Chapter 5, are able to detect and operate instantaneously at fault current levels well below the rated current of the unit protected by them, say 0.2 pu or less. This feature is very important in applications associated with very large machines. This can be appreciated by considering the machines being installed in Daya Bay Power Station in China. It was pointed out in the introduction to this chapter that the rated stator current of these machines is 25.7 kA. Clearly currents of these levels could not be allowed to flow in fault paths because the consequential damage would be unacceptable and very sensitive protection settings in per-unit or percentage terms are essential.

7.2.6 Current-balance schemes

Current-differential schemes are not capable of detecting interturn faults in a machine winding because such faults do not cause differences in the currents at the two ends of a winding. Unacceptably high currents may nevertheless flow in the fault paths. In machines with high current ratings, each winding may have two or more parallel paths in it, i.e. a complete winding may be made up of two or more separate windings connected in parallel. In such cases, each of the parallel paths should normally carry almost the same current at every instant. If an interturn fault developed on one of the paths however, the parallel-connected windings would share the total current unequally and the presence of the fault could be detected by comparing pairs of winding currents. As an example, if two windings were connected in parallel, and current transformers were connected in series with each of them, a fault could be detected using the current-balance arrangement shown in Fig. 7.1(a). Alterna-

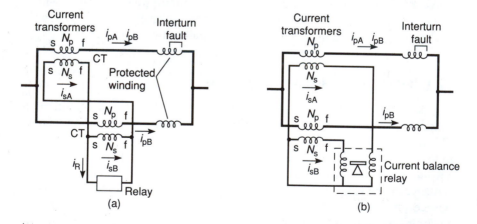

Fig. 7.1 Current balance arrangements. (a) $i_R \neq 0$ when $i_{pA} \neq i_{pB}$. (b) Relay does not balance when $i_{pA} \neq i_{pB}$.

tively, a current-balance relay with two windings energized from the two current transformers as shown in Fig. 7.1(b) would be unbalanced as a result of an interturn fault.

If a winding had three parallel paths (1, 2 and 3) then two current-balance relays would be needed to compare the currents in, say, paths 1 and 2 and 1 and 3 respectively.

7.2.7 Phase-unbalance relays

It is obviously desirable that three-phase machines should operate under balanced conditions and even relatively small amounts of unbalance may cause operating efficiencies to be significantly reduced.

A negative-sequence component in the supply voltages of a motor may cause large losses in its rotor and it is therefore usually essential that a three-phase machine should not operate with an open-circuited phase.

Several types of relays are available to detect unbalance in voltages or currents. Clearly voltage- or current-fed sequence networks may be used to feed outputs to operate electromagnetic relays when unacceptable levels of unbalance are present.

7.2.8 Voltage-operated relays

Attracted-armature or other types of relays capable of being set to operate at particular voltage levels may be used to detect when supply voltages to motors or outputs from generators are outside acceptable levels.

7.2.9 Control equipment

Many motors must be provided with control equipment to ensure that the correct starting sequences are used. As examples, the main windings of three-phase motors may be connected in star during starting periods and then connected in delta when the speeds are approaching the running value or resistors may be connected in either the stator or rotor winding circuits during starting. This equipment, which is often supplied by the manufacturers of machines, is usually housed in a cabinet or case which is mounted near to the machine it controls. In these situations protective equipment, such as overcurrent and field-failure relays, is often also housed with the control equipment.

7.2.10 Applications of protective schemes to machines

It will be clear from the above that there is a range of relays and schemes available for protecting machines and whilst only one device, such as a fuse, may provide sufficient cover for a small machine, a combination of devices and/or schemes is usually needed to protect a large machine. Guidance on the practices currently adopted is provided in the following sections.

7.3 THE PROTECTION OF MOTORS

As stated earlier, motors represent loads in power systems and therefore their protective equipments should be selected taking account of the withstand abilities of their supply networks and the behaviours of the motors themselves. A number of applications are considered below.

7.3.1 The protection of small motors

There must be many millions of relatively small, single-phase a.c. motors incorporating commutators and with their armature and field windings connected in series, i.e. series-commutator machines, in service in domestic appliances, such as vacuum cleaners and washing machines. Their power ratings are of the order of a few hundred watts and so they draw currents of 3 A or less when operating from the voltages normally available in homes and offices (i.e. 240 V).

These motors are switched directly to their supply voltages and draw surges of current during starting, but because of the relatively low inertias of their rotors, the starting periods tend to be very short. Standard cartridge-type domestic fuselinks, rated at 13 A or less, will not operate during these surges and they are therefore suitable for protecting such motors. They will operate if a fault occurs in a machine but relatively slow clearance may be obtained if the fault current is not greatly in excess of the fuselink rating. Further damage to a motor, which may result from slow clearance, will probably be unimportant because it will not be likely to greatly increase the cost of the repair above that which would have been necessary in any event, and even if complete replacement of the motor is necessary it is not likely that the cost will be unacceptably high.

The fuses would certainly ensure that cables would not be damaged by faults on the motors supplied by them. It must be recognized, however, that the fuses will not operate at low-overcurrent levels, but this is usually acceptable as such conditions do not often occur with motors used in domestic appliances because their mechanical loads tend to be limited.

Some applicances, such as refrigerators, incorporate single-phase induction motors. These usually have a main winding which is continuously rated and a short-time rated starting winding. These motors are protected by domestic-type cartridge fuselinks, rated at 13 A in the UK, and these ensure that the supply circuits will not be damaged by faults within a motor. They will not, however, protect the starting winding if for any reason the accelerating period is unacceptably long. To cater for this condition, some manufacturers incorporate thermal devices containing bimetal strips to open the starting circuit when necessary.

7.3.2 The protection of large induction motors

Large induction motors invariably have three-phase stator windings suitable for operation at voltage levels up to 11 kV (line). Power ratings up to 10 MW are available. Such machines may suffer damage from a range of causes and these are considered separately below.

Stator-winding insulation

All insulating materials deteriorate with age and the speed of this process increases with the temperature of the insulation. As a result prolonged operation at high temperature is undesirable. In the case of a motor, heat is generated as a result of the power loss in the stator windings. Clearly the flow of high currents for short periods produces relatively small amounts of energy and only small temperature rises because of the mass of the conductors being heated. Such conditions, which can occur during starting, can therefore be accepted. On the other hand, currents above the rated level flowing for long periods because of overloading can cause the insulation to reach high temperatures which will be maintained until the overloading ceases. Such conditions, especially if repeated regularly, will significantly damage the insulation, which could then break down.

The temperature levels at which insulation may be operated safely are dependent on the insulants being used. The motor industry has established standards, those associated with induction motors being given in NEMA MG1–1987 [3].

It will be appreciated that the withstand time/current curve for any motor is inverse and protective equipment must ensure that the limits are not exceeded.

As stated earlier, unacceptably high-temperature levels may be reached in the stator of a machine due to defects in the ventilating system, such as blocked ducts or a faulty fan, and these could arise even though the current levels were at rated values or less. It is therefore desirable that stator windings be protected by three-phase overcurrent relays and thermal devices.

Guidance on suitable relay settings is provided in manufacturers' literature and standards such as ANSI/NFPA 70–1987 [4]. Depending on the design and duty of a motor, its relay should initiate tripping at currents in the range 115–125% of the rated value of the motor. These relays may have either thermal elements or induction-type movements.

Relatively small machines operating at voltage levels such as 415 V (line) and currents up to 20 A may incorporate thermal protectors containing bimetal elements. These are capable of directly interrupting currents somewhat above the rated values of their motors when overheating occurs. Larger machines incorporate temperature detectors embedded in their stators. These detectors, forms of which were described in section 7.2.3, initiate the opening of their motor circuit-breakers when necessary.

Faults on stator windings

Conditions which may affect the insulation of the stator windings of induction motors have been considered and protective arrangements, which may be used to try to ensure that consequential damage will not result from them, have been examined. Nevertheless, breakdowns of the insulation of stator windings, although not common, are experienced and these can be quite serious in machines operating at relatively high voltages. It is desirable that they should be cleared quickly and for this reason current-differential protective schemes are applied to large machines.

As explained in section 5.2.5 (page 154), these schemes may be set to detect fault currents of magnitudes much lower than the rated current of the protected equipment. They do not operate if faults occur external to their zone or when high currents are flowing as a result of overloading and they must therefore be used in conjunction with overcurrent protective relays.

The application of current-differential schemes to three-phase motors is clearly affected by the connections of the stator windings, i.e. star or delta.

Star-connected windings may be protected using the arrangement shown in Fig. 7.2(a). All six current transformers, which will have the same ratios, should ideally be produced by the same manufacturer and have identical characteristics. The three transformers on the supply side of the machine should be mounted on the incoming connections to the switchgear so that the motor supply cables and the stator windings are in the protected zone and the other three transformers should be mounted in the machine as near as possible to the neutral or star-point connection.

The protection of delta-connected windings is not so simple and in practice several arrangements are used. Six current transformers may be connected as shown in Fig. 7.2(b) and (c). With the arrangement shown in Fig. 7.2(b) the six transformers have the same ratios and should ideally be identical. Each phase winding of the motor is protected by its group of two transformers and relay, but no protection is provided for the supply cables to the motor or the connections between the phase windings of the stator.

With the arrangement shown in Fig. 7.2(c), three current transformers are mounted in the supply circuit to the motor on the incoming side of the switchgear. Their secondary windings are connected in star. The other three transformers are connected in the phases of the stator windings with their secondary windings connected in delta. All the current transformers have the same step-down ratio and the secondary output current of each of the transformers in the supply circuits balances with the difference of the output currents of two of the transformers mounted in the motor under healthy conditions, e.g. $I_{R2} = N_p/N_s(I_a - I_{ab} + I_{ca})$. This arrangement has the advantage that the protected zone includes the supply cables and switchgear, but the achievement of the necessary balance is more difficult than with the scheme shown in Fig. 7.2(b) because the current transformers in each balancing group will normally be

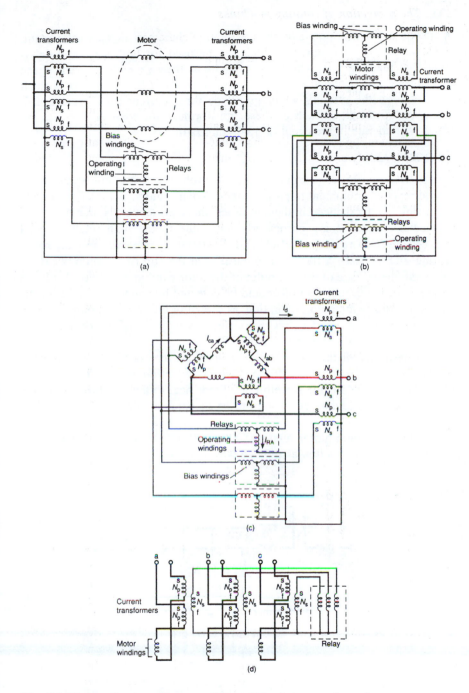

Fig. 7.2 The application of current-differential schemes to three-phase motor stator windings. (a) Star-connected winding; (b) and (c) delta-connected winding; (d) arrangement suitable for both star- and delta-connected windings.

operating at different points on their excitation characteristics at each instant.

An alternative arrangement, which can be used for both star- and delta-connected machines, in which the currents at the two ends of each phase winding are compared directly is shown in Fig. 7.2(d). Both ends of each winding form the primary windings of the three current transformers. The secondary windings of each of these transformers are connected to a relay which will operate in the event of a winding fault.

Rotor windings

Rotors with squirrel-cage windings are very robust and it is not necessary to provide special equipment to protect them. Wound rotors, however, may have quite high e.m.f.s induced in them at times, those present at the beginnings of starting operations being particularly large due to the high relative speed between the rotating magnetic field and the stationary or slowly-moving rotor. During such periods, a flashover to earth or at the slip rings could occur. In this event the voltage of the star point of the rotor winding would be displaced from that of the ground and this could be detected by connecting a star/open-delta transformer to the rotor winding, as shown in Fig. 7.3. The output of this transformer may be used to operate either a voltage-sensitive or a current-sensitive relay to disconnect the machine.

The sensitivity of this arrangement reduces as the speed of the protected motor increases because of the consequent reduction in the rotor e.m.f.s. Operation is not attainable therefore at speeds near to the synchronous value. This behaviour is usually considered to be acceptable, because flashovers are less likely when the induced e.m.f.s are low. Should a serious rotor-flashover occur, however, on a motor running near synchronous speed, reliance has to be placed on the protection associated with the stator windings.

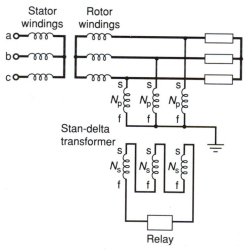

Fig. 7.3 The detection of rotor flashovers.

Failure to start

An induction motor may fail to start for several reasons including those below:

1. A mechanical fault such as a seized bearing.
2. Excessive load torque.
3. A low supply voltage.
4. An open circuit in a winding.

When a motor is started, the motor is stationary initially and its windings are cut by the rotating magnetic field set up by the stator. A large e.m.f. of the same frequency as that of the voltage supplied to the stator is induced in the rotor, the action being the same as that in a transformer, i.e.

$$E_s = \frac{N_s}{N_p} E_p = -\frac{N_s}{N_p} V_p$$

in which N_p and N_s are the turns in the stator and rotor windings respectively, E_p and E_s are the induced e.m.f.s in the stator and rotor windings and V_p is the supply voltage. Relatively large currents (I_p and I_s) flow in the stator and rotor windings, their relationship being given by:

$$I_s = -\frac{N_p}{N_s} I_p$$

The actual current values in any given application are dependent on the arrangements used to start the motor, such as star connection of the stator during starting and delta connection for running, or the progressive removal of resistance included in the rotor circuit during accelerating periods. In all cases, the initial current levels are significantly above the rated current, multiples of three to seven being common. If a motor does not accelerate from standstill for any reason, these above normal currents will persist and cause unacceptable heating. This situation is also aggravated by the fact that the cooling effect produced by the rotation of the rotor is not provided.

It will be evident that these stalled conditions can be detected by a relay which operates when the initial motor starting current persists for more than a certain time. This time must be great enough to allow normal starting to take place but it must, in addition, be short enough to ensure that conductors and insulation do not suffer consequential damage. Alternatively, or in addition, the condition may be detected by thermal devices set in the stator and/or in the proximity of the rotor and these can initiate the disconnection of the machine before unacceptable temperature rises are produced. In practice, the stator of a machine may reach its limiting temperature before its rotor, during locked-rotor conditions. This is designated stator limiting and thermal devices should clearly be embedded in the stator of such a machine.

It is necessary that the time for which a motor can remain connected to its supply with its rotor locked be determined from its manufacturers, to enable the necessary time delays associated with overcurrent and thermal devices to be selected. In cases when the permissible locked-rotor time is shorter than the normal acceleration time, shaft speed sensors may be used to detect this condition.

Unbalanced operation

Unbalanced operation may occur in several ways. The supply voltages may be unbalanced and in these circumstances they will contain significant negative- and zero-sequence components. An extreme condition could occur if one phase of the supply became open circuited, as shown in Fig. 7.4.

The application of negative-sequence voltages to a motor causes a magnetic field to be set up rotating in the opposite direction to that set up by the positive-sequence voltages. As is well known, such a field induces very large e.m.f.s in the rotor windings of a machine rotating in its normal direction and consequently large negative-sequence currents flow in both the stator and rotor windings. In addition, e.m.f.s of nearly double the supply frequency are induced in the magnetic material of the rotor. These conditions can cause quite rapid heating of machines even though the total input currents are not very high and, because of this, the operating times of overcurrent relays may be too long to afford the necessary protection.

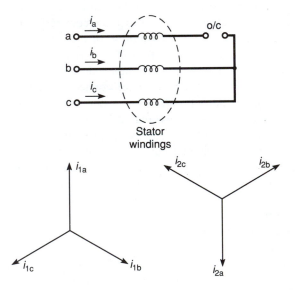

Fig. 7.4 Operation with one phase on open circuit. $i_a = 0$, $i_b = -i_c$, $|i_{1a}| = |i_{1b}| = |i_b|/\sqrt{3}$.

Similar unbalanced effects can be produced by faults such as short-circuited turns in a winding. In such cases, negative-sequence currents flow because of physical unbalance and will do so even though the supply voltages to the machine are balanced.

It is evident that the above conditions cannot all be detected by fitting relays which detect the negative-sequence components in the supply voltages to machines and therefore relays which monitor the stator-winding currents and detect the negative-sequence components in them should be used. Because the damage which results from unbalanced operation is caused by overheating, the period for which unbalance may be allowed to persist is dependent on the magnitude of the negative-sequence current in the windings, the relationship clearly being inverse. To ensure that relays with suitable time/current characteristics are chosen for particular applications, it is necessary that the appropriate withstand-abilities of motors are obtained from the manufacturers.

Overvoltage and undervoltage operation

Standard specifications require that motors be designed to operate satisfactorily at their rated outputs over a range of voltages. As an example, NEMA MG1–1987 quotes a limit of ± 10% of rated voltage. Clearly variations within such a range will not overstress the insulation of a machine but they do affect the heating which occurs because different levels of current will flow when the rated output is being provided.

Outside the specified operating voltage limits a machine could be damaged if the condition persisted for a significant period. If desired, voltage-sensitive relays could be provided to monitor the supply to a motor and disconnect it if necessary. Because excessive heating is caused, however, by the flow of above-normal currents, the overcurrent relays provided to detect other conditions should also give adequate protection when abnormal voltages exist.

The application of a very high surge-voltage to a motor may cause severe damage to the insulation of its windings. It must be recognized that overvoltage relays are not able to provide protection against such an event, because they could not operate and cause a motor to be disconnected from its supply before damage had been caused. It is therefore desirable that suitable surge-suppression devices or voltage limiters be fitted to the supply network if such conditions may arise.

Under- or over-frequency operation

The frequencies of the voltages supplied by major networks are closely controlled and the deviations are small, typically less than 1% of the nominal value. In more localized networks, however, greater variations may occur and therefore motors are usually designed so that they will operate satisfactorily from supplies with frequencies in the range of the nominal ± 5%. Such variations will affect the speed of an induction motor and therefore its power output when its load requires a fixed torque. Corresponding changes will occur in the currents supplied to the motor and the heating within it.

As with the overvoltage condition considered above, the overcurrent relays should detect this abnormal behaviour and ensure that the motor is not allowed to operate in a way which may cause it to deteriorate. Although the system frequency could readily be monitored and motors could be disconnected when values outside the normal limits were detected, this is not normally considered necessary.

Co-ordination of the protective equipment applied to induction motors

It will be appreciated that relatively small machines are normally only provided with simple and cheap protective equipment whereas large machines operating from relatively high voltages, such as 11 kV, are protected by several devices and schemes including one operating on the current-differential principle.

Several of the protective devices have a single setting and duty. For example, a temperature sensor, which will probably be installed by the motor manufacturer, must operate at a level below that at which damage will occur. Similarly a current-differential scheme should be set to operate at a level well below the rated current of the windings being protected, the only constraint being that it should not be so sensitive that it may operate incorrectly, say as a result of a starting-current surge. Voltage- and frequency-sensitive relays have settings which may be selected quite simply.

Overcurrent relays with inverse time/current characteristics have to be set, however, to perform satisfactorily under several conditions. Firstly they must not operate when a motor is carrying its rated current and indeed must allow it to run when slightly overloaded. They should have a minimum operating current corresponding to the maximum current which the motor being protected can carry continuously without being damaged, a typical value being 1.2 times the rated current of the motor. At higher current levels, the relays should operate in times slightly below those for which the motor can carry them. To ensure that this requirement will be met withstand time/current curves, of the form shown in Fig. 7.5, should be obtained from motor manufacturers if they are not already available. A further requirement is that the relays should operate in a time less than that for which the motor protected by them can withstand the locked-rotor condition. It is not possible to directly check that this latter requirement will be met by relays with a particular time/current characteristic because the starting currents of motors are not constant, but reduce from an initial high value in a manner dictated by the starting arrangements. In practice, therefore, it is usual to assume that the initial starting current will continue throughout the whole acceleration period and to check that this would not cause relay operation. This clearly ensures that a factor of safety will exist.

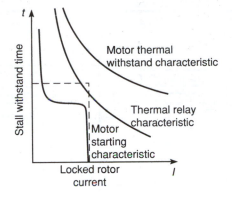

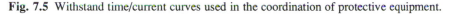

Fig. 7.5 Withstand time/current curves used in the coordination of protective equipment.

A further requirement is that the relays must not operate during normal starting periods. This latter requirement must always be achievable because motors must be so designed that they will not be damaged as a result of carrying their starting currents for the periods associated with load torques up to the rated values.

Thermal relays are particularly suitable for protecting motors because their operation is caused by heating and therefore their operating time/current characteristics tend to be similar in form to the withstand time/current characteristics of motors. When using such relays, it can be advantageous to omit compensation for ambient temperature variations because the times for which motors can withstand particular overcurrents are affected by the temperature of their surroundings, as also are the operating times of uncompensated relays.

It can also be advantageous to omit instantaneous-reset features as relay operation is then dependent on the current levels over a significant period and because of this a motor could be prevented from making several normal starts in a short period, a condition which cannot normally be allowed because of the progressively higher temperatures it would produce in a motor. Clearly, a relay which reset instantaneously would not provide the necessary protection in these circumstances because each motor start would be sensed independently and allowed to take place.

A typical arrangement of the protective equipment applied to an induction motor is shown in Fig. 7.6.

In the past such arrangements were implemented by providing separate relays to operate for each of the various forms of protection being applied, i.e. inverse-time overcurrent, instantaneous overcurrent, etc. Whilst this practice is still adopted today, electronic relays, which process information and are capable of performing several functions simultaneously, are being produced [5]. As a result, although the same forms of protection are being applied to motors, the number of separate relay units is reducing.

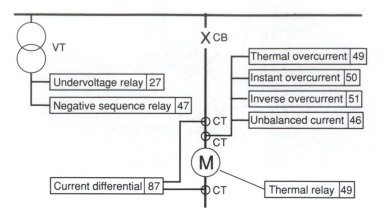

Fig. 7.6 Schematic arrangement of induction motor protective equipment. (System device function numbers according to IEEE C37.2–1987.)

Further information about relaying equipment and other protective arrangements are provided in references [6–9].

7.3.3 The protection of synchronous motors

The stators of three-phase synchronous motors are similar to those of induction motors and they have the same limitations. Clearly the same current-differential schemes and thermal overcurrent relays may be used to protect them.

The rotors of the two types of motor are, however, dissimilar, the rotors of synchronous machines having a winding fed with direct durrent.

Some of the conditions considered in section 7.3.2, which can cause induction motors to suffer consequential damage, can also adversely affect synchronous motors. There are, however, some conditions which are solely associated with synchronous machines and these are examined below.

Loss of synchronism

Non-synchronous running, which is clearly unacceptable, may be caused by the factors considered below.

Mechanical overloading of a machine   When a machine is running synchronously on no load, its excitation must be such that the e.m.f.s induced in each of the phases are of almost the same magnitudes and phases as those of the supply. Slight differences must exist however to drive small currents through the phase impedances of the stator windings and there must be a small power input equal to the electrical and mechanical losses. This condition is illustrated in Fig. 7.7(a). When load is applied to the motor shaft, the output power

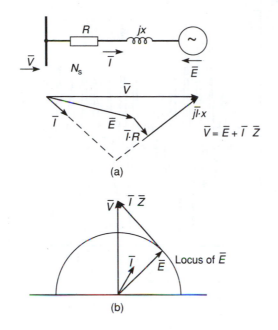

Fig. 7.7 Operation of a synchronous motor.

temporarily exceeds the input power and the rotor decelerates for a period. As a result, the phases of the rotor e.m.f.s fall further behind those of the supply voltages, causing the input currents and power to rise until equilibrium is established and synchronous running is re-established. Such a condition is shown in Fig. 7.7(b). At a particular load, which causes the phases of the stator e.m.f.s to be about $\pi/2$ radians behind those of the supply voltages, the limit of synchronous running is reached and any further increase in load will cause the motor to fall out of step. In these circumstances, slip-frequency e.m.f.s are induced in the rotor winding and the deceleration of the rotor increases. The stator-winding currents vary as shown in Fig. 7.8 and when the rotor reaches standstill they are very high, the conditions being similar to those in a stalled induction motor. As the conditions during deceleration cannot be allowed to persist, it is necessary that they be detected so that the motor can be disconnected from its supply.

Faults on the supply network   Faults occurring on the supply networks to which motors are connected probably cause most of the incidents of loss of synchronism. During faults, the supply voltages are often depressed, and under such conditions the load torque at which a motor may begin to decelerate may fall below its rated value. In such circumstances, loss of synchronism may occur unless rapid fault clearance is effected.

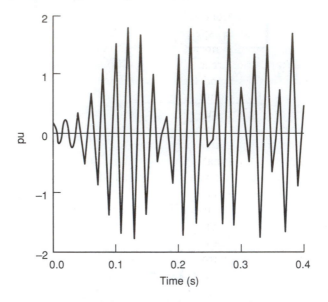

Fig. 7.8 Current variation during asynchronous operation.

Loss of excitation   Another common cause of loss of synchronism is under-excitation of a machine. This may arise because of the opening or short-circuiting of its field-winding circuit. It is clearly necessary that such conditions be detected and that the stator windings be disconnected from the supply. In addition, the appropriate action should be implemented to protect the field circuit. For example, if a short-circuit or an interturn fault is present in a field winding it should be disconnected from the supply and connected to its discharge resistor.

Starting

Synchronous motors, as shown above, can only run satisfactorily at synchronous speed and arrangements must be provided to enable them to be started.

One arrangement employs a small starting motor, often called a pony motor, which is coupled to the synchronous machine. During starts the synchronous motor is not connected to its supply and provides no driving power. The pony motor therefore provides the torque to accelerate the combination up to synchronous speed. When this speed is reached and the stator-winding e.m.f.s of the synchronous motor are close enough in magnitude and phase to the voltages of the supply sytem, it is connected to its supply. It is subsequently controlled to provide the required output. The stator-winding currents of the synchronous motor should not be abnormal during this process and should not affect the protective equipment. The pony motor must of course be provided with its own protective equipment.

An alternative starting method which is often adopted is to run up a motor in the induction-motor mode, special rotor windings being provided for the purpose, or alternatively the normal rotor windings may be short-circuited. When the speed has almost reached the synchronous value either the extra windings are open-circuited and the normal rotor winding is energized with direct current or the short-circuit is removed from the normal rotor winding and then it is fed with direct current. In either case, the motor will pull into step and run synchronously. During such starts, a large current surge flows initially as the motor accelerates from standstill, this surge clearly being of the form associated with normal induction motors and then a second smaller current surge flows as the machine is pulled into the synchronous mode.

In recent years, developments in the field of power electronics have enabled power-supply units with variable frequency outputs to be produced. Such units, with relatively low power outputs, are now being used to supply some synchronous motors during starting periods. The frequency output of a unit is increased gradually in a controlled way until the motor supplied by it is running at its synchronous speed and thereafter the motor is disconnected from the supply unit and connected to the main power-supply network.

In such cases, no high current surges should flow in the motors during starting periods.

Co-ordination of the protective equipment applied to synchronous motors

As with induction motors, the larger the rating of a synchronous motor, the more justified are relatively complex and expensive protective arrangements. Several of the available relays, devices and schemes, described in section 7.3.2, are applied to both types of machine.

Current-differential schemes with low fault-settings may be applied to the stator windings of large machines to detect short-circuits and faults to ground. Overcurrent relays with appropriate time/current characteristics are commonly used to ensure that damage is not caused by prolonged operation under overload conditions. Thermal devices are incorporated in the stators of machines to directly detect overheating and sensors to detect a failure to rotate from standstill may be used.

In addition, protective equipment to detect asynchronous running and field-circuit faults is essential on large machines. As stated earlier, asynchronous operation causes the phase angle between the stator-winding currents and voltages to vary to levels outside the normal operating range and indeed it changes continuously as the rotor speed falls. All three phases vary similarly and therefore a single relay, fed with the voltage and current of one phase, may be made to operate when the power factor or phase angle is in a range not encountered under synchronous conditions, say when the power factor is negative, i.e. when the phase angle is in the range $\pi/2$ to $3\pi/2$ rad. Further information is provided on this topic in section 7.4.2 (page 261).

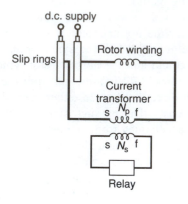

Fig. 7.9 Method of detecting asynchronous operation.

Alternatively, or in addition, the field-winding current can be monitored to detect the alternating-current components which are present when machines are running asynchronously. This is done by including a current transformer in the field circuit, its secondary winding being connected to a sensitive current-operated relay as shown in Fig. 7.9. It must be recognized that one cycle of alternating current flows in the field winding for a slip of one pole-pair pitch, i.e. for one revolution of slip in a two-pole machine. The frequency of the alternating current is therefore very low when non-synchronous running begins and the current transformer must be able to reproduce these slowly changing or low-frequency currents so that relay operation will occur quickly after synchronous operation commences.

It is vital that adequate current should flow in the field winding of a motor to maintain synchronous operation and therefore either the supply voltage to the field circuit or the current in it should be monitored by a relay which can disconnect the stator windings from their supply if the excitation becomes inadequate. Current-sensing relays, which must be fed from a shunt in the field circuit, have the advantage that they will operate in the event of an open circuit in the motor field circuit as well as for a loss of the supply voltage.

Protection should also be provided to detect faults associated with the rotor windings of relatively large synchronous motors. Rotor circuits are not normally earthed and therefore a single fault to ground does not affect the operation or cause further damage. The condition may be allowed to persist but it is desirable that an indication be provided to the operator so that action may be taken as soon as the motor may be conveniently disconnected. One method which is employed is shown in Fig. 7.10. It employs two lamps connected in series between the supply connections. Under healthy conditions, both lamps have the same voltages across them and they should therefore be equally bright. Should a fault occur on the field winding at any point other than the

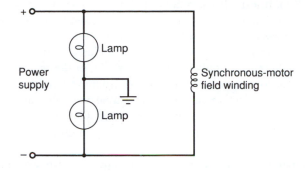

Fig. 7.10 The detection of earth faults on a field winding.

centre, unbalance will occur causing one lamp to be brighter than the other, because of the unequal voltages across them. As an alternative, the lamps may be replaced by two equal resistors and a voltmeter may be connected from their junction to earth.

Should it be considered desirable to disconnect a motor in the event of a single earth fault on its rotor winding then the second arrangement described above may be used, a voltage-operated relay being included rather than a voltmeter. This scheme has the slight disadvantage, already referred to, that operation cannot be obtained for faults near the mid-point of a winding. Alternative schemes have been developed to eliminate this limitation. One simple arrangement, shown in Fig. 7.11, includes a varistor in series with one of a pair of divider resistors. Again there is a point on the field winding at which an earth fault would not cause the relay to operate. Should the supply voltage subsequently change, however, the fault would then be detected, because the varistor resistance would have altered and the fault would not then be at the point on the rotor winding at which operation would not now be obtained. These and other possible arrangements are discussed later in section

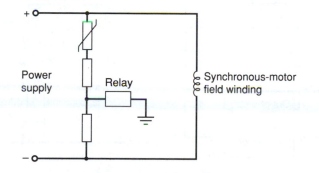

Fig. 7.11 The detection of earth faults on a field winding.

7.4.2 (page 261). If a single earth fault is allowed to persist on a rotor winding and then a second fault occurs, a short-circuit condition will exist. Such a condition could cause a high current to flow in the rotor circuit and this should be detected by the protective equipment associated with the exciter. This equipment should then not only cause the opening of the rotor circuit but also that of the stator as otherwise asynchronous running would ensue. It is also probable that the protective relaying associated with the stator circuit will operate when two faults are present on the rotor winding because the excitation is likely to be significantly reduced causing high currents in the stator windings.

Whilst temperature sensors can be fitted into the stators of machines and their outputs can readily be connected to monitoring equipment, it is not practicable to detect overheating of rotors in this way because extra slip rings would have to be provided to enable the necessary connection to be made. Consequently rotor temperatures were not monitored in the past. Recently, however, a relatively simple method, based on the resistances of rotor windings, has been developed to detect rotor overheating. Clearly the resistance of a winding is dependent on its surroundings and the current it carries. It will therefore rise above the value it would have when carrying a given current, if the adjacent rotor material is at an above-normal temperature. Relay operation can therefore be initiated either when the resistance of a winding exceeds a particular value or when it exceeds a value dependent on the current in it.

The protective equipment associated with the stator windings of synchronous motors is generally similar to that applied to induction motors, as stated earlier. Details of this were provided in section 7.3.2.

An extra feature which is required with synchronous motors is an out-of-step relay to detect loss of synchronism. Such relays compare the phase of one of the stator currents with a supply voltage, an example being shown in Fig. 7.12(a).

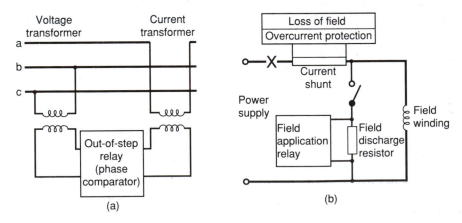

Fig. 7.12 The detection of asynchronous operation.

A typical arrangement of the equipment used to control and protect the field windings of large synchronous motors is shown in Fig. 7.12(b). Included in this arrangement is a field application relay, the duty of which is to initiate the closing of the field circuit at the appropriate instant towards the end of starting periods. This relay monitors the slip-frequency voltage across the field discharge resistor and when its frequency is low enough and its instantaneous value is appropriate the main field circuit is closed, after which the motor is pulled into synchronism.

7.3.4 The protection of d.c. motors

Direct-current motors have stationary magnetic fields produced by field windings mounted within their stators. Armature windings mounted in the rotors are fed via commutators. They may, depending on their duties, have their armature windings connected in series with their field windings (series-connected) or the windings may be connected in parallel (shunt-connected). In addition, shunt-connected machines may have extra field windings connected either in series with the armature windings or in series with the supply, to provide compounding.

Normal operation

Alternating e.m.f.s are induced in the conductors of an armature winding as they rotate in the constant magnetic field set up by m.m.f.s provided by the field windings. As a result of the commutator action, a direct e.m.f. is effectively present between the brushes, its magnitude being dependent on the current in the field windings.

This so-called back e.m.f. is almost equal to the applied voltage in a well-designed machine. As a result, the speed of a shunt-connected machine with a fixed field current remains almost constant for all normal loads. Lowering of the field current causes such a machine to run at a higher speed, so that the necessary back e.m.f. can be produced.

The field currents of series-connected machines cannot be separately adjusted and they are governed by the mechanical load applied to them, because the greater the load, the greater the motor current. Increase of load therefore increases the field and for a constant e.m.f. to be produced the machine speed must fall.

Compounding enables intermediate load speed characteristics to be obtained.

More detailed information on the normal behaviour of these machines may be obtained from reference [10].

Causes of failures in d.c. motors

Breakdowns and failures may occur from a range of causes and these are considered separately below.

Overspeeding   Operation at very high speeds can cause very high stresses in the rotating parts of a machine and it may therefore result in mechanical failures. It is clear from section 7.3.4 (page 251) that this undesirable condition will arise if the mechanical load on a series-connected machine is reduced to a very low level. In the case of a shunt-connected machine, overspeeding will occur if the current in its field winding is reduced below a certain level.

Overloading   Mechanical overloading of all d.c. motors causes them to draw currents above their rated values and this results in the temperatures of their armature windings and any series-connected field windings rising. Should such conditions be allowed to persist for long periods then temperature levels will be reached which will cause the insulating material around conductors to deteriorate.

The permissible periods for which particular degrees of overloading may be permitted depend on the designs of the machines and the operating conditions prior to the overloading.

The extreme condition of stalling, during which very high currents flow, must not be allowed to persist.

Undervoltage operation   A motor operating at its rated load will run at reduced speed and draw currents above the rated value if the supply voltage is below its normal level. Such a condition clearly may not be allowed to persist.

Overheating   Overheating may not only occur as a result of overloading but also because of inadequate cooling due to factors such as blocked ventilating ducts. Clearly mechanical defects of these types do not affect the electrical behaviour initially, and input currents do not rise, but if the conditions are allowed to persist insulating materials could then be damaged.

Commutator faults   Severe sparking may occur due to unsatisfactory commutation and this can lead to arcs forming between the positive and negative brushes. In these circumstances the armature is short-circuited and very high currents will be drawn from the supply. Should the condition be allowed to persist, severe physical damage would be caused, particularly to the commutator.

Unsatisfactory starting   Manually-operated starters are used with many relatively small machines and automatic starters may be used with large machines. Should the resistance be reduced too rapidly during starts, excessive currents will flow leading to possible physical damage.

The application of protective equipment

Relatively small machines may be protected by fuselinks in the main supply connection. As stated in section 1.9.8 (page 28), special fuselinks capable of withstanding starting-current surges are produced for this application. Generally, separate fuselinks are not included in the field-winding circuits of shunt-connected machines.

Overcurrent and inadequate field-current protection is provided on many machines, it being commonly incorporated within the starting equipment. Overcurrent relays or devices are usually set to operate if the input current rises above 1.3 times its rated value and, if necessary, operation may be delayed to ensure that it will not occur during starting-current surges. Relays or devices to detect inadequate field-current conditions are arranged to disconnect machines from their supplies.

Relays are often provided to monitor supply voltages and initiate the disconnection of machines should the voltages fall below acceptable levels. Speed-measuring devices may also be provided to detect both overspeeding and operation at unacceptably low speeds, including stalled conditions. Operation of these devices also causes their machines to be disconnected.

It is essential that circuits should be so arranged that starting equipment will be reset whenever machines are disconnected from their supplies, so that the correct sequence will be followed when machines are restarted.

Direct-current motors fed from rectifiers

As a result of the development of power-electronic equipment, electronically-controlled rectifiers are being increasingly used to supply direct-current motors. Rectifier/motor units in a range of ratings are produced and it is necessary that each unit be adequately protected. A typical arrangement of the protective equipment applied to such units is shown in Fig. 7.13.

Overcurrent and overvoltage protective devices are provided within rectifier housings. A fuse may be included in the supply circuit to a unit or alternatively a circuit-breaker which can be tripped by an overcurrent relay may be provided. Fault clearance must be effected rapidly enough to ensure that the

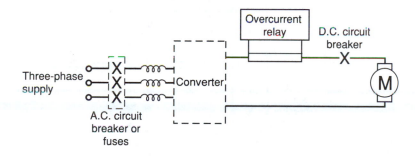

Fig. 7.13 The protection of a motor fed from a converter.

let-through energy (I^2t) does not exceed that which the equipment can withstand, typical maximum operating times being in the range 5–100 ms. Inductors are often included in units to limit fault currents to acceptable levels.

Signals from current sensors and overcurrent devices are fed to the electronic equipment in the rectifiers to control their output currents.

Should it be possible for a drive to operate in a regenerative mode at times, a d.c. circuit-breaker must be provided which can operate in the event of a short-circuit occurring in the rectifier.

7.3.5 The protection of variable-speed drives

In recent years, both induction motors and reluctance motors fed from electronically-controlled supply units have been used to provide variable-speed drives. Because of the diversity of the various arrangements which have been developed it is not possible to consider the protection of them in a general way in this work.

Clearly, in all cases, relays and other devices must be provided to detect abnormal conditions in the motors, such as overheating and faults within windings. Protective equipment for the supply units, however, may vary greatly, being dependent on their individual designs. To provide an illustration, the protection of the supply unit shown in Fig. 7.14 is considered below.

During normal operation, the incoming three-phase supply is rectified and fed to the inverter to produce a variable-frequency alternating supply to the motor. A large inductive reactor (L) is included between the rectifier and the inverter to ensure that the current fed to the inverter (I_{dc}) cannot change rapidly.

During faults in the inverter or the motor, the current can be brought to zero quickly by adjusting the firing pulses to the rectifier and by opening the circuit-breaker.

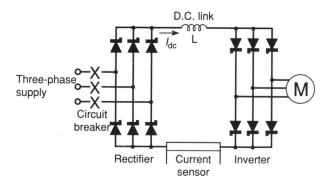

Fig. 7.14 The protection of a supply unit of a variable-speed drive.

If the drive operates in the regenerative mode at times or a fault occurs in the rectifier, the current fed back by the motor can be limited by adjusting the firing pulses supplied to the thyristors in the inverter.

Current sensors may be included in the connections from the a.c. supply, in the link between the rectifier and the inverter or in the motor circuit, but for those applications where regeneration may occur satisfactory performance will only be obtained using sensors in the last two positions.

More detailed information about particular electronic-drive systems and their protection is provided in reference [11].

7.4 THE PROTECTION OF ALTERNATORS

Alternators are basically similar to synchronous motors but they have much more other plant directly associated with them and they are produced with power ratings ranging up to much higher levels. All alternators are driven by prime movers, which at the lower power levels may be reciprocating diesel or petrol engines. At the high power ratings, however, they are invariably driven by water or steam turbines. In the past it was common to install a group of boilers which together supplied a number of turbines, this arrangement being referred to as the range system. In most installations today, however, complete generating units are produced consisting of the water supplies or steam-raising plant, the turbine and alternator. In addition, as stated in the introduction to this chapter, it is the practice in the large generating stations to have a power transformer directly connected to each alternator to step up its output voltage to that of the transmission system. There are many auxiliaries, such as fans and pumps, associated with each generating unit and the supplies needed by them are usually supplied by a second transformer which is also connected directly to the alternator, this transformer being referred to as a unit transformer. A typical arrangement is shown in Fig. 6.1 (page 180).

It will be appreciated that the items of plant in a generating unit are interdependent and therefore they must all be monitored to provide a co-ordinated protective scheme which will operate appropriately when any abnormality occurs. Important factors which must be taken into account when selecting protective equipment are examined in the following section.

7.4.1 Alternator construction and behaviour

Alternator construction

Alternators, except those of very low ratings, are invariably three-phase. The main output windings of a machine are mounted in slots in the stator core and the field windings, which are fed with direct current supplied by an exciter, are mounted in the rotor slots. The exciter may in turn be fed by a pilot exciter, the main machine rotor and the exciters being mounted on the common shaft driven by the turbine. To assist in maintaining synchronous running, damper bars or windings are also fitted in the rotor.

The rated stator currents of large attenuators are very high; for example, the rated current of the 500 MW, 22 kV machines installed in several generating stations in Britain is 22 kA. The stator windings of such machines, which may be made up of two or more windings in parallel, are invariably connected in star.

Earthing of stator windings

The neutral points of alternators are usually earthed to ensure that excessive voltages are not likely to be present between their stator windings and the surrounding core. Should a fault nevertheless occur between a winding and the core, and a large current flow as a result, then severe arcing could melt the core locally and cause laminations to become welded together. The conductors affected might be replaced at a relatively low cost but the necessary repairs to the core, to ensure that excessive heating would not occur in it during subsequent normal operation, could be extremely costly.

To avoid this situation it is usual to include impedance in the neutral connection to earth to limit the magnitudes of the earth fault currents which may flow. In some installations, earthing resistors, which will limit earth fault currents to the rated current of the machine, are used. On very large machines, such as those referred to earlier, however, the rated currents are very high and earth faults of these magnitudes would certainly cause extensive damage to a stator core. It is the standard practice in such cases to include earthing resistors which limit earth fault currents to maximum levels, 200 A being a commonly-used value. In such instances even this value has been considered to be excessive and higher values of earthing resistance have been used with a few machines.

It will be obvious that limiting earth fault currents to very low values in the above way requires the use of resistors of high ohmic value in the neutral

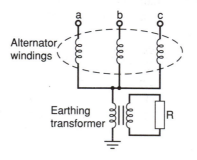

Fig. 7.15 Method of earthing an alternator.

connections, say about 4000 Ω to limit currents to 5 A and such resistors may not be very robust. As an alternative, earthing transformers of 5–100 kVA rating are used in some installations, the connections being as shown in Fig. 7.15. In the extreme case of an earth fault on one of the alternator output terminals, its phase voltage would be present across the primary winding of the earthing transformer. Under this condition its secondary winding output voltage would be typically in the range 100–500 V and this is applied across a resistor. The fault current can thus be limited to a low value using a robust resistor of relatively low ohmic value, i.e. reduced from the value needed with direct connection by the square of the earthing-transformer ratio. Further information on this method of earthing is provided in reference [12].

Interphase and interturn faults on stator windings

In the previous section, methods used to limit the magnitudes of earth faults on stator windings were considered. Such techniques, however, would quite obviously have no effect on interphase and interturn faults which could cause large currents to flow.

Because of the physical distribution of the phase windings around the stator cores of machines, interphase faults are unlikely and can only occur on the end connections of stator coils and in slots which contain the sides of two coils belonging to different phases. Interturn faults are also quite rare. Should any of these faults occur, however, it is likely that the resulting damage will cause the faults to develop and quickly become earth faults.

Faults on rotor windings

The field circuits of alternators, like those of synchronous motors, are normally unearthed and therefore a single earth fault on a rotor winding has no effect on the performance of the machine. It may be allowed to persist and indeed some machines have been allowed to continue operating in this condition for lengthy periods.

However, it is clearly desirable that an indication should be given of the presence of such a fault, so that remedial work may be undertaken at the earliest convenient opportunity.

It must be realized, when a decision is taken to allow a large alternator to remain in service with an earth fault on its rotor winding, that considerable consequential damage could occur if a second earth fault then developed on the winding. In such a situation a parallel path would be created in which a large current could flow and this could damage both the conductors and the rotor core material. In addition, the magnetic field then produced by the rotor might no longer be sinusoidally distributed around the periphery and it might also be asymmetric, the flux densities at corresponding points on the poles being different. This could have serious consequences because the attractive forces existing between the poles and the stator core could be unequal and vibrations could be set up during each revolution of the rotor. This behaviour could cause severe physical damage, particularly to components such as bearings.

Interturn faults rarely occur on rotor windings, but should they do so it is very probable that they will develop rapidly and become earth faults.

Inadequate excitation

The output of an alternator connected to busbars is dependent on its excitation and the power supplied by its prime mover. For a given power input the power output per phase must be almost constant, i.e. $I \cos \varphi$ must be almost constant and therefore the current locus may be taken to be a straight line vertical to the phase voltage (V), as shown in Fig. 7.16. With relatively weak excitation, the phase e.m.f. (E) is lower than the terminal voltage (V) and as a result the machine operates at a leading power factor. As the excitation is varied, the locus of the e.m.f. must be the straight line shown in Fig. 7.16. It is evident that synchronous running will not be maintained if the excitation is reduced below that required to produce the limiting e.m.f. (E_L) at point A on the locus.

Such situations should not normally occur, but they could arise because of faults in exciters or excitation circuits.

Asynchronous operation

Insufficient excitation considered above can lead to a machine running at a constant speed, above the synchronous value as an induction generator, the

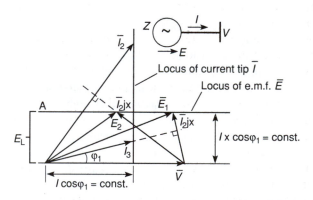

Fig. 7.16 The operation of an alternator. (Note: The resistances (R) of the windings of alternators are very much lower than the inductive reactances (X) and they have been neglected in the figure. In practice the locus of E slopes slightly from the vertical. Two operating conditions are shown, designated by the suffixes 1 and 2.

main field then being produced by the reactive components of the stator-winding currents. In this mode, e.m.f.s are induced in the rotor core, and in the damper bars and the slot wedges, causing significant heating. Such operation cannot be allowed to continue indefinitely, the permissible periods being of the order of a few minutes. The precise withstand times, which are dependent on the machine design, should be obtained if necessary from the machine manufacturer.

An alternator with a healthy excitation system may also fall out of step and operate asynchronously because of faults on the network being supplied by it. For example, an interphase fault on a transmission line may cause the terminal voltages of a machine to be depressed and the power transferred to the loads may then be severely affected. As a result of the e.m.f. phasors moving relative to the terminal voltages, the magnitudes and phases of the stator-winding currents may vary widely and unacceptably. The machine may not return to synchronous operation when the fault has been cleared, in which event it must be disconnected and then reconnected after completion of the synchronizing procedures.

Unbalanced loading

In the case of motors, unbalanced operation only normally occurs if the supply voltages are unbalanced, whereas with alternators it is normally caused by unbalanced loading of the network supplied by the machine or by faults.

An alternator, like a synchronous motor, overheats quite rapidly if it carries even relatively low currents of negative sequence, because such currents set up a magnetic field rotating in the opposite direction to that of the rotor. They therefore induce significant e.m.f.s at double the operating frequency in both the rotor core and winding.

The amount of negative-sequence current which a machine can carry depends on its construction and factors such as the cooling methods it employs. Because unbalanced loading may continue for long periods, machines are assigned continuous negative-sequence ratings which are expressed as percentages of their continuous ratings. These values, which should be obtained from machine manufacturers when selecting protective schemes, are typically in the range from 10% to 15% for turbine-driven alternators with cylindrical rotors.

During fault conditions, large negative-sequence currents may be provided by an alternator. As examples, an interphase fault current, say phase a to phase b, has a negative-sequence component with a magnitude of 57.7% of that of the fault current and a single-phase to earth fault has a negative-sequence component equal to one third of the fault current. The magnitudes of these component currents, which can be quite large, may be calculated using the sequence impedances of machines which are to be protected.

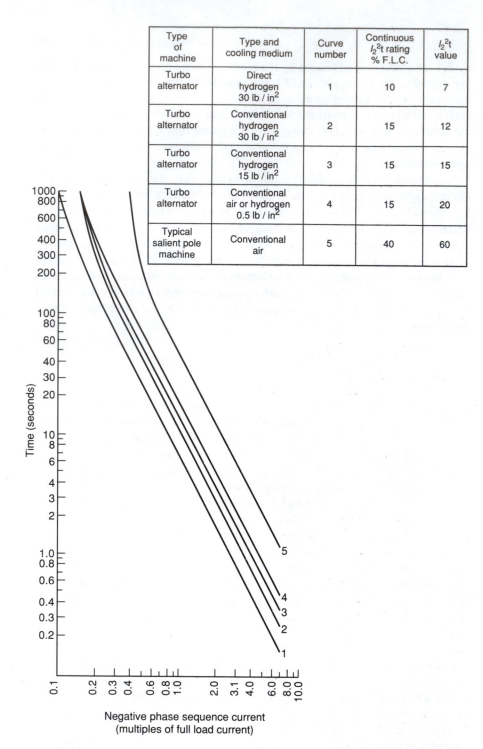

Type of machine	Type and cooling medium	Curve number	Continuous I_2^2t rating % F.L.C.	I_2^2t value
Turbo alternator	Direct hydrogen 30 lb / in²	1	10	7
Turbo alternator	Conventional hydrogen 30 lb / in²	2	15	12
Turbo alternator	Conventional hydrogen 15 lb / in²	3	15	15
Turbo alternator	Conventional air or hydrogen 0.5 lb / in²	4	15	20
Typical salient pole machine	Conventional air	5	40	60

Time (seconds)

Negative phase sequence current
(multiples of full load current)

Fig. 7.17 Typical negative sequence current withstand levels of alternators with different forms of cooling. (Reproduced from *Protective Relays – Application Guide, 3rd edn*, GEC Measurements, 1987 with the permission of GEC Alsthom Protection and Control Ltd).

Because fault durations are short, it may be assumed that little heat is lost by a machine while they are present. In addition, it is the heating caused by these currents which may cause damage and therefore it is the input energy which must be limited. As a result, the time integral of the square of the negative-sequence current must be kept below a particular value, i.e. $I_s^2 t \leq$ constant. A typical withstand time/current characteristic for a hydrogen-cooled machine is shown in Fig. 7.17.

Mechanical equipment

Alternators are electro-mechanical structures subject to mechanical stresses. Large machines are reliant on gas and/or water cooling systems which must perform correctly, and action must be taken quickly if they become defective in any way. Similarly prime movers, such as turbines, and other plant which form part of an integrated generating unit are individually complex and may not operate satisfactorily or safely because of a range of defects within themselves or their auxiliary equipment. Consequently, a large number of monitoring or protective devices are incorporated in such plant during manufacture and erection and they must function in conjunction with the protective equipment associated with the alternators.

7.4.2 The application of protective equipment to alternators

Small single-phase alternators, such as those fitted in motor-cars, may be protected by fuses fitted in the field and main winding circuits. The fuse in the main winding circuit will operate for overloads or faults to earth on the main winding whilst the fuse in the field circuit would operate for faults in its circuit. Machines of this type do not operate synchronously and loss of excitation merely causes loss of output voltage and therefore a reduction, possibly to zero, of the output current. Thermal devices may also be embedded in such machines to disconnect them in the event of overheating.

With larger machines and certainly for those which are to run in synchronism on a network and have three-phase outputs, fuses do not represent a satisfactory form of protection. The operation of one of the output fuselinks would produce unacceptable unbalanced operation and operation of a fuselink in the excitation circuit would result in asynchronous running. In these applications co-ordinated protected equipment should be provided to cover the whole unit, including the alternator, driving machine and excitation circuit. The cost of such units is usually very considerable and therefore the use of several forms of protection is justifiable and, of course, very complex schemes are not only justifiable but essential on very large machines such as those installed in major generating stations. The various forms of protective equipment available for application to alternators are considered below.

Current-differential schemes

In some installations, alternators are connected directly to busbars feeding distribution circuits, a typical arrangement being shown in Fig. 7.18(a). In such cases, six current transformers, which should ideally be identical, should be installed, three in the connections between the windings and the neutral point and the other three in the output connections. Biased relays may be used, as shown in Fig. 7.18, to enable the sensitive settings, needed to detect small

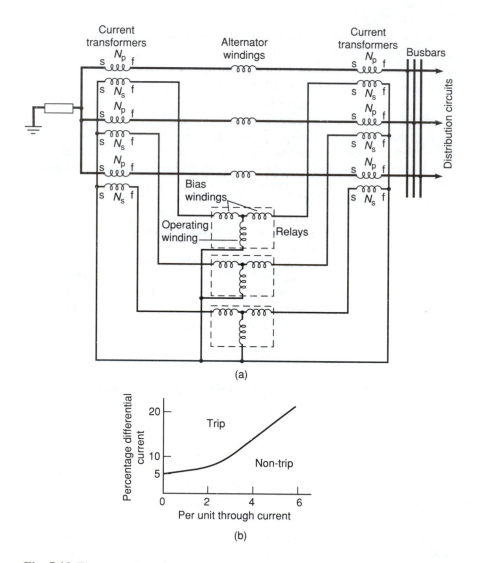

Fig. 7.18 The protection of an alternator supplying distribution circuits.

earth fault currents to be obtained. The relays, which may be of the induction pattern, usually have a relatively low bias, say 5%, i.e. as shown in Fig. 7.18(b).

Alternatively, high-impedance relays may be used, their operating voltages being set high enough to ensure that operation would not occur even if the core of one of a pair of the balancing current transformers was continuously saturated whilst the other remained unsaturated. This condition, which was examined in section 5.6 (page 171), becomes more difficult to satisfy as the resistances of the interconnecting conductors increase. In applications associated with machines, however, the interconnections are usually quite short and acceptable settings can readily be obtained. As mentioned earlier, it is usually necessary to connect a non-linear resistor in parallel across the operating windings and the series resistors of these relays to limit the voltages which might otherwise be present across them during internal faults.

These schemes do not detect interturn faults but they will certainly operate in the unlikely event of interphase faults on a stator winding because there is usually a very low impedance in the paths of such faults. As stated earlier, earth fault currents are limited by resistors included in the neutral connections. These usually have high ohmic values and therefore the current magnitude during an earth fault on a winding will vary almost linearly with the voltage at the fault position and thus its distance from the neutral point. It must be accepted therefore that faults near the neutral point will not be detected. As an example, if the current which will flow when an earth fault is present at the output terminal of a phase winding is limited to 200 A by the earthing resistor and the relays are set to operate for currents of 20 A in their current transformer primary windings, then the 10% of the winding adjacent to the neutral point would not be protected. This limitation has not been considered to be serious in the past but recently schemes capable of protecting all of the stator windings have been investigated. One type detects earth faults by examining the harmonic voltages produced by alternators whilst another is based on the injection of additional relaying signals [13–15].

Earth fault protection

Earth fault protection may be applied to supplement that provided by a current-differential scheme. This can be provided by mounting a low-ratio current transformer over the neutral connection of a machine and feeding its output to a relay, as shown in Fig. 7.19. Because this arrangement does not involve balancing of currents, the relay may be given a very sensitive setting.

If this form of protection is applied to an alternator which is directly connected to its loads, operation will be obtained for faults at any point on the network and discrimination must then be achieved with other protective equipments by introducing an adequate time-delay before the alternator is disconnected or shut down.

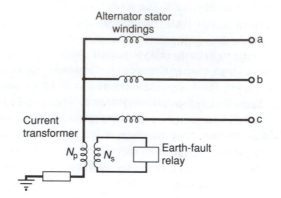

Fig. 7.19 Sensitive earth fault protection.

In applications associated with alternator–transformers, the earth fault relay may, however, initiate immediate action when the transformer windings connected to the alternator are delta-connected. In this, the usual arrangement, relay operation will only occur for earth faults associated with the alternator and the primary windings of the transformers.

Some machines have only four bushings, rather than six, three being used for the output connections from the phases of the stator winding and one for the neutral connection, the neutral point being within the casing. In such cases, four current transformers interconnected as shown in Fig. 7.20 may be installed to provide earth fault protection. High-impedance or biased relays are used in this application and again it must be accepted that faults near the neutral point will not cause operation but nevertheless with sensitive settings most of each phase winding can be covered. Clearly this protective arrangement will not detect interphase or interturn faults, but this may not be regarded as unacceptable because such faults occur so infrequently.

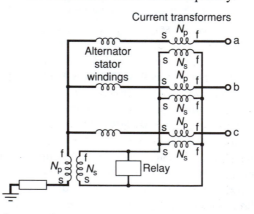

Fig. 7.20 Earth fault protection of the stator windings.

Many alternators, particularly those with high output ratings, operate in conjunction with transformers. As stated in the introduction to this chapter, it is the practice in large generating stations to directly connect each alternator to a step-up transformer which then feeds busbars operating at the transmission voltage level, for example 400 kV. The transformers almost invariably have their low voltage windings connected in delta and their high voltage windings in star with the neutral point solidly earthed. It is also usual to connect a second transformer with a lower rating to the output connections of each alternator. This transformer steps down the voltage to a level suitable for the supply of the other plant associated with the alternator. Figure 6.1 (page 180) shows the electrical layout of an alternator and its two associated transformers which is common in large power stations. The step-up and step-down transformers are usually referred to as the 'main' and 'unit' transformers respectively.

Because the three items, the alternator and the two transformers are interdependent, circuit-breakers are not installed in the connections between them and all three are usually included in the zone of the current-differential scheme which is invariably installed to protect them. The scheme therefore operates from nine current transformers, as shown in Fig. 7.21, and balance has to be effected in each phase, during healthy conditions, between the outputs of three transformers of different designs and ratios. For the rated values in the example shown in Fig. 7.21, the current transformers associated with the alternator, main and unit transformers could have ratios of $5000/5$, $500/5/\sqrt{3}$ and $10\,000/5/\sqrt{3}$ respectively and it will be noted that they must be connected in star and delta configurations to allow for the connections of main and unit transformers. This situation is clearly similar to that examined in section 6.3.2 (page 198) of the previous chapter except that in the case of power transformer protection there are normally only two sets of balancing current transformers. As in those applications, either biased or high-impedance relays may be used in schemes protecting alternator–transformer units and, because of the importance of these units, it is vital that their protective schemes should perform correctly at all times.

It is essential, first, that the settings of the relays should be such that they will not operate in the event of faults on the networks connected to the secondary (output) windings of the main and unit transformers. To enable this to be achieved, fault calculations should be done to determine the most extreme conditions which might occur. It should be recognized that the current transformers in the balancing groups will normally all be operating at different points on their excitation characteristics at each instant and therefore their transformation errors will not balance out. If high-impedance relays are to be used, their settings should be determined assuming that the core of one transformer will be continually saturated during external faults while the others are producing their ideal secondary currents.

Having selected the relay settings, the percentages of the various windings on which earth faults will not cause relay operation should be determined, in

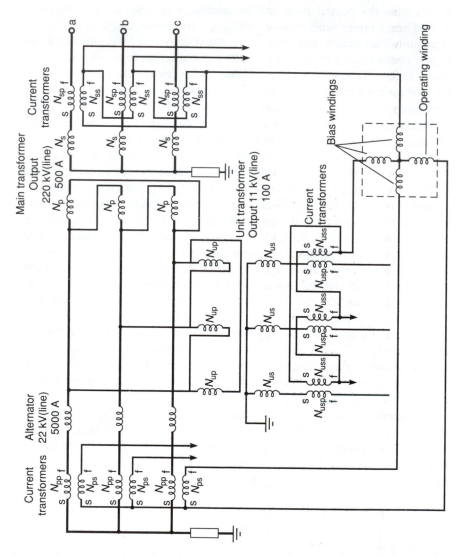

Fig. 7.21 The protection of an alternator and its associated transformers (the relay for one phase only is shown).

To obtain balance $\dfrac{N_{pp}}{N_{ps}} = \dfrac{N_p}{N_s} \dfrac{N_{sp}}{N_{ss}} = \dfrac{N_{up}}{N_{us}} \dfrac{N_{usp}}{N_{uss}}$.

a similar way to that described above, to ascertain that they are sufficiently low to be acceptable.

Similar checks should be made if biased relays are to be used, to determine that the optimum percentage bias is selected.

In the previous chapter it was pointed out that current-differential schemes should incorporate features such as harmonic-biasing, to prevent them operating during the exciting current surges which may occur when a power transformer is initially energized. Exciting current surges are less likely to occur, however, when alternators and transformers are directly connected together because during run-up the voltage applied to the primary windings of a transformer increases as the exciting current of the alternator increases, i.e. there is no sudden application of a high voltage. When a unit is subsequently synchronized to the busbars, the transformer secondary voltages are made almost equal in magnitude and phase to those of the busbars and therefore, when the circuit is closed, any associated current surge will normally be very small.

It is, however, possible that significant exciting current surges may occur as a result of faults on the network connected to a transformer. For example, should a short-circuit occur on the network fed by a transformer at a point close to it, the secondary voltage could collapse to zero. If this happened at an instant when the voltage had been small, the transformer core flux would have been near its maximum value. If the fault was subsequently cleared at a time when the voltage would have been at its maximum value, i.e. a current zero in an inductive circuit, then a further flux variation could be required which would take the core into saturation, thus causing an exciting current surge. This condition is illustrated in Fig. 7.22. Because of the possibility of such events occurring it is desirable that the current-differential protective scheme should have the same features as those used to protect large power transformers which are not associated directly with alternators.

A further limitation of overall current-differential protective schemes is that they may not detect faults in the unit transformers. Such transformers have relatively low ratings and therefore have impedances which restrict the currents which can be fed to faults on their windings. Because of this, unit transformers are often protected by their own schemes.

Schemes to detect interturn faults

As with large synchronous motors, it is common for alternators with high volt-ampere ratings to have stator windings with two or more paths in parallel. In such applications it is desirable that a current-balance scheme be included as part of the overall protective equipment. This scheme, which is illustrated in Fig. 7.1, operates when the total current in a winding is not being carried equally by the parallel paths. It can therefore detect faults, such as interturn faults, to which current-differential schemes are insensitive.

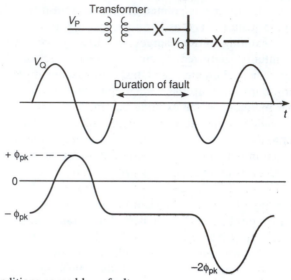

Fig. 7.22 Conditions caused by a fault.

Alternators with single-path stator windings clearly cannot be protected in the above manner and, in the past, equipment was not provided to detect interturn faults on such machines. This practice was considered acceptable because these faults occur very infrequently and many of them develop quickly into faults involving earth. Serious damage can, however, result from an uncleared interturn fault and it is certainly desirable that they should be detected and cleared quickly. A scheme which is now widely used to do this determines the zero-sequence component present in the voltages across the three phases of a machine. This component will be present when an interturn fault is present, but it is normally insignificant during healthy conditions. The scheme may therefore be implemented by connecting the primary windings of three voltage transformers on a five-limb three-phase transformer across the phases of a machine as shown in Fig. 7.23. The zero-sequence voltage is then derived by connecting the three secondary windings in series, i.e. in open delta.

It must be recognized that a zero-sequence component may be present in the voltages of a machine when earth faults are present on the network connected to it and therefore the voltage-operated relay used to monitor the output in the above scheme should be arranged to operate after a time delay sufficiently long to enable faults on the network to be cleared before the machine is disconnected.

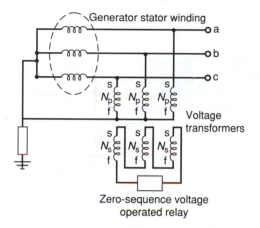

Fig. 7.23 Method of detecting interturn faults in a winding.

Schemes to detect unbalanced operation

Alternators, as stated earlier in section 7.4.1 (page 255), are not capable of operating continuously under unbalanced conditions which cause their currents to contain significant negative-sequence components and therefore all large machines should be provided with negative-sequence relays which have inverse time/current characteristics coordinated with their withstand time/current capabilities.

Schemes to detect asynchronous operation caused by low excitation

Asynchronous running of an alternator may be caused by its excitation falling below critical levels which are dependent on the power it is supplying. A machine may, however, continue to run synchronously at a low leading power factor, even with very low excitation, if it is lightly loaded, i.e. if its prime mover is supplying only a small amount of power. This can be seen from Fig. 7.16 in which the value of the limiting e.m.f. (E_L) is low if $I \cos \varphi$ is low.

A voltage-operated relay may be connected across a shunt in the excitation circuit to monitor the current flowing in it. The relay would operate in the event of a complete loss of excitation, due to an open circuit or failure of the exciter. It must, however, have a low setting, say 5% of the continuous rating of the circuit, to ensure that its machine will not be disconnected at a time when it could continue to run synchronously. This is not completely satisfactory, however, as synchronism might be lost at higher field-current levels if the power output was greater.

A superior method which is now widely used is to monitor the impedance presented at the output terminals of an alternator. As explained earlier in section 7.4.1 (page 255), the phase e.m.f.s produced by an alternator rotate relative to its output voltages when it runs asynchronously, i.e. at both super- and sub-synchronous speeds. As a result, the current varies during each slip cycle.

This behaviour may be illustrated by considering the simple circuit shown in single-phase form in Fig. 7.24(a) in which E and V are the e.m.f.s provided by the asynchronous machine and the synchronous source respectively, X_a is the inductive reactance of the machine and Z_s is the reactance of the synchronous source and the connecting circuit to the machine terminal at point A. To simplify the treatment the alternator resistance is neglected. The current I and the terminal voltage V_a are given by:

$$\bar{I} = \frac{\bar{E} - \bar{V}}{j X_a + \bar{Z}_s} \quad \text{and} \quad \bar{V}_a = \bar{E} - \frac{j X_a}{j X_a + \bar{Z}_s}(\bar{E} - \bar{V})$$

The apparent input impedance (Z_{ap}), seen at terminal A, is:

$$Z_{ap} = \bar{V}_a / \bar{I} = \frac{E\bar{Z}_s + j V X_a}{\bar{E} - \bar{V}}$$

With zero excitation the machine e.m.f (E) would be zero, in which event the apparent impedance (Z_{ap}) would be:

$$\bar{Z}_{ap} = -j X_a$$

With weak excitation, the magnitude of the e.m.f. E would be less than that of the voltage V. As an example, if $E = 0.5\ V$, i.e. the e.m.f. has a magnitude half that of the voltage and they are of the same phase. In this case the apparent impedance (Z_{ap}) would be given by:

$$\bar{Z}_{ap} = -(2 j X_a + \bar{Z}_s)\ \Omega$$

Half a slip cycle later, when $E = -0.5\ V$, the value of Z_{ap} would be:

$$\bar{Z} = -\frac{j 2 X_a - \bar{Z}_s}{3}\ \Omega$$

In practice the impedance Z_s will be smaller than $2X_a$ and therefore the above impedance will always appear to be capacitive when the e.m.f. E is less than the voltage V. It can be shown that the loci of the apparent impedances as the e.m.f. E rotates relative to the voltage V will always be circular, examples being shown in Fig. 7.24(b).

Clearly, therefore, a relay energized with a current and voltage proportional to those at the stator terminals of a machine, and set to operate when the apparent impedance is abnormal, may be used to detect asynchronous operation caused by low excitation. It will be clear that only one relay, associated with one phase of a machine, is needed for this purpose because the behaviour of each of the phases is the same for this condition.

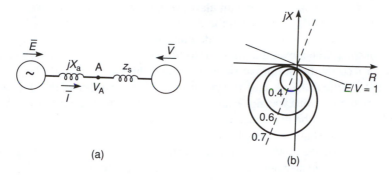

Fig. 7.24 Apparent impedance (Z_{ap}) at point A.

Relays with circular operating zones in the impedance plane, the zone centres being displaced from the origin as shown in Fig. 7.25, are commonly used to detect asynchronous operation. These relays which are described as mho type, are considered in some detail in section 11.3.2. To enable them to be correctly set for this application, the apparent impedance loci for protected machines during asynchronous and both healthy and other abnormal conditions should be determined to ensure that the relays will discriminate correctly. It will generally be found that satisfactory performance will be provided by a relay with a mho characteristic centred at a point $R = 0$ and $X = -(0.75 X_{d'} + 0.25 X_d)$ and with a radius of $0.25 X_d$, in which $X_{d'}$ and X_d are respectively the direct axis transient and synchronous reactances of the protected machine.

Asynchronous operation caused by low excitation may be allowed to persist for periods ranging from a few seconds to several minutes, the permissible times being lower the greater the rating of a machine. In attended stations it

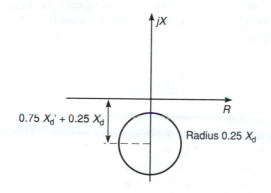

Fig. 7.25 Operating zone of a relay set to detect asynchronous operation.

is therefore quite common for a relay, which has detected that a machine is running asynchronously, to give an alarm so that the operators may assess the situation. They could firstly open the excitation circuit to enable the machine to run in the induction-generator mode and then they could reduce the power output of the prime mover and thus possibly re-establish synchronous running. If this procedure did not prove successful then the machine could be disconnected from the network and re-synchronized after checking the excitation circuit. If the network was not heavily loaded, and it was thought preferable, the machine could be disconnected on receipt of the alarm.

In unattended stations, the relay could initiate the opening of the excitation circuit and cause the power output of the prime mover to be reduced to attempt to re-establish synchronous running and, should this not occur, then the machine could be disconnected from the network.

Schemes to detect pole slipping

Pole slipping may occur as a result of a disturbance or fault on the network connected to a machine. In such cases, the speed of the machine rotor changes slightly for a short period during which high currents will flow and large mismatches will exist between the input and output powers if full excitation is maintained. As a result, the mechanical stresses and vibrations produced will be much higher than those present when asynchronous operation occurs because of a loss of excitation. If the disturbance or fault is removed quickly, synchronous running may be re-established when the rotor has moved forwards or backwards by two pole pitches, i.e. one revolution in a two-pole machine, and no action need be taken. If this does not occur, however, the condition cannot be allowed to continue because of the damage which will result.

Pole slipping, because of its similarity to asynchronous running, may also be detected by monitoring the apparent impedance at the output terminals of a machine. Again a relay with an appropriate characteristic in the impedance plane must be selected and it must be different from that chosen to detect asynchronous running produced by low excitation, the impedances for this condition being inductive rather than capacitive. Relays with mho-type characteristics may prove satisfactory but it may be found necessary, after examining the various impedance loci, to employ relays with quadrilateral characteristics of the form shown in Fig. 7.26.

In attended stations, the relay could initiate the opening of the field circuit-breaker to enable the machine to establish asynchronous running and thus remove the damaging stresses. The power input to the prime mover could then be reduced to a level where the machine might resume synchronous operation. If this did not occur, the field circuit-breaker could be reclosed to allow a low exciting current to flow. This should normally cause the machine to become synchronous. The alternative is to arrange that relay operation will initiate the

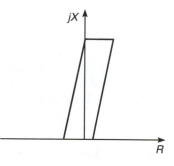

Fig. 7.26 Relay characteristic used to detect pole-slipping.

opening of the main circuit-breaker to disconnect the machine from the network, after which it could be resynchronized.

In unattended stations, the latter practice must usually be adopted, unless the control systems are highly automated and capable of initiating the procedures outlined above.

Overloading

Alternators in attended stations are not likely to be overloaded. Firstly the power they can provide is limited to the maximum rating of their prime movers and the operators will ensure that their excitations do not allow them to provide excessive VAr outputs. It is not usual therefore to provide protective equipment to detect the overloading of such machines.

Alternators in unattended stations may, however, supply excessive currents at times due to failures of control equipment, such as voltage regulators, and overload relays with suitable operating-time characteristics are often fitted to such machines.

Overcurrent protection

Although the synchronous positive-sequence impedances of alternators are usually high, values of 1 pu or more being normal on large machines, the sub-transient and transient values are much lower, as also are the negative and zero-sequence impedances. The currents which will flow in them to faults on the networks they supply can therefore be quite high initially. They will, however, decay to the normal rated values or less if a fault persists, the actual rate of decay, in a given case, being dependent on the decrement characteristic of the machine and the behaviour of its voltage regulator.

IDMT relays may be used to provide a back-up feature to the protective equipment on a network. The time and current settings of such relays must be selected after studying the characteristics of the machines to which they are to be applied. When these relays are to be used in a network fed by a single machine, the current transformers feeding the relays should be in the neutral connections of the machine so that operation will occur for faults, both internal and external to it. In those situations where several machines operate in parallel, the current transformers should be in the output connections of the machines, operation for both external and internal faults then being obtained because of the currents that will be fed into a faulted alternator from the other machines.

The detection of earth faults on field circuits

The field windings of alternators are normally unearthed and therefore the incidence of a single earth fault will not cause current to flow. It is desirable, however, that an indication of the condition should be given so that remedial work may be done at the earliest convenient opportunity. Several methods are used in practice to detect this condition. A simple scheme incorporates a potentiometer connected across the field winding. The centre point of the potentiometer may be connected to earth through a high resistance and a sensitive relay. This scheme, which is also applied to synchronous motors, has the disadvantage that it is not sensitive to faults near the centre of the field winding. An alternative arrangement, described in section 7.3.2 (page 235), in which a varistor is introduced does ensure that such faults will be detected when the excitation changes significantly. Such a change may not occur for a long period on large base-load alternators and therefore another arrangement which may be adopted in attended stations is to include a switch, as shown in Fig. 7.27, so that its operation will cause the relay to be connected to a different point on the potentiometer which would then enable a fault near the centre of the field winding to be detected. This check would be made manually every few hours.

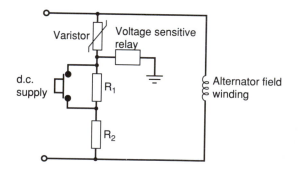

Fig. 7.27 Scheme to detect earth faults on field windings.

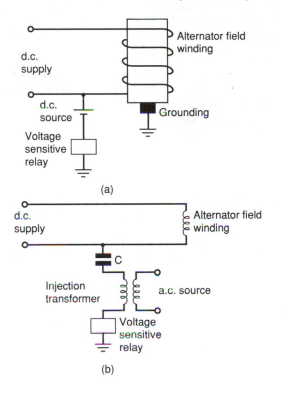

Fig. 7.28 Schemes for detecting earth faults on field windings.

More complex arrangements involving auxiliary supplies are used on large motors and alternators, two being shown in Fig. 7.28 (a) and (b). In each, current will flow through the relay in the event of an earth fault on the field-winding circuit and an immediate indication of its presence can be provided.

Other monitoring and protective schemes

The protective schemes applied to alternators have been considered in the preceding sections. Associated with them must be devices to detect other abnormal conditions which may arise in the prime movers and auxiliary plant. As an illustration, conditions such as low steam pressure, loss of vacuum and loss of boiler water should be detected when driving power is provided by stream turbines. In addition, the supply of lubricating oil and the performance of governors should be monitored.

When transformers form units with alternators they are included in the zones of the overall current-differential schemes, as stated earlier. In addition, however, the transformers should have earth fault protection to cover their second-ary windings, Buchholz relays and devices such as temperature sensors.

Overall protective arrangements

For any particular application, the appropriate protective schemes should be provided to ensure that all unacceptable behaviour which occurs in the protected plant will be detected. Each of the protective schemes or devices should, when necessary, provide an indication of the abnormality it has detected and also initiate the appropriate actions to ensure that any damage which might have occurred is limited to the lowest possible level. As an example, the operation of a current-differential scheme should cause the appropriate circuit-

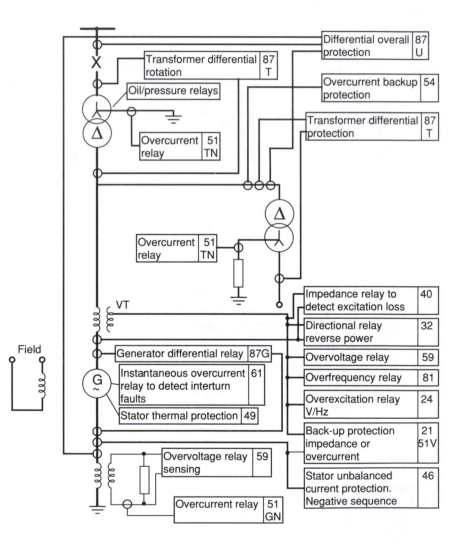

Fig. 7.29 Schematic of alternator–transformer unit protection.

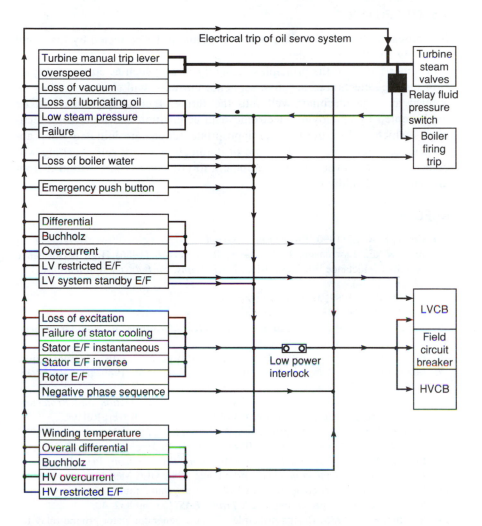

Fig. 7.30 Typical tripping arrangements for alternator–transformer unit. (Reproduced from *Protective Relays – Application Guide, 3rd edn*, GEC Measurements, 1987 with the permission of GEC Alsthom Protection and Control Ltd).

breakers to be opened to ensure that the unit is disconnected from the network and also that its excitation is removed. In addition, the steam or water supply to the driving turbine should be cut off or reduced by operating the appropriate control valves.

An arrangement suitable for the complete protection of a large alternator–transformer unit is illustrated in Fig. 7.29. More examples may be found in references [6, 7, 16, 17]. A typical tripping arrangement for an alternator–transformer unit is shown in Fig. 7.30.

7.5 THE FUTURE

It is probable that small machines will continue to be protected by fuses and relatively simple thermal devices and relays.

Schemes based on the principles presently in use, such as current comparison and the detection of negative-sequence currents, will almost certainly be applied to large machines well into the future. Increasingly they will be implemented by employing electronic data acquisition and processing techniques which will ensure that the appropriate actions are initiated when abnormal conditions occur. A measure of duplication and/or self checking will be necessary, however, in such equipment to prevent a failure within it from disconnecting healthy plant.

REFERENCES

1. Creck, F. R. L. (1990) *Power Eng. Journal.*
2. Ray, W. F., Lawrenson, P. J., Davis, R. M. *et al.* (1986) High performance switched reluctance brushless drive, *IEEE Trans. on Industry Applications*, IA–22, pp 722–30.
3. NEMA MG1–1987, *Motors and Generators.*
4. ANSI/NFPA 70–1987, *National electrical code.*
5. Type MCHN02 Motor protection relays, in *GEC Measurements Modular Protective Relays*, **1**, 155–64.
6. Protective Relays Application Guide, Third Edition, 1987, GEC Measurements.
7. Motor Protective Relays in *Protective Relays*, Catalog R. 1989, pp 14/1–14/18, published by Siemens Actiengesellschaft.
8. Generator and Motor Protection, Section 4 in *HV Protection and Protection Systems*, Buyer's Guide 1989–1990, Vol I, Asea Brown Boveri Relays.
9. IEEE Guide for AC motor protection, ANSI/IEEE C37.96-1988.
10. Say, M. G. and Taylor, E. O. (1980) *Direct Current Machines*, Pitman.
11. Finney, D. (1988) *Variable frequency AC motor drive systems*, Peter Peregrinus.
12. *Protective relays application guide*, 3rd edition, 1987, GEC Measurements, p. 301.
13. Pope, J. W. (1984) A comparison of 100% stator ground fault protection schemes for generator stator protection, *IEEE Trans. PAS-103*, pp 832–40.
14. Marttila, R. J. (1986) Design principle of a new generator stator ground relay for 100% coverage of the stator winding, *IEEE Trans, PWRD-1*, pp 41–51.
15. Type RAGEA 100% generator stator ground-fault relay in *Asea Brown Boveri Buyer's Guide B03-4012E.*
16. Generator Protection: Application Guide, Asea Relays AG03–4005 Dec 1986.
17. IEEE Guide for AC Generator Protection, *IEEE C37.102*–1987.

FURTHER READING

IEEE Relaying Committee Report (1988) Survey of experience with generator protection and prospects for improvements using digital computers, *IEEE Trans. on Power Delivery*, **3**, 1511–22.

GEC Measurements, *Industrial generator protection application guide*, Publ R-4016B.

8

The protection of busbars

Busbars are vital parts of power networks because they link incoming circuits connected to sources, to outgoing circuits which feed loads. In the event of a fault on a section of busbar all the incoming circuits connected to it must be opened to clear the fault. In practice, because of the amount of interconnection of circuits and the possibility of back feeds from load circuits, all the circuits connected to a faulted section of busbar are disconnected. Such disconnection clearly causes considerable disruption and the greater the operating voltage and current levels of a busbar, the greater will be the loss of supply resulting from a fault. It is therefore necessary that busbars should be so designed and constructed that the incidence of faults occurring on them is reduced to a very low level and it is also essential that the protective schemes applied to busbars are highly discriminative so that they will not wrongly cause a busbar, or a section of it which they are protecting, to be disconnected when faults occur on circuits external to it. In addition, because of the severe and extensive damage which may result from busbar faults it is necessary that they should be detected by protective schemes and then cleared very rapidly.

The following section provides some historical background and then information on busbar layouts and constructions and protective arrangements is provided in the later sections of this chapter.

8.1 HISTORICAL BACKGROUND

During the early decades of this century both the ratings of generating plants and the degrees of interconnection between networks were relatively low and therefore the fault levels, in terms of volt-amperes, were fairly limited. It was also felt that the designs of switchgear were such that the likelihood of faults occurring on or adjacent to busbars was very small. As a result of these factors busbar-protective schemes based on the then well-known Merz–Price principle were not applied, it being thought that they were not sufficiently reliable and that they might maloperate and cause unnecessary disconnection of busbars. Reliance was usually placed on the definite or inverse-time overcurrent and earth fault relays covering the circuits connected to the busbars. This practice was not wholly satisfactory because of the delays involved in clearing busbar faults and the fact that such relaying systems cannot always discriminate correctly.

In some installations, therefore, the above arrangements were supplemented by applying frame-leakage protection. This simple form of protection, which was available before 1915, was also applied to other power equipment, such as power transformers. Its principle, which was outlined in section 6.3.6 (page 212), is to lightly insulate the casing or enclosure of a piece of equipment from earth and to then connect the primary winding of a current transformer between earth and the enclosure. A relay connected to the secondary winding of the current transformer can be set to operate at a low current, say 20% or less of the secondary rating of the current transformer, and thereby internal faults to the enclosure can be detected. As faults external to the protected equipment would not cause currents to flow to earth through the casing, these could not cause relay operation and therefore discrimination could be obtained without the need for time delays. It will be appreciated, however, that this scheme cannot detect interphase faults which do not cause breakdowns to the enclosure.

In spite of the steady growth in generator outputs and networks in the first decades of this century it was not until about 1935 that the necessity for introducing and applying more advanced busbar-protective schemes was accepted. Two basic forms were developed, namely interlock and current-differential.

The interlock schemes basically employed directional relays associated with each of the circuits connected to the sections of busbars being protected. In the event of an external fault one of these relays would detect current leaving the busbar and it would inhibit the tripping of the circuit-breakers. Such schemes clearly required quite large numbers of relays and both current and voltage transformers had to be provided to supply their windings. In addition, starting relays, which operated when currents were above normal levels, were included so that the tripping of circuit-breakers could only be initiated when faults were present. In this way the possibility of incorrect operation was reduced.

Current-differential schemes, which operated on the Merz–Price principle, were introduced and biased relays of various forms were developed. By 1950 devices such as transductors were introduced to reduce the complexity of the electro-mechanical relays required, it being felt that this would increase the reliability of the schemes. A scheme of this form, designated 'Monobias' [1], was produced by A. Reyrolle and Co Ltd.

Because of the initial reluctance of electricity supply authorities to install busbar protection, which was caused by their fear that healthy busbars might be disconnected incorrectly, the practice of applying two independent schemes to major busbars was adopted and this has continued up to the present time. As an example, A. Reyrolle and Co Ltd produced Dualock protection which had an interlock scheme and a current-differential scheme and, of course, both had to operate before the tripping of circuit-breakers could be initiated.

Because of the rarity of busbar faults, protective schemes are seldom required to operate and this leads to the possibility that defects might have arisen of which operating staff could be unaware. For this reason, manufacturers have always incorporated equipment into their schemes to enable operators to perform tests regularly and in many cases equipment is provided to initiate tests automatically.

8.2 BUSBARS

Busbars, because they interconnect several circuits, are associated with circuit-breakers and therefore they are mounted within switchgear units produced for systems operating up to medium voltage levels. When circuits operating at very high transmission-voltage levels were introduced, the above practice was not possible because of the spacings needed between the conductors, and in these applications the busbars were normally mounted in air, connections then being taken to the circuit-breakers. This practice is still used extensively but recently fully-enclosed switchgear containing gaseous insulants has been produced for use at very high voltage levels and the busbars are housed within these units, as described later.

Constructional features associated with these arrangements are examined in the following section and then busbar configurations are considered.

8.2.1 The construction of enclosed switchgear units

In switchgear produced at the beginning of this century, conductors, supported on insulators, were housed in air-filled compartments, usually at the tops of the units and they ran along to interconnect the units and thus formed busbars. The voltage levels were low and small spacings between the phases were acceptable. The amounts of generation and the load levels were small and therefore the damage which could result from a busbar fault was not great and the consequences of loss of supply were not serious.

As demand increased, larger machines were installed and operating voltages increased. Because of the increased consequences of busbar faults, switchgear was produced in which the busbars were mounted in separate metal chambers filled with bituminous compound. The spacings between adjacent busbars were great enough to ensure that the breakdown voltages between the phases were higher than those on the circuits connected to them.

Because faults on the conductors connecting the busbars to the circuit-breakers would have the same consequences as faults on the busbars themselves, these connections, which were within metal enclosures, were also insulated to high levels and surrounded by bituminous compound to ensure that their breakdown voltages were as high as those of the busbars. In addition, the circuit-breakers had not only to be capable of interrupting the fault currents which could flow in the event of faults on the busbars or on the circuits

Fig. 8.1 800 kV GIS substation produced by ABB, in operation since February 1988 in the 800 kV system of ESCOM in South Africa. (Reproduced from *ABB Review 6/91*, with the permission of ABB Relays AG, Baden, Switzerland).

connected to them, but also to be so insulated that the likelihood of faults occurring within them was reduced to a minimum.

The above principle has been adhered to in all the forms of enclosed switchgear produced subsequently.

Air, oil, synthetic resins and gases have been and still are used as the main insulants in different types of enclosed switchgear. In the past such switchgear was only suitable for use in the lower voltage sections of networks, the open-type switchgear layouts described below being used in sections operating at transmission-voltage levels. In recent years however, very compact fully-enclosed switchgear which uses the gas sulphur hexafluoride (SF_6) as the main insulant has been developed and designs are available for use at all voltages and current levels. An exterior view of an installation operating at 800 kV is shown in Fig. 8.1.

8.2.2 Open-type switching sites

Busbars which operate at high voltages may be made of either uninsulated flexible cable or rigid conductors. They are either suspended in air from string insulators or supported on post-type insulators. In all cases, the insulators are longer than those used on the rest of the associated networks and the spacings between the phases and to earth are great enough to ensure that the breakdown voltages are very high.

The busbars are linked to air-blast, small-oil volume, or bulk oil circuit-breakers, which are insulated to high levels by connectors with large spacings between them and to earth.

Because of the sizes of these sites they are usually outdoors and allowance must be made for the effects of the weather and possible pollution on the withstand-voltage levels of the insulation. A further important factor is that exposed conductors could be struck by lightning, thus initiating busbar faults. To eliminate this possibility, a grid of earthed conductors is usually provided above the busbars.

In a number of installations, for example at the Ratcliffe-on-Soar Generating Station of Powergen in Britain, it was felt that outdoor switching sites were too vulnerable to climate conditions and therefore the costs of suitable buildings to house the busbars and switchgear were considered to be justified and they were provided. Indoor switching sites are also used near coasts to avoid them being polluted by salt.

8.2.3 Further methods to reduce the incidence of busbar faults

Whilst direct lightning strikes to busbars can be prevented, extremely high voltages can nevertheless be produced between phases or between phases and earth on transmission lines as a result of lightning strikes on or near them. If such a situation arises, voltage waves are set up, travelling in both directions from the point where the disturbance has occurred, and should they reach busbars to which the line is connected they could cause a breakdown of the busbar insulation.

To try to eliminate this possibility and also to prevent damage to other vulnerable equipment, such as power transformers, an earthed conductor is provided above the phase conductors of each transmission line to reduce the likelihood of high voltages being produced on the phase conductors by lightning discharges and, in addition, horn gaps are fitted across each of the string insulators to try to prevent any high voltage surges which do occur from reaching the ends of the line.

As a further measure, two earthed conductors are often run in parallel over the end sections of transmission lines to try to ensure that high voltage surges will not be induced, by lightning, on the phase conductors at points near terminal equipment or busbars, because there are few opportunities for such surges to be attenuated.

In addition to these measures, surge absorbers, i.e. voltage limiters, are often fitted near the terminals of overhead lines to prevent excessively high voltages reaching busbars or other equipment.

8.3 SECTIONALIZATION

Because of the practices described above, the number of faults which occur on busbars and the connections to them represent only a tiny fraction of the total number of faults which occur on power systems in given periods. Nevertheless it must be accepted that some busbar faults will occur and therefore it is the general practice to sectionalize busbars to reduce the number of circuits which must be opened to clear a fault and at the same time enable a significant part of the network to continue in service.

The simplest and cheapest arrangement is to include one bus-section circuit-breaker to divide a busbar into two sections as shown in Fig. 8.2(a). With this arrangement it is usual to operate with the bus-section circuit-breaker closed and should a fault occur on one of the sections all the circuit-breakers associated with the faulted section and also the bus-section circuit-breaker must be opened to clear the fault. For this arrangement to be effective each of the sections should contain about equal numbers of incoming and outgoing circuits and where possible important loads should be fed by two circuits, one connected to each of the busbar sections.

Where it is thought to be desirable, busbars are provided with two or more bus-section circuit-breakers, thus creating three or more sections. These practices, of course, increase the initial cost of an installation but they reduce the amount of disconnection which will occur in the event of busbar faults.

Busbar faults can cause severe damage which may necessitate extensive repairs and, as a consequence, consumers connected to a faulted section of a busbar could be left without supplies for a considerable time. To alleviate such situations, it is usual to have duplicate busbars in important installations, a typical arrangement being shown in Fig. 8.2(b). The primary purpose of this practice is to allow circuits, which become disconnected because of a fault on

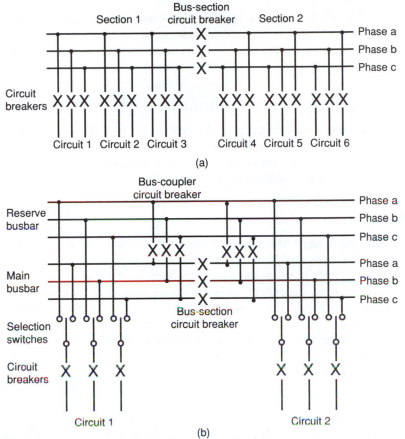

Fig. 8.2 Busbar arrangements.

a busbar section, to be transferred to the other busbar. This requires that selector switches, not capable of interrupting fault currents, be provided for each circuit to enable the necessary transfers to be effected.

If the second set of busbars is used solely as a reserve and is not normally energized, the transfer procedures may be quite lengthy because supply and load circuits cannot be transferred simultaneously. It is therefore usual, but more costly, to include one or more bus-coupler circuit-breakers, which are normally closed, so that both busbars are energized, thus enabling transfers to be effected relatively quickly in the event of a fault on any busbar. Two three-phase bus-coupler circuit-breakers are shown in Fig. 8.2(b).

In some installations bus-section circuit-breakers are included in both sets of busbars thus providing symmetry and complete duplication. This arrangement makes it possible to isolate individual sections of busbars for short periods during which maintenance work may be done; for example, the insulators in outdoor installations may be cleaned and inspected.

It will be evident that the clearance of a fault on a section of busbar will require that all the circuit-breakers associated with incoming and outgoing circuits connected to the section be opened, as must also bus-section and bus-coupler circuit-breakers associated with the section.

8.4 FAULTS ON OR NEAR BUSBARS

It is essential when considering the designs of busbars and the protective equipment which will be applied to them that the levels of the currents which may flow when faults are present be known. Behaviour during both internal and external fault conditions is examined below.

8.4.1 Internal faults

Busbars are unlike transformers and machines in that the currents which will flow in the events of faults on them are not dependent on their positions because at any instant, on any phase, the voltage is the same at all points.

Fault levels are dependent, however, on the number of circuits connected at any time, which can feed currents to a fault. Clearly if all the circuit-breakers in circuits connected either directly or indirectly to sources are closed at the time of a short-circuit then the total fault current will be at the maximum possible level. Should some circuit-breakers be open, however, then lower fault currents will flow and, of course, similar effects will be produced if there is significant impedance in a fault path. In addition, earth fault currents will be affected by the earthing arrangements on the various circuits.

Clearly, the range of fault currents which may be experienced in a particular application should be determined using one of the well-established techniques to ensure that satisfactory performance will be obtained from the proposed protective equipment.

In general, fault levels are normally well above the rated currents of individual circuits and very sensitive fault settings are not usually needed.

Whilst all types of faults, i.e. phase to earth and interphase faults, could possibly occur on open-type outdoor busbar installations and in many other enclosed switchgear arrangements, there are some switchgear designs in which the phases are separated by earthed barriers and of course in such cases only phase to earth faults are possible.

8.4.2 External faults

In the event of a fault on a circuit at a point near a circuit-breaker, as shown in Fig. 8.3, the fault current would be the same as that for a busbar fault and, if all the circuits were connected, the maximum possible fault level would be reached. If the faulted circuit was feeding only a load, its circuit-breaker would have to clear the total fault current and would have to have the appropriate

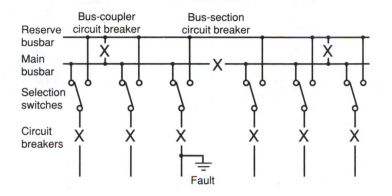

Fig. 8.3 Fault on an outgoing circuit. Circuit shown in single line form for simplicity.

breaking capacity. If, however, the faulted circuit was connected to a source, its circuit-breaker would have to clear a current somewhat less than the total fault current. In practice, however, it is usual, in the interests of standardization, for all the circuit-breakers associated with a set of busbars to have the same breaking capacity, namely that needed to clear the maximum-possible fault current.

It will be appreciated that current-differential protective schemes associated with busbars have to be set sensitively enough to detect internal faults and yet not operate in the event of external faults involving currents up to the highest levels possible.

8.5 POSITIONING OF CURRENT TRANSFORMERS AND INCORRECTLY PROTECTED ZONES

As pointed out in section 8.4.2, large currents flow in the event of faults on circuits near the circuit-breakers which connect them to busbars. Such faults should be detected by the protective equipments associated with the circuits on which they occur and only the circuit-breakers in a faulted circuit should be opened. In practice, however, there must be connections between circuit-breaker contacts and current transformers and these connections are protected incorrectly as explained below.

Busbar-protection current transformers are invariably mounted on the outgoing sides of the circuit-breakers and should a fault occur on the connections between the circuit-breaker contacts and the current transformers feeding a current-differential protective scheme covering a section of the busbars, as shown in Fig. 8.4, all the circuit-breakers associated with the section would be opened incorrectly. This possibility cannot be eliminated, but to reduce it to a minimum the connections between the current transformers and the circuit-breaker contacts should be both highly insulated and short.

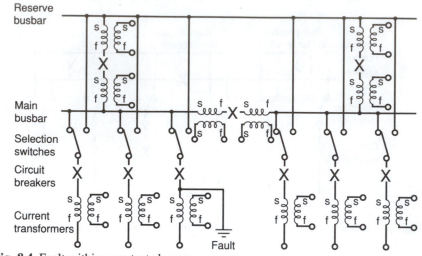

Fig. 8.4 Fault within a protected zone.

This topic was examined in more detail in section 2.2.7 (page 71), as was the positioning of the current transformers feeding the protective schemes associated with the outgoing circuits.

8.6 PROTECTIVE ARRANGEMENTS FOR BUSBARS

Simple busbars used in direct-current distribution networks operating at relatively low voltages may be protected by fuses. Correct discrimination in the event of faults either on the busbars or the circuits connected to them can be achieved by selecting fuselinks with the appropriate current ratings and operating times, when there are few circuits connected to sources and the other circuits, supplying loads, cannot feed current back into the busbars. Fuses may also be used to protect simple busbars used in single-phase distribution networks operating at relatively low voltages and, again, satisfactory operation should be obtainable. Fuses are not suitable for application to three-phase busbars, however, because complete circuits would not be cleared in the event of single-phase faults and, as a result, equipment such as motors could be supplied with imbalanced voltages.

IDMT relays are used to protect the busbars of some single-phase and three-phase distribution networks. In these applications, the current and time settings of the relays should be selected in the manner described in section 4.4 (page 124) and satisfactory grading should be achievable.

With these arrangements, fault clearance times may be quite long, a situation which cannot be accepted on major busbars where the fault levels may be very high and the consequential damage could be very great. It will be clear, therefore, that high-speed, current-differential schemes should be used in these applications and this has been the practice for many years.

Depending on the busbar layouts and the designs of the switchgear, several different current-differential schemes are used and these are now considered.

8.6.1 Application to simple single-phase, unsectionalized busbars

Such busbars probably would not be protected by current-differential schemes, but if they were, then a current transformer would be included in each of the circuits on the outgoing side of the circuit-breakers as shown in Fig. 8.5(a).

The current transformers, which would each have the same ratio, would have their secondary windings connected in parallel and the sum of their output currents would flow in the operating winding of the single relay.

The protected zone would, as shown in Fig. 8.5(a), include the connections on the busbar sides of the current transformers, the circuit-breakers and the busbar.

Unless there was a fault present within the protected zone, the primary currents of the transformers would sum to zero at every instant, i.e.

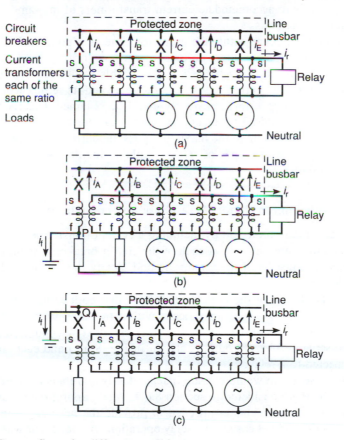

Fig. 8.5 Current flows for different conditions.

$i_A + i_B + i_C + i_D + i_E = 0$. Ideally, the secondary winding currents should also sum to zero causing the current in the operating winding of the relay to be zero. In practice, because of the non-linearity of the excitation characteristics of the transformers and the fact that their primary currents would probably all be different, the secondary currents would not sum to zero and some current would flow in the relay operating winding.

A particularly onerous condition would be a fault just outside the protected zone, at a point such as P in Fig. 8.5(b). In this event, the fault current (i_f) could have a very large magnitude as would the primary current i_A. The other currents ($i_B - i_E$) would probably all have different magnitudes but all of them would be lower than that of the current i_A. In these circumstances quite considerable mismatching could occur and a significant current (i_r) could flow in the operating winding of the relay. To assist in minimizing this current, the current transformers in a scheme should all be of the same design. There are circumstances, however, where this may not be practicable, because the rated currents of the individual switchgear units may vary considerably and it may not be possible to accommodate current transformers of the same dimensions in each of the circuits.

Should a fault occur within the protected zone, at a point such as Q in Fig. 8.5(c), the sum of the primary currents of the transformers would be equal to the fault current, i.e. $i_A + i_B + i_C + i_D + i_E = i_f$. This would cause a current, dependent on the fault current, to flow in the relay and its setting would have to be such that the necessary sensitivity could be obtained. Clearly biasing or other features would be necessary to ensure that operation would not occur when external faults up to the maximum possible current levels were present. These factors and the types of relays available for this duty are considered later in section 8.7.

8.6.2 Application to simple three-phase unsectionalized busbars

These busbars are protected by including a current transformer in each of the phases on the outgoing side of each of the circuit-breakers. Again the current transformers must be of the same ratio and ideally they should all be of the same design.

When the design of the switchgear is such that both earth faults and interphase faults clear of earth may occur within the zone to be protected, the secondary windings of the current transformers must be connected in parallel in three separate phase groups as shown in Fig. 8.6(a) and each group must be connected to a relay.

Each phase group would operate independently and its behaviour would be the same as that considered in section 8.6.1. Again appropriate relay settings and biasing arrangements are needed to provide the necessary sensitivity to internal faults and yet ensure that relay operation will not occur when external faults are present.

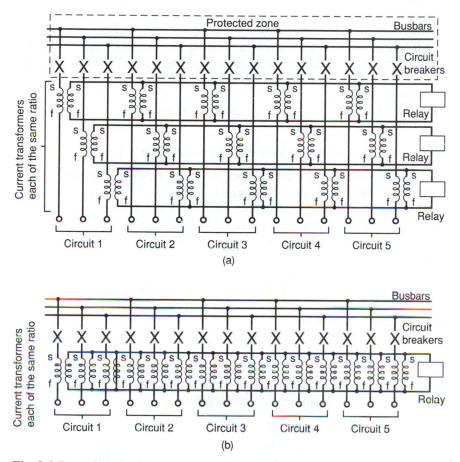

Fig. 8.6 Protective-circuit arrangements.

It will be evident that only one relay will operate in the event of a single phase to earth fault but two or all three relays may operate when interphase faults occur. In each case, however, relay operation will initiate opening of the circuit-breakers associated with the faulty section which will thus be completely isolated.

Should the switchgear be so designed that there are earthed barriers between the phases, thus eliminating the possibility of interphase faults clear of earth, then the secondary windings of all the current transformers could be connected in parallel and connected to a single relay as shown in Fig. 8.6(b). The protected zone would again include the connections on the busbar side of the current transformers, the circuit-breakers and the busbars.

With this arrangement the number of current transformers operating in parallel is three times the number of circuits connected to the busbars. It will be appreciated that greater imbalance is likely to be caused by mismatching when the number of balancing current transformers is increased. In addition, obtaining a given fault setting may require the operating current of a relay to be reduced because of the increased total of the exciting currents of the transformers. Such schemes may therefore require greater biasing to ensure that correct discrimination will be achieved. In some circumstances it may be preferable to segregate the current transformers into phase groups and use three relays, i.e. use the arrangement shown in Fig. 8.6(a).

8.6.3 Application to complex three-phase sectionalized busbars

As explained earlier in section 8.3, important busbars are sectionalized, so that internal faults will not cause all the circuits linked to a busbar to be disconnected. In addition, such installations usually have duplicate busbars, a main set of busbars to which the circuits are normally connected and a reserve set of busbars to which the circuits may be connected when a fault has occurred on the main busbars or when maintenance work has to be done on them. In such installations, busbar section and coupler circuit-breakers are needed and selector switches, incapable of clearing load and fault currents, must be provided to enable individual circuits to be connected to either the main or reserve busbars. A typical arrangement is shown in Fig. 8.7.

Clearly a current-differential scheme providing a single protective zone, covering all the busbars in an installation, would not be satisfactory as a single fault anywhere on the busbars would cause the disconnection of all the circuits. In practice, separate and independent schemes must be used to cover each of the sections of busbar. For the arrangement shown in Fig. 8.7, there

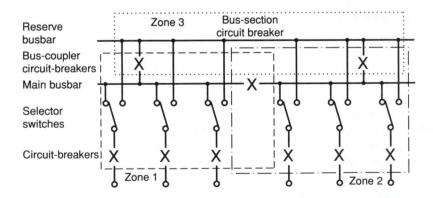

Fig. 8.7 Protected zones.

would be schemes covering three zones, i.e. the two sections of the main busbar and the reserve busbar.

As in other applications, all the current transformers would have the same ratio and those associated with the incoming and outgoing circuits would be mounted on the circuit sides of the circuit-breakers in positions as close to the contacts as possible to minimize the incorrectly protected zones. In addition, current transformers could be mounted on each side of the busbar section and busbar coupler circuit-breakers. Ideally all the current transformers should be of the same design to assist in achieving balance when the busbars are healthy.

Should the switchgear and busbars be of designs where both earth faults and interphase faults could occur, then the phases would be protected separately and for the arrangement shown in Fig. 8.7 the current transformers would be connected to form nine groups each with its own relay, i.e. three zones.

Because each of the incoming and outgoing circuits could be connected either to a section of the main busbar or the reserve busbar, the selector switches must have auxiliary switches which will connect the secondary wind-ings of the current transformers into the appropriate zone groups and thus ensure that correct discrimination will be obtained at all times. An important point which should be noted is that the selector switches are not capable of breaking current and they are therefore only operated when their associated circuit-breakers are open. There is therefore no possibility of the protective schemes being imbalanced when any of the current transformers are trans-ferred from one group to another because they will not be carrying current when the auxiliary switches operate. The detailed current transformer circuitry is shown in Fig. 8.8.

It is evident that the numbers of current transformers in the groups associ-ated with the zones will vary and at any time they might be quite different, the number in the phase groups of one zone being low while there may be a large number in the groups of another zone. As an example, if none of the circuits were connected to the reserve busbar, then there would only be the two bus-coupler current transformers in its phase groups whereas there would be large numbers of current transformers in the groups associated with the sections of the main busbar.

For given relay current settings the minimum fault currents needed to cause operation may be considerably affected by the numbers of current transformers in the zone groups; for example, if high-impedance relays are used, the mini-mum fault current for operation is approximately proportional to the number of current transformers in a group. A further factor which must be recognized is that the amount of mismatching and therefore the imbalance during external fault conditions may be affected by the numbers of transformers in groups. Studies should therefore be done to determine that the proposed relay settings and biasing will provide acceptable performance under all operating condi-tions. This topic is considered further in section 8.7.

The inclusion of auxiliary switches in the secondary circuits of current

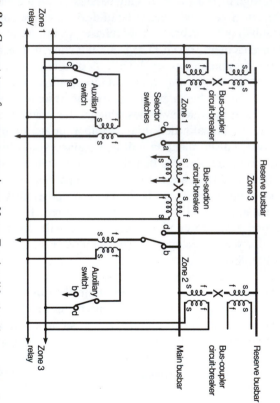

Fig. 8.8 Current-transformer connections. Note: To simplify the diagram it is shown single-phase and only the connections for zones 1 and 3 are shown. All the current transformers have the same ratio.

transformers has always been accepted with some reluctance and this is particularly so when they are feeding busbar protective schemes. The switches could be eliminated if the zones were completed by including current transformers in both the connections between each selector switch and the busbars, as shown in Fig. 8.9. This arrangement would increase the total number of current transformers required and also make it necessary to provide further protective equipment and current transformers to enable faults on the selector switches, the circuit-breakers and the connections between them to be detected. This extra complexity has never been considered to be justified, because of the high performance record of the current schemes incorporating auxiliary switches.

Whilst it is the common practice to have main and reserve busbars and to only have circuits connected to the reserve busbar after a fault has been present on a section of the main busbar or when maintenance work is being done, it can be argued that it is better to have circuits connected to each of the busbars so that fewer circuits will be disconnected in the event of any busbar fault. If this procedure is adopted it does not affect the protective arrangements in any way because they must be such that satisfactory performance will be obtained with all the possible connections.

As stated earlier, some switchgear units have the phase connections completely segregated by incorporating earthed barriers between them. In such cases, all the current transformers associated with a protected zone may be connected in parallel and only one relay is provided per zone. This clearly reduces the number of relays needed but it may affect the sensitivity of a scheme and increase the imbalances during healthy and external fault conditions as stated earlier in section 8.6.2.

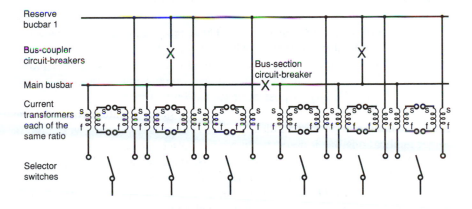

Fig. 8.9 Current-transformer connections without auxiliary switches.

8.6.4 Interconnections in current-differential schemes applied to busbars

Switchgear produced for relatively low voltage and current operation tends to be compact and the interconnections between the current transformers associated with current-differential protective schemes are usually quite short and therefore of low resistances. In addition, the primary current ratings of the current transformers are not high and therefore low secondary current ratings, e.g. 1 A, can readily be provided. As a result, the voltage drops on the connections, even during external faults involving high currents, are not great and the imbalances caused by mismatching and current transformer errors will be limited.

On large outdoor switching sites, however, the connections between the current transformers may be quite lengthy and the current ratings are high, e.g. current transformers with primary windings rated at 2000 A are used in large British generating stations. In such applications current transformers with 5 A secondary ratings would be used if possible to limit the number of secondary winding turns, but the voltage drops that would be produced in busbar protective schemes with such transformers might not be acceptable and the extra expense of producing 1 A secondary windings is often considered to be justified.

8.7 RELAYS USED IN CURRENT-DIFFERENTIAL SCHEMES

As explained in Chapter 5, both high- and low-impedance relays may be used in current-differential schemes and their use in busbar protective schemes is considered below.

8.7.1 High-impedance relays

High-impedance relays have proved very satisfactory in several different applications over many years, a particular example being in the restricted earth fault protection of power transformers. These relays have the advantage that the minimum operating voltage, at which they must be set to ensure that they will not operate during external faults, can be determined easily. It is shown in section 5.6 (page 171) that this is the voltage which would be produced across the relay during the largest possible external fault, if the core of the current transformer carrying the total current was continuously saturated whilst the others in the group performed ideally. This voltage is not usually very high in other applications but it could be in busbar protective schemes, because the current levels in the event of faults near busbars may be great and the resistances of the secondary windings of high-ratio current transformers are high.

Whilst a relay could be set to operate at any voltage level, this voltage would have to be produced at the fault-setting level and each current transformer in a balancing group would require the necessary exciting current (I_e) in its primary windings, i.e. the fault-setting current would be approximately nI_e A in which n is the number of current transformers in a balancing group. Clearly in applications to busbar sections to which a large number of circuits are connected, the numbers of current transformers in balancing groups may be high and therefore fault settings will be high unless the exciting currents (I_e), needed to induce the relay operating voltage, are low. In practice exciting currents can be made low by using current transformers with cores of large cross-sectional area, but obviously there is a practical limit and it must be recognized that an increase in core dimensions is accompanied by an increase in secondary winding resistance which would increase the relay setting voltage needed. This could be counteracted by using conductors of larger cross-sectional area for the secondary windings which would cause a further increase in the cost and size of the current transformers.

There are usually several circuits on a busbar which are connected to sources. As they are connected in parallel their effective impedance is relatively low. In addition, in the case of very high voltage installations there may be several solidly-earthed neutral points. As a consequence, fault currents on such busbars will usually be high and therefore, unlike the situations which can arise on other equipment such as power transformers, very sensitive fault settings are not necessary.

It will be clear that fault studies must be done when protection is to be applied to a busbar installation to determine the levels of the internal and external fault currents which may flow. The behaviour of the proposed scheme must then be examined to determine the relay voltage settings which must be used to ensure that operation will not occur when external faults are present. Thereafter, the minimum internal fault current which will cause operation must be determined and the acceptability of this value must be considered.

8.7.2 Low-impedance relays

It will be apparent from the preceding section that satisfactory performance may not always be obtainable from busbar protective schemes which include high-impedance relays and indeed many schemes produced in the last 40 years have used biased low-impedance relays.

Again it is desirable that the possible causes of imbalance be reduced to a minimum by using both interconnecting conductors of relatively large cross-sectional area and therefore of low resistance and current transformers with low secondary current ratings, e.g. 1 A. The current transformers should be of the same design and well matched.

A particular fault-current setting can be provided by setting the relays in the zones to the appropriate values; for example, in a scheme using current transformers of ratio 1000/1 and relays set to operate at 0.2 A, operation would be obtained at a little over 200 A, the exciting currents of the transformers being low at these current and voltage levels.

Relays with such low settings and without any form of biasing would almost certainly maloperate when large currents were flowing to faults outside their protected zones. Biasing must therefore be provided and it must clearly be effective under all possible conditions.

It is evident that the greatest current will flow in the circuit on which a fault is present and bias provided by this current would be effective. Clearly, however, no advance indication is given of the circuit on which a fault will occur and therefore bias must be derived from all the incoming and outgoing circuits of a protected zone. It would not be practicable to provide relays with a bias winding for each circuit of an extensive busbar zone and therefore the individual currents must be summed to produce an input to, preferably, a single bias winding.

A direct summation of the instantaneous currents in the individual circuits would be zero at all times when external faults were present and therefore a satisfactory bias input could not be derived in this way. In practice as was shown in section 5.4.5 (page 167), no form of direct summation of the alternating currents can provide satisfactory bias inputs which are independent of the locations of external faults, i.e. the circuits on which they occur.

A satisfactory method which has been in use for over 40 years is to rectify the secondary currents of the current transformers in the individual circuits of a protected zone and then sum them. In this way a bias input, proportional to the external fault current but independent of the circuit on which it is present, is obtained. This bias input is also produced when internal faults occur and therefore a fixed percentage bias is effective for all types of faults and this clearly prohibits the use of biasing of 100% or more.

A typical protective scheme of the above form was shown in Fig. 5.18.

A major difficulty, which arises when applying schemes of this type, is to determine the amount of biasing which must be used to ensure that operation will be prevented from occurring when external faults occur. It will be appreciated that the imbalances which could cause such incorrect behaviour can be caused by several factors, all or some of which may be present on any occasion. Some factors such as the resistances of the interconnections between the current transformers in a balancing group and their relay could be determined and their effects could be assessed. Other factors such as the possible residual fluxes in current transformer cores and various non-linearities make it impossible to make relatively simple assessments of behaviour. Accurate allowances for the effects of the saturation of current transformer cores are also difficult to make.

This situation, together with the vital need to ensure that busbar protective schemes will perform correctly, led to the introduction of the conjunctive testing of schemes on the premises of manufacturers. In these tests the actual current transformers to be used on a site were tested with the protective scheme to be used with them. High test currents, provided at low voltages by laboratory machines, were supplied to the primary windings of the current transformers to simulate the conditions which could arise in service and in this way the suitability of the scheme and its settings was ascertained.

In more recent years, the development of simulation techniques and accurate mathematical modelling have enabled theoretical studies of schemes to be undertaken. It must be recognized, however, that there are no simple methods of assessing the performance which will be obtained from a particular protective scheme incorporating biased low-impedance relays. It must be added, nevertheless, that many such schemes have been installed and that they have performed excellently over many years.

8.7.3 Low-impedance relays with compensation for current-transformer saturation

It will be clear from the two preceding sections that it can be difficult to apply current-differential protective schemes to busbars. When high-impedance relays are used very large current transformers may be needed so that the exciting currents required to provide the e.m.f.s needed to operate the relays are not so high that acceptable fault settings cannot be obtained. Alternatively, when low-impedance relays are used lengthy conjunctive testing may be needed to assess the suitability of protective schemes for particular applications and again large current transformers may be needed.

Because of the introduction of modern electronic equipment and acceptance of its use in important protective schemes, other features can now be provided in busbar protective schemes to enable the necessary performance to be obtained without the need for very large current transformers.

It has always been apparent that the imbalances caused during external fault conditions by mismatching of the current transformers and the resistances of their secondary-circuit resistances are small compared with those which can result if the core of one of the current transformers saturates whilst those of the others in its balancing group do not. It will be evident that the current transformer in the circuit on which the fault is present will be the one in which saturation will occur because it will have the greatest primary current and loss of secondary output from it will cause very significant imbalance.

It was therefore realized that current-differential schemes could incorporate equipment which would monitor the current transformer outputs and provide signals to inhibit relay operation during periods when the core of any current transformer is saturated.

This technique has been developed and used in a scheme [2] produced by GEC Measurements of Stafford, England. The basic feature on which the technique is based is that the secondary current of a current transformer with a resistive burden collapses to zero when saturation occurs and it remains at zero until the time when the next zero crossing would have occurred. Thereafter normal behaviour is obtained until saturation occurs again and the process is then repeated. This behaviour, which was discussed earlier in section

2.3.1 (page 78), would not be obtained if the secondary circuit was highly inductive, but this is not the case in current-differential schemes in which the resistances of the current transformer secondary windings and the interconnections form the burdens.

The detection of the sudden collapse of the secondary current of a transformer during either a positive or negative excursion could be detected in several ways. The method developed by GEC Measurements, referred to above, is shown in simplified form in Fig. 8.10(a). The primary winding of an auxiliary current transformer is connected in the secondary circuit of each main current transformer. The output of the auxiliary transformer is full-wave rectified and the resultant current passes through a resistor (R), connected in

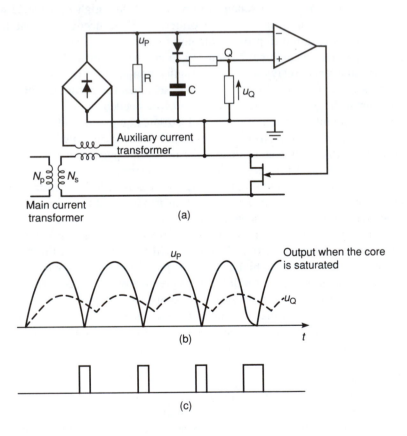

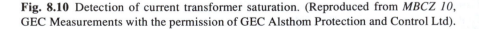

Fig. 8.10 Detection of current transformer saturation. (Reproduced from *MBCZ 10*, GEC Measurements with the permission of GEC Alsthom Protection and Control Ltd).

parallel with a rectifier-capacitor circuit. A voltage (v_P) proportional to the rectified current is produced across resistor R as shown in Fig. 8.10(b). A voltage (v_Q) derived from the rectifier capacitor circuit is provided at point Q in the circuit. This voltage is approximately equal to half the voltage (v_P) during the earlier part of each half cycle as shown in Fig. 8.10(b). The voltages v_P and v_Q are applied to the inputs of a comparator which provides an output when the voltage v_Q exceeds the voltage v_P and this output is used to close a switch which short circuits the interconnecting conductors and thus the operating winding of the relay.

It will be appreciated that inhibiting pulses are produced at each zero crossing of the current waveform but these will be of short duration when saturation of the transformer core does not occur. However, when significant core saturation occurs longer inhibit pulses will be obtained as shown in Fig. 8.10(c). Clearly this will ensure that operation will not be caused by the effects of core saturation when faults are present outside a protected zone. Similar action would occur should a current transformer saturate during a fault within a protected zone, and in these circumstances inhibit signals would also be generated. However, a large current would have to be flowing to cause the saturation of a core and therefore the current levels during the period when there were no inhibit signals would be great enough to ensure that relay operation would be obtained.

Further details of this particular scheme and other modern busbar protective schemes can be obtained from reference [2].

An alternative scheme, which will not be affected by the saturation of current transformer cores, is currently being studied. In this scheme, the currents in the circuits of a protected zone would be sampled many times during each half cycle. Each group of samples would be summed algebraically and clearly under healthy conditions the sums should always be zero. If the core of a transformer saturated, however, its secondary current would be affected and during such periods the groups of samples would not sum to zero. This is somewhat similar to the scheme described above but in this case relay operation would not be initiated if more than a certain percentage of the sums was zero. Clearly it is not likely that the sums would be zero if a fault was present within the protected zone.

8.7.4 The duplication of current-differential protective schemes

The numbers of faults which occur on busbars are very low because of the levels of insulation associated with busbars and the spacing between adjacent phase conductors and to earth and because of other arrangements such as the installation of voltage limiters at the incoming ends of overhead lines.

Current-differential schemes applied to the zones of busbars are therefore required to operate very infrequently but they must nevertheless balance and not initiate the tripping of circuit-breakers for a wide range of healthy conditions which may involve the individual circuits carrying different load currents and this process must continue through all of every day. They must also cope with more onerous conditions when faults occur on the circuits connected to their busbars, at points outside the zones covered by the busbar protective equipment and again balance of the current-differential schemes must be maintained. Because the numbers of faults of this type will always exceed the small number of busbar faults it is imperative that steps be taken to reduce the possibility of zones of busbars being disconnected unnecessarily because of the maloperation of a busbar protective scheme.

In practice this is achieved by applying two separate and independent protective schemes to cover each section of a set of busbars and it is so arranged that both schemes must operate before the tripping of the appropriate circuit-breakers may be initiated.

Clearly, complete duplication of all the current transformers and the connections and relaying equipment would be very costly on major busbar installations in which there are main and reserve busbars and several bus-section and coupler circuit-breakers. In such applications, therefore, a compromise solution is usually accepted in which current transformers and relaying equipment are provided to protect the various zones as described earlier. Extra current transformers and relaying equipment are then provided to form an overall zone covering the complete busbar installation. With such an arrangement, which is shown in Fig. 8.11, current transformers must be included in all the circuits connected to the complete installation but they are not needed adjacent to the busbar section and busbar coupler circuit-breakers. In addition, extra auxiliary switches do not need to be provided on the busbar-selector switches in each of the circuits, because the current transformers do not have to be switched into different zones when the positions of selector switches are changed.

Should a busbar installation be so constructed that both earth faults and interphase faults may occur then, as before, the extra current transformers must be divided to form three separate balancing groups, i.e. one associated with each of the phases, and three relays must be provided. In those installations, however, where only faults to earth are possible then, again, all the extra current transformers can be connected to form one overall group and, of course, only one extra relay is needed.

It will be recognized that this compromise arrangement has the slight disadvantage that the extra scheme will operate should a fault occur anywhere on a busbar installation and therefore if a fault occurred in a zone and not only the relays associated with that zone operated but those in another zone also operated wrongly, then two zones of the busbars would be disconnected instead of one. Another possibility is that the relays associated with a faulted zone might not operate whilst those associated with a healthy zone might do so incorrectly. In these circumstances the fault would not be cleared, a result which would have occurred even if a duplicate scheme had not been fitted,

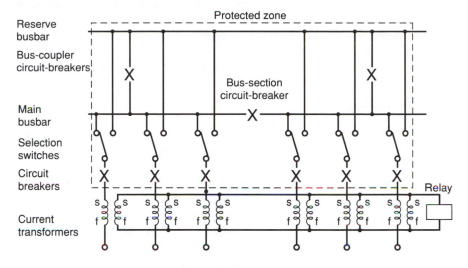

Fig. 8.11 Overall protective zone.

but, in addition, a healthy zone would have been disconnected unnecessarily.

The possibilities of such happenings are considered to be so small that the advantages of the compromise arrangement justify its usage.

8.8 MANUAL AND AUTOMATIC TESTING

As stated earlier, switchgear installations are so designed and constructed that faults occur only rarely on their busbars, but nevertheless, when they do so, large fault currents will flow and considerable damage may result unless they are cleared quickly. It was to limit such damage that busbar protection was developed and applied in spite of fears that it might operate incorrectly on occasions and that large and unnecessary disruptions of supply could thus be caused.

To reduce the risks of circuit-breakers being opened incorrectly or failing to operate on the infrequent occasions when it is necessary, manually-operated or automatically-operated testing equipment has been produced for use with busbar protective schemes. Such equipment enables checks to be made by operators or automatically at regular intervals. During such checks it is necessary that the protective equipments be inhibited from opening circuit-breakers and thus for short periods the protection is ineffective. During the checks, signals are injected to simulate both internal and external faults and the behaviour of the protective equipment is monitored to determine whether it is functioning correctly.

The inclusion of checking equipment does increase the complexity of a protective scheme and there are those who feel that its reliability is thereby reduced. As a consequence some supply authorities do not allow testing equipment to be incorporated in busbar protective schemes.

Clearly the desirability of including such equipment should be considered when schemes are to be installed and it must be recognized that the advantages and disadvantages of particular arrangements must be studied.

8.9 THE FUTURE

Current-differential protective schemes applied to busbar installations have performed extremely well over many years. Their basic principle, namely Kirchhoff's law, that the algebraic sum of the currents entering a healthy network must be zero at all times, is so simple that it must continue to be used for such applications. These schemes have the advantages that they can be set to operate rapidly when low-current internal faults occur and that they can discriminate readily between faults internal and external to their protected zones.

When conventional electro-mechanical relays are used in these schemes, quite large current transformers and connecting cables may be required to ensure that unacceptable imbalances will not occur when high fault currents flow to points outside the protected zones. Electronic relays have already been produced which enable smaller current transformers to be used and it seems certain that further developments will use digitally-processed signals obtained from relatively small current transducers to determine whether the circuit currents sum to zero or not. It has been stressed earlier that busbars perform such a vital role in the operation of a power system that their protective equipment must be extremely reliable. Because of this, it will be necessary that electronic relaying be produced to very high standards. Very secure power supplies and even duplicate supplies will be necessary and built-in checking features must be incorporated in all the equipment. In addition, the present practice of applying duplicate sets of protective equipment to busbars will almost certainly continue.

An area where the introduction of microprocessor technology is expected to lead to further developments, is to provide busbar protection as part of an integrated protection system. In such schemes signals needed for busbar protection are shared with other protective and control functions. Further details may be found in references [3], [4], [5] and [6].

REFERENCES

1. Leyburn, H. and Lackey, C. H. W. (1952) The protection of electric power systems – a critical review of present-day practice and recent progress, *Proc. IEE*, **99**, (II), 47–66.
2. Low impedance busbar protection MBCZ10, *Publication R-4026A*, GEC Measurements.

3. Cory, B. J. and Moont, J. F. (1970) Application of digital computers to busbar protection, *IEE Conference on the Application of Computers to Power System Protection and Metering*, Bournemouth, England, May 70, pp 201–209.
4. Udren, E. A. (1985) An integrated microprocessor based system for relaying and control of substations—design features and testing program, *12th Annual Western Protective Relaying Conference*, Spokane, Washington, October 24.
5. Udren, E. A. Protection function, in IEEE Tutorial course on *Microprocessor Relays and Protection Systems* Course Text 88 EH0269–1–PWR, pp. 43–45.
6. Phadke, A. G. and Thorp, J. S. (1988) Computer relaying for power systems, *RSP*, pp. 182–186.

FURTHER READING

Guide for protective relay applications to power buses, *ANSI/IEEE Standard C37.97–1979*.

9

The protection of overhead lines and cables by current-differential schemes

Current-differential protective schemes, as stated earlier, have the great advantages that they can discriminate between faults within and external to their protected zones and can operate very rapidly when required to do so. Their basic principle, namely Kirchhoff's first law, that the algebraic sum of the currents flowing into a healthy circuit is zero, is very simple.

The behaviour of these schemes was considered in a general manner in Chapter 5 and their application to transformers, rotating machines and busbars was examined in Chapters 6, 7 and 8. They have been widely applied to relatively short cables and overhead transmission and distribution lines, which are often referred to as feeders, since the beginning of this century and are still being installed today.

Their behaviour and limitations are examined in detail in this chapter after the following section in which some historical information is provided.

9.1 HISTORICAL BACKGROUND

As stated in Chapter 5, C. H. Merz and B. Price submitted an application for a British patent on a current-differential protective scheme in February 1904. This was granted and numbered 3896. It included arrangements to cover both feeder circuits and transformers and the proposed scheme employed the balanced-voltage principle, the current transformers being so connected that their outputs opposed each other during healthy and external fault conditions. As a result, the secondary winding currents were zero under these conditions and the secondary winding e.m.f.s were of equal magnitudes and opposite polarities. One end of each secondary winding was earthed and hence only one interconnecting conductor was needed to protect a single-phase circuit. The inventors referred to the interconnecting conductors as pilot wires and this term, which has remained in use, will be used in this chapter. During internal faults a current flowed in the pilot wire and also in the relay windings, which were connected in series with the pilot wire, at each end of a protected line.

Merz and Price submitted two further patent applications in 1904 and were granted patents numbered 11364 and 15796. These covered improvements to the initial scheme. One proposal enabled the number of pilot wires needed in three-phase applications to be reduced. The use of auxiliary transformers which would allow several currents to be fed to separate windings mounted on a common magnetic core was also proposed. An output could be obtained from a further winding. These were forerunners of summation and core-balance transformers and indeed Merz and Price were granted a patent, numbered 28186 in 1908, on the latter devices.

Schemes based on the above patents were applied to lines and cables in the first decade of this century. In an article [1] published in the *Electrical Review* in 1908, it was stated that about 200 miles of high-tension network were protected in the North of England at that time by the Merz–Price system. It was further stated that the scheme had not just been perfected but had passed the perfected development stage for some 18 months or more. Most importantly, numerous faults of the most severe description had been isolated without visible shock on the system, and, better still, without opening healthy sections.

It was further stated that the relays were of the simplest design. They consisted of an ordinary electromagnet with a very light armature having, for the sake of rapidity in action, a small air-gap. They clearly did not possess any biasing windings.

It was presumably found, as the lengths of the lines and cables being protected increased, that incorrect operations of Merz–Price schemes were occurring because of the capacitance currents which were flowing between the pilot wires and earth during external faults. This situation is illustrated in Fig. 9.1(a), from which it can be seen that half the capacitance current to earth flows in each of the relays and clearly this current is proportional both to the length of the protected line and the current flowing in it.

To overcome this situation, special compensated pilot wires, covered by a British patent, were introduced. These cables incorporated metallic sheaths which surrounded the interconnecting conductors, there being a break in each sheath at the mid-point of a cable run. The protective arrangement with such cables was as shown in Fig. 9.1(b), from which it can be seen that the capacitance currents between a pilot wire and its sheath do not return through the relays.

As line operating voltages and fault current levels increased further, more complex current-differential schemes were produced. One of these, which was designated as 'split-pilot' protection, was described in a paper [2] presented at the International Conference on Large Electrical High-Tension Systems, which was held in Paris in 1931. This scheme, which incorporated summation transformers and had three pilot wires, was connected as shown in Fig. 9.2. The use of the parallel-connected pair of pilot wires ensured that imbalance currents did not flow in the relays during healthy conditions and the inclusion of the diverter reactors reduced the currents which flowed in the relays be-

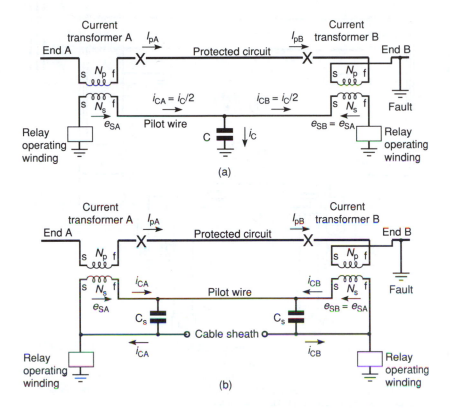

Fig. 9.1 Operation when an external fault is present. C is the pilot wire capacitance to earth in (a), and C_s is the pilot wire capacitance to the sheath in (b).

cause of mismatching. This scheme was used on three-phase lines operating at voltages up to 220 kV (line).

A range of special relays was developed for use in Merz–Price schemes. One, which was described as an 'anti-surge' relay, was covered by British patent No. 235605. This relay had two movements, one being electromagnetic and the other purely mechanical. The two movements were mechanically coupled together. Because of the construction, the effective restraining torque on the electromagnetic movement varied with the current in its operating winding, being large when a high impulse current flowed as a result of an external fault and low when a comparatively low current flowed.

Protection schemes which were developed subsequently are described in the later sections of this chapter.

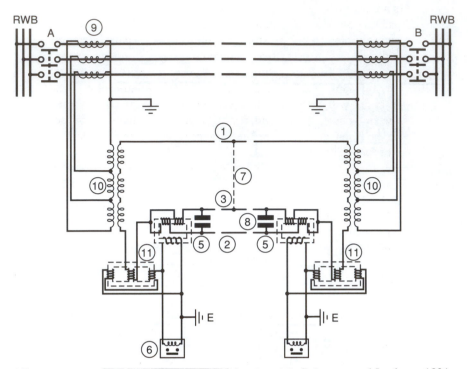

Fig. 9.2 Split-pilot protective scheme. (Reproduced from Leason and Leyburn, 1931, *CIGRE Conference, Paper 106* with the permission of CIGRE.)
(1) Common pilot No. 1. (2) Split-pilot No. 2. (3) Split-pilot No.3. (5) Split-pilot transformer.
(6) Relay. (7) Mid-point tripping connection. (8) Small tuning condenser. (9) Current transformers.
(10 Air-gap summation transformer. (11) Diverter reactor.

9.2 CABLES AND OVERHEAD TRANSMISSION AND DISTRIBUTION LINES

Cables are much more expensive per unit length than overhead lines and therefore they tend to be used primarily for distribution circuits in built-up areas where overhead lines would be considered to be unacceptable. They are usually relatively short, typically 20 km or less. They are also used in a few extra high voltage transmission circuits; for example, in areas of great scenic beauty where overhead lines would be too obtrusive. Again they are usually relatively short, transmission being continued by overhead lines when areas of less beauty are reached.

Because cables incorporate dielectric materials with breakdown voltages well above their operating levels and they are buried in the ground, the incidence of faults on them tends to be low. They may, however, be damaged as a result of subsidence or when excavations are taking place.

Overhead lines are engineered to high standards but it must be accepted that breakdowns will occur when extreme weather conditions are experienced. The conductors are under tension and the stresses may increase greatly when thick ice coatings build up on them and, as a result, fractures may occur. Conductors may also swing violently during stormy conditions and again breakages may result. In addition, flashovers between conductors or to ground or across string insulators may occur during lightning storms. These factors, coupled with the great lengths of many transmission lines, make it inevitable that significant numbers of faults will be experienced on them. Whilst many of these faults will be between a phase and ground, it must be accepted that interphase faults will also be experienced and therefore protective equipments should be able to detect all types of faults.

Whilst large currents flow in the event of many faults on cables and over-head lines, there may be occasions when quite low currents will flow. As an example, a broken conductor may make poor contact with rocky ground, causing little fault current to flow. It is therefore desirable that protective schemes should have sensitive fault settings.

A factor which should be recognized is that arcs will be present in most of the faults which occur on overhead lines. It is therefore desirable that faults be cleared very rapidly so that the resultant burning of vital parts is minimized and so that the production of ionized gases is limited. In many cases, if rapid fault clearance has been effected, it will be possible to reclose a line after a dead time of say 0.2 s, during which period the ionized gases will have dispersed, and normal operation will then be resumed. This practice, which is termed 'auto reclosing', is common on overhead lines.

The presence of capacitance between pilot wires was referred to in section 9.1 and, of course, there is also capacitance between the conductors of over-head lines and cables. Clearly their capacitances are dependent on the effective diameters of the conductors and the spacings between them and also on their lengths. The presence of capacitance currents causes the input and output currents of the individual phase conductors of the lines and cables to be unequal and therefore exact balance cannot be expected to occur in current-differential protective schemes applied to them.

In practice, however, the capacitance currents of relatively short lines or cables, say up to 20 miles in length, are very low relative to the rated currents. As an example, the capacitance current per phase of a 20-mile long, three-phase, 132 kV overhead line rated at 440 A is about 7 A, i.e. only 2%, a level well below the lowest fault settings likely to be used. During limiting conditions when a short circuit is present just outside a protected zone, the voltage at the faulted end of the circuit will be almost zero and therefore the capacitance current will be roughly half its normal value and extremely small relative to the fault current.

It is usual, therefore, when considering the behaviour of protective schemes to be applied to relatively short lines or cables to neglect the capacitance currents and assume that the input and output currents are equal and this practice will be assumed in the following sections.

9.3 THE APPLICATION OF CURRENT-DIFFERENTIAL PROTECTIVE SCHEMES

As explained in Chapter 5, these schemes may be implemented in two basic forms, i.e. circulating current and balanced voltage. Their application to overhead lines and cables was considered in Chapter 5 but more detailed information is provided below.

9.3.1 Comparison arrangements

All current-differential schemes must compare the currents flowing at the two ends of their protected zones. In applications to equipment such as three-phase transformers and rotating machines, the individual phase currents are compared using four interconnecting conductors, as was indicated in the earlier chapters. This practice is acceptable because the lengths of the interconnecting conductors are relatively short and their cost is therefore low.

In applications to overhead lines and cables, however, the interconnecting connections may be quite long, say 15 miles, and in such cases the cost of providing four pilot wires could be very significant. It is therefore the practice in these applications to perform a single comparison between quantities derived from the three-phase currents at each of the ends of protected zones so that only two pilot wires are needed.

Two possible methods of producing single quantities from sets of three-phase currents for comparison purposes are examined below.

Sequence networks

Sequence networks of various forms may be used to determine the positive- and negative-sequence components of sets of three-phase currents and the zero-sequence component may be derived by simply summing the three-phase currents.

Zero-sequence and negative-sequence components are not present in the currents associated with balanced three-phase faults and therefore neither of these components could be used on their own or together for comparison purposes. Positive-sequence components are present however in the currents associated with all types of faults. They represent only one third of the magnitude of the current which flows in the event of a single phase to earth fault and therefore it would be difficult to provide sensitive earth fault settings if only this component was used. In practice, therefore, comparison quantities obtained from at least two of the sequence outputs are needed to provide acceptable performance.

A factor which should be appreciated is that the magnitudes of the output signals derived from two sequences are dependent on the phases on which faults are present and this would clearly cause the sensitivity to such faults to

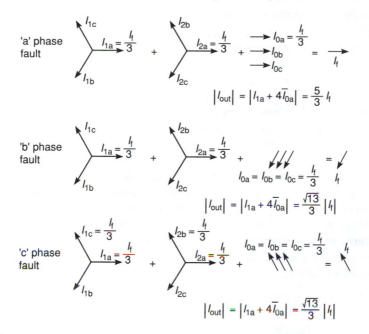

Fig. 9.3 Output current derived from positive- and zero-sequence currents in the ratio 1 to 4, for single phase faults on phases 'a', 'b' and 'c'.

vary. As an example, if a signal were derived from the 'a' phase positive-sequence current and the zero-sequence current, in the ratio of 1 to 4, then the ratios of the magnitudes of the outputs for single phase to earth faults of a given magnitude on phases a, b and c would be 5, $\sqrt{13}$, $\sqrt{13}$, respectively as shown in Fig. 9.3.

A comprehensive study of the possible quantities which can be used for relaying purposes was undertaken by Adamson and Talkan [3].

Variations of sensitivity can be accepted, but the networks needed are quite complex when signals are derived from symmetrical components because phase shifting of two of the input currents must be performed to determine the positive and/or negative sequence components.

This situation has been accepted and such networks have been used in differential protective schemes in the United States. In Britain, however, the use of summation transformers or summation techniques has been preferred and they are still used in differential schemes.

Summation transformers

A summation transformer may have three input windings supplied with the currents from the secondary windings of the current transformers associated with the three phases of a protected system or network. These input windings, which each have a different number of turns, would be mounted together with an output winding on a magnetic core.

The usual, and more economical, arrangement of summation transformers, which may be used when the current transformer secondary windings are connected in star, has a single, tapped input winding, and one or more output windings. Such a transformer is shown in Fig. 9.4. It will be seen that the common or neutral connection of the secondary windings of the current transformers is connected to one of the tappings (x, y, z) on the primary winding of the summation transformer. The other ends of the secondary windings of the current transformers are connected to the terminals a, b and c of the primary winding of the summation transformer. Because of the asymmetry of the arrangement, different sensitivities are obtained for the various current combinations which may be experienced during power-system faults.

The input m.m.f.s to a summation transformer are proportional to the number of turns of its primary winding through which the input currents flow.

When phase to phase faults of a given magnitude (I) are present, the input m.m.f.s are as follows:

IN_{ab} for a fault between phases 'a' and 'b'.
$I(N_{ab} + N_{bc})$ for a fault between phases 'a' and 'c'.
IN_{bc} for a fault between phases 'b' and 'c'.

It is usual for the turns N_{ab} and N_{bc} to be the same, in which case the sensitivity to faults between phases a and c will be twice as great as it is for the other interphase faults.

For a three-phase balanced fault of magnitude I, i.e. $|I_a| = |I_b| = |I_c| = I$, the input m.m.f. is $|I_a N_{ab} - I_c N_{bc}|$. For a transformer in which $N_{ab} = N_{bc}$ this m.m.f. would have a magnitude of $\sqrt{3} I N_{ab}$, i.e. the sensitivity to three-phase faults would be $\sqrt{3}$ times as great as that to faults between phases a and b and b and c.

The sensitivities to phase to earth faults, which are normally higher than

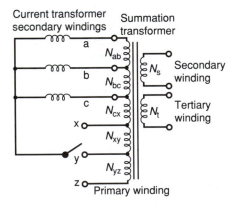

Fig. 9.4 A summation transformer.

those for interphase faults because N_{cx} is greater than N_{ab} and N_{bc} also vary with the phase on which a fault occurs being greatest for faults on phase a and lowest for faults on phase c. The desired sensitivities may be selected by connecting the star point of the secondary windings of the current transformers to the appropriate tapping (x, y, z) on the primary winding of the summation transformer.

Whilst it is the usual practice, as stated above, to use summation transformers in which the turns N_{ab} and N_{bc} are the same, there are applications where it could be advantageous to make them different.

As an example, if a line is supplying a transformer which has delta-connected windings on the line side and star-connected windings on its remote side, then the phase currents on the line side will be in the ratio $-2 : 1 : 1$ in the event of an interphase fault on the remote side of the transformer. This situation was discussed earlier in Chapter 6. It will be seen that should it be the 'b' phase current which has double the magnitude of the others and because it will be in anti-phase with the others there would be zero m.m.f. applied to summation transformers in which the turns N_{ab} and N_{bc} were equal. If this occurred in a current-differential scheme applied to a line the quantities to be compared would both be zero and relay operation would, quite correctly, not occur, the fault being outside the protected zone. If, however, a current transformer core at one end of the scheme saturated, imbalance would occur and currents could flow in the operating windings of the relays at each end of the scheme. No biasing current would be available however at the end of the scheme at which the current transformers were behaving correctly because there would be no output from the associated summation transformer and incorrect operation of the relay at this end could occur.

It will be found, therefore, in some cases, that the primary winding turns N_{ab} and N_{bc} on summation transformers are not the same and as a result the sensitivities to each of the types of interphase faults are different.

9.3.2 Circulating-current schemes

The basic form of a single-phase circulating current scheme is shown in Fig. 9.5(a). To achieve the ideal situation in which no current will flow in the relay unless there is a fault within the protected zone, i.e. between the current transformers, the following conditions must be met:

(a) the circuit must be physically symmetrical,
(b) the current transformers must be physically identical,
and
(c) the current transformers must produce identical secondary currents at each instant when they have the same currents in their primary windings.

To meet the first condition the resistances of the pilot wires on each side of the relay must be the same, i.e. $R_{a1} + R_{a2} = R_{b1} + R_{b2}$. If this were achieved by

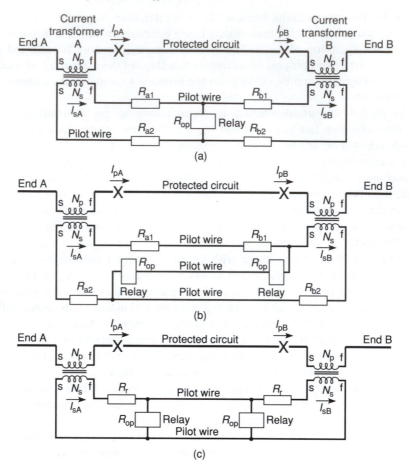

Fig. 9.5 Circulating-current schemes.

connecting the relay between the mid points of the interconnecting cables, so that $R_{a1} = R_{a2} = R_{b1} = R_{b2}$, it would be remote from both ends of the protected zone and in a position where it might not be readily accommodated. In addition cables would have to be run to enable the relay to initiate the tripping of the circuit-breakers. The alternative arrangement, shown in Fig. 9.5(b), in which two relays are connected in series with each other and in which $R_{a1} = R_{b2}$ and $R_{b1} = R_{a2} = 0$, would provide both the necessary balance and position the relays near the circuit-breakers. It has the disadvantage, however, that an extra interconnecting conductor is needed.

An alternative arrangement in which a measure of imbalance is accepted is shown in Fig. 9.5(c). The relays, which are mounted at the ends of the protected zone, have their operating windings connected across the interconnecting conductors at points where the voltages are not zero during healthy or external fault conditions. The application of restraint to the relays by biasing windings carrying the current which circulates during such conditions can ensure that the relays will not operate. The necessary performance may nevertheless be obtained when faults occur within the protected zone.

In general, circulating-current schemes have not been widely applied to overhead lines or cables in the past, the balanced-voltage schemes described in the following section having been preferred. Some relatively modern schemes, however, do operate on the circulating-current principle.

9.3.3 Balanced-voltage schemes

In these schemes the outputs of the current transformers at the two ends of a protected zone are effectively opposed to each other and therefore under ideal conditions no current flows in the interconnecting cables when the protected circuit is healthy. This was the arrangement patented by Merz and Price in 1904. When a fault occurs within the protected zone, balance is not achieved and a current circulates in the interconnecting cables. This may be detected by connecting the operating windings of the relays in series with the interconnecting cables at each end of the protected zone. These relays are then ideally situated to initiate the tripping of the circuit-breakers.

As stated in section 9.1, such schemes were widely applied to transmission and distribution lines and cables in the first decades of this century. As the current and voltage levels rose, the need for biasing features became evident and more advanced schemes were developed, two of these being 'Solkor' and 'Translay' which were produced in Britain by A. Reyrolle and Co Ltd and Metropolitan-Vickers Electrical Co Ltd respectively. These schemes are described in some detail below to illustrate features needed to provide both the necessary fault settings and discrimination between faults within and external to their protected zones.

Solkor protective scheme

The circuit of the original Solkor scheme [4] which was applied to three-phase transmission or distribution circuits is shown in Fig. 9.6, from which it will be seen that the current transformers, which were connected in star, all had the same ratio. Ideally they should all have been of the same design and been well matched. Because of the various features within the protected scheme, current transformers with solid cores could be used rather than the large transformers with cores having distributed air gaps (DAGs) needed by earlier schemes, such as split-pilot protection, which was referred to earlier in section 9.1. In fact, the name 'Solkor' was derived from reference to the 'solid cores' of the transformers used with it.

To enable comparison to be achieved using only two interconnecting conductors, summation transformers were included.

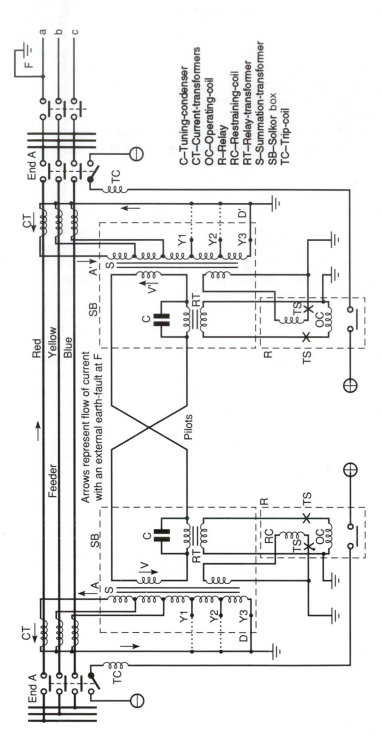

C—Tuning-condenser
CT—Current-transformers
OC—Operating-coil
R—Relay
RC—Restraining-coil
RT—Relay-transformer
S—Summation-transformer
SB—Solkor box
TC—Trip-coil

Fig. 9.6 Connections of Solkor plain feeder protective scheme. (Reproduced from *Solkor protective schemes, Pamphlet 806/3–54*, with the permission of Reyrolle Protection.)

During healthy conditions or when faults were present external to the protected zone, balance had to be maintained between the e.m.f.s induced in the secondary windings of the summation transformers. To ensure that this would be achieved, the summation transformers had to be well matched. For conditions such as earth faults of high current levels, large m.m.f.s were applied to the summation transformers. To prevent the r.m.s. values of the secondary winding e.m.f.s of the summation transformers being very high in these circumstances and very large voltages thus being present between the pilot wires, the summation transformers were so designed that their cores saturated at these levels. As a result, the higher secondary e.m.f.s were not sinusoidal but contained harmonic components and a limited fundamental component. The relationship of r.m.s. voltage to current was as shown in Fig. 9.7.

Ideally, no current should have circulated in the interconnecting conductors during the above conditions but in practice, because of inevitable mismatching, a current containing fundamental and harmonic components did flow. In addition, the presence of a large balancing e.m.f. caused significant current to flow in the capacitance between the interconnecting conductors and half of this current was fed from each end of the scheme. Clearly the magnitudes of these capacitance currents were proportional to the length of the protected zone, i.e. the length of the pilot wires. A further factor which had to be recognized was that the presence of harmonic components in the balancing e.m.f.s caused corresponding components to be present in the capacitance current and indeed, because of the inverse relationship of the reactance associated with the capacitance between the pilot wires to frequency, the resulting harmonic components of the capacitance current increased with their order.

To cope with the above situation, the scheme incorporated relay transformers at the ends of the balancing loop and capacitors were connected across the primary windings of these transformers to form tuned circuits (RT, C in

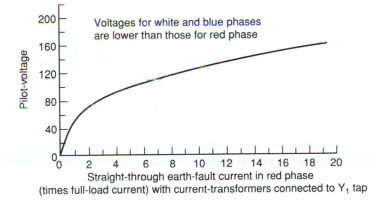

Fig. 9.7 Relation between pilot voltage and primary current. (Reproduced from *Solkor protective schemes, Pamphlet 806/3–54*, with the permission of Reyrolle Protection.)

Fig. 9.6). Because of this feature large proportions of the harmonic components of the imbalance currents flowed through the parallel-connected capacitors and as a result the currents fed to the operating windings of the relays were predominantly produced by the limited fundamental components of the summation transformer e.m.f.s.

It nevertheless had to be accepted that currents would flow in the operating windings of the relays when faults were present outside a protected zone and, to counteract this effect, the relays were provided with biasing windings connected to tertiary windings on the summation transformers. The amount of restraint was dependent on the currents flowing in the primary windings of the summation transformers, but it was not a fixed percentage bias because of the core saturation which occurred at high current levels.

When a fault was present in the protected zone, the e.m.f.s produced by the summation transformers did not balance and a current circulated in the pilot wires. As an example, in the event of an earth fault on one of the phases, currents would have flowed to it from each end. E.m.f.s were induced in the secondary windings of the summation transformers and these e.m.f.s caused a current to circulate in the pilot wires. As a result, currents flowed in the operating windings of the relays. Clearly these currents and those in the pilot wires were dependent on the impedance of the path around which the current circulated and therefore the fault currents needed to cause operation increased with the length of circuit to be protected and in practice there was an upper limit to the impedance which could be present. It was usual to use conductors with seven strands of 0.029 inch diameter copper and these had a resistance of about $5 \, \Omega$ per 1000 yards. They enabled satisfactory operation to be obtained in applications in which the length of the protected zone did not exceed 15 miles. Whilst the impedance of longer pilot wires could be kept down to the necessary levels by using cables with conductors of greater cross-sectional areas, their costs tended to be unacceptably high and, in addition, the effects of the high capacitances between such conductors would have made it necessary to use very high percentage biasing to prevent incorrect operation.

To enable the necessary sensitivity to internal faults to be obtained in Solkor schemes, rotary-sensitive relays of the form described earlier in section 5.7 (page 174) were employed. These relays could be set to operate with currents in the range 35–80 mA in their operating windings, the required value being obtained by adjusting the travel and thus the initial air gaps between their armatures and electromagnets. Restraining windings, supplied from the tertiary windings of the summation transformers, enabled the percentage biasing needed to ensure that correct discrimination would be obtained up to the maximum fault-current levels likely to be encountered. To further assist in this process, weights could be placed on pans mounted on the relay spindles. These increased the inertias of the movements and thereby slightly increased the operating times and reduced the possibility of operation due to transient imbalances.

It will be clear from the above that the fault settings obtainable with this scheme were dependent on both the minimum current at which the relay would operate and the interconnecting conductors. As an example, if the scheme were applied to an overhead line of length 6 miles, with interconnecting conductors of seven strands of 0.029 inch conductor, and the relays were set to operate at 55 mA and if the least sensitive tappings on the summation transformers (i.e. N_{cx} = minimum value) were used, the fault settings were as shown in Table 9.1.

Figures 9.8 and 9.9 show the effects of the lengths of the pilot wires and the operating currents of the relays on the fault settings.

Because the summation transformers included in Solkor schemes saturated when their input currents were high, the demands on the main current transformers were limited.

To ensure that acceptable balance would be obtained in the event of high-current faults outside a protected zone, it was nevertheless necessary that the current transformers installed at the two ends of a scheme were able to provide adequate secondary winding e.m.f.s. A large number of the original Solkor schemes and variants of it were installed around the world and many are still in use and performing satisfactorily.

Translay protective scheme

This scheme [5] also employed the biased balanced-voltage principle, its general arrangement being as shown in Fig. 9.10(a).

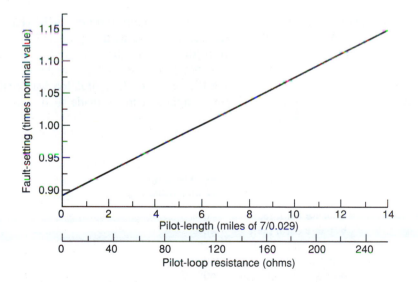

Fig. 9.8 Variation of fault setting with pilot length. (Reproduced from *Solkor protective schemes, Pamphlet 806/3–54*, with the permission of Reyrolle Protection.)

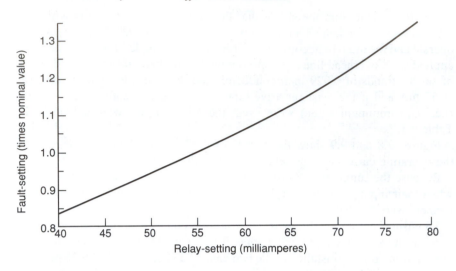

Fig. 9.9 Variation of fault setting with relay setting. (Reproduced from *Solkor protective schemes, Pamphlet 806/3–54*, with the permission of Reyrolle Protection.)

It will be seen that the secondary windings of the current transformers at each end of a protected zone were again connected in star. They were all of the same ratio and should ideally have been of the same design and well matched.

At each end, the secondary windings of the current transformers were connected to a summation winding mounted on the centre limb of an 'E'-shaped electromagnet. As a result, the effect was the same as that produced by the summation transformers incorporated in the Solkor scheme described above, namely that the m.m.f.s applied were not solely dependent on the current levels but also on the phases in which they flowed. This situation was again accepted because it enabled a single comparison to be made to determine

Table 9.1

Type of fault	Fault setting (percentage of current-transformer primary rating)
'a' phase earth fault	40%
'b' phase earth fault	48%
'c' phase earth fault	60%
'a–b' phase-fault	240%
'b–c' phase-fault	240%
'a–c' phase-fault	120%
Three-phase fault	140%

whether a fault was within a protected zone or not and, most importantly, it reduced the number of pilot wires needed to a minimum, namely two.

In the Translay scheme, however, induction-type relays were employed and the secondary windings on their main E-shaped electromagnets and the windings on the lower C-shaped electromagnets were connected in series via the pilot wires, as shown in Fig. 9.10(a).

During healthy conditions or when faults were present external to the protected zone, the m.m.f.s applied to the main electromagnets of the two relays by the currents in the summation windings, would ideally have been of the same magnitude at each instant, as would the e.m.f.s (e_{sr}) induced in the secondary windings. Because the e.m.f.s were connected in opposition, no current should have flowed in the pilot wires or the windings of the lower electromagnets of the relays and therefore no torque should have been produced on the relay discs.

In practice, significant currents could circulate in the comparison circuit because of mismatching when external faults involving high currents were present. To ensure that these did not cause the relays to operate, the relays were biased by copper shading loops fitted to the main E-shaped electromagnets. These provided restraining torques proportional to the square of the flux magnitudes in the main magnets. In addition, permanent magnets were fitted to provide eddy-current damping of the discs and this feature further improved the performance of the relays.

Again, capacitance currents (i_c) flowed between the interconnecting conductors during healthy conditions and they would be quite high when external faults involving high currents were present. These currents flowed in both the

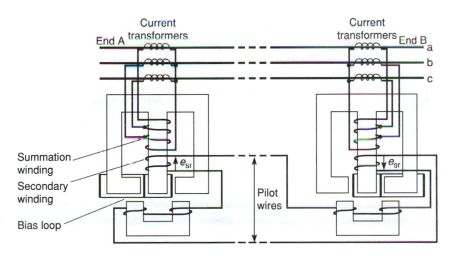

Fig. 9.10 (a) Translay differential protective scheme. (Reproduced from *Protective Relays – Application Guide, 3rd edn*, GEC Measurements, 1987 with the permission of GEC Alsthom Protection and Control Ltd.)

secondary windings of the main electromagnets and the windings on the lower electromagnets as shown in Fig. 9.10(b). In the absence of resistance in the comparison circuit, the capacitance currents would have lead the e.m.f.s (e_{sr}) by $\pi/2$ rad and they would thus have been in phase with the fluxes in the centre limbs of the main electromagnets of the relays. In these circumstances, the fluxes set up by the lower electromagnets would have been in phase with those in the main electromagnets and the resultant torques produced on the discs would have been zero. The presence of significant resistance in the pilot wires affected the above conditions because it caused the capacitance currents to lead the e.m.f.s (e_{sr}) by less than $\pi/2$ rad and caused the fluxes set up by the relay electromagnets to be out of phase with each other, the displacement increasing with increase of the resistance of the pilot wires. This situation was counteracted by the shading loops fitted to the main electromagnets, which displaced the phase of the pole-tip fluxes from that of the main fluxes in the centre limbs of the main electromagnets. The phase adjustment so produced increased the capacitance current which could be allowed and so enabled the scheme to be applied to relatively long lines and cables.

In the event of a fault occurring within the protected zone in which currents were fed to the fault from each end of the line or cable, as shown in Fig. 9.10(c), fluxes would have been set up in the centre limbs of the main electromagnets of both relays and a current would have circulated in the comparison circuit and in the windings of the lower electromagnets of both relays. Provided that the faults currents at both ends of the protected zone were above the fault-setting levels, both relays would have operated and initiated the opening of their associated circuit-breakers. If, however, a fault occurred

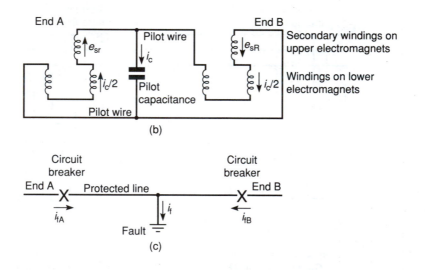

Fig. 9.10 (b), (c)

within the protected zone but current was fed to it from only one end of the line or cable, then flux would only have been produced in the centre limb of the main electromagnet of the relay at the end from which the fault was fed. Current would nevertheless have circulated around the comparison circuit, but only the relay at the end from which the fault was fed would have operated. This situation, for which the current i_{fB} in Fig. 9.10(c) would have been zero, might not have been acceptable, in which case arrangements would have been made to inter-trip the second circuit-breaker whenever one of them was opened by the protective equipment.

This scheme has been widely applied to distribution circuits. In the majority of applications the following settings were found to be satisfactory:

Least sensitive earth fault	40% of rating
Least sensitive interphase fault	90% of rating
Three-phase fault	52% of rating

Relays with more sensitive earth fault settings were available for applications in which earth fault currents could have been very low and phase-fault settings above the rated-current level could be selected in some instances to ensure that relay operation would not be caused by full-load currents if a fault occurred on the pilot wires.

The e.m.f.s provided by the secondary windings on the main electromagnets of the Translay relays were not directly proportional to the currents in the summation windings even though there were significant air gaps in the magnetic circuits. Saturation of the cores, which occurred at the higher current levels, was allowable because of the close matching achieved during manufacture. As a result, the r.m.s. voltages applied between the pilot wires did not exceed about 180 V. As with the Solkor scheme, the voltage waveform did become distorted and high peak values could be produced when external faults of high current levels were present. To cope with such situations the relay windings were insulated to withstand test voltages of 5 kV.

It will be appreciated that the currents which flowed around the comparison circuit when external faults occurred were affected by the resistance of the interconnecting conductors and could be quite low when the protected line or cable was long. Because the relay torque was dependent on the fluxes produced by both its electromagnets, operation could be achieved with limited current in the winding of the lower electromagnet, but then higher currents were needed in the summation winding, i.e. the fault settings increased with the resistance of the pilot wires. In practice it was found that the scheme could operate satisfactorily with resistances up to 1000 Ω in the comparison circuit.

Solkor Rf protective scheme

The original Solkor and Translay schemes have been described in detail in the preceding sections because they incorporated the features which are necessary

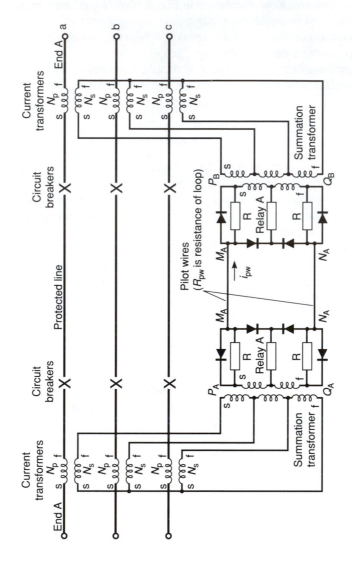

Fig. 9.11 (a) Solkor Rf protective scheme.

when lines are to be protected. They did, however, operate relatively slowly when internal faults occurred and the relays used in them were quite complex. As a consequence several other schemes, which employ different techniques, have been introduced. One of these is the Solkor Rf scheme [6] produced by NEI Reyrolle, which incorporates simple polarized relays and several rectifiers. The basic arrangement of this scheme, which employs the circulating-current principle, is shown in Fig. 9.11(a).

The resistors R_A are of greater resistance then half the total resistance of the two pilot wires (R_{pw}). When an external fault is present the output windings of the summation transformers produce equal e.m.f.s which assist each other and thereby cause a current to circulate around the loop formed by the pilot wires, causing the voltage distributions during the alternate half cycles to be as shown in Fig. 9.11(b). Because of the arrangement of the rectifiers the voltages across the relays are always of the opposite polarites to those needed to cause current to flow in them. In effect the relays are negatively biased, the bias being dependent on the amount by which the resistance R_A exceeds that of the pilot wires (R_{pw}).

When faults are present on a protected line the above conditions do not obtain and the relays at both ends of the scheme operate when the fault-setting level is exceeded if current is being fed into only one end of the line. In the event of a current being fed from each end of a line to a fault, operation occurs at levels below the setting values.

Typical fault settings are given below for the various types of fault:

Phase a to earth	25%
Phase b to earth	32%
Phase c to earth	42%

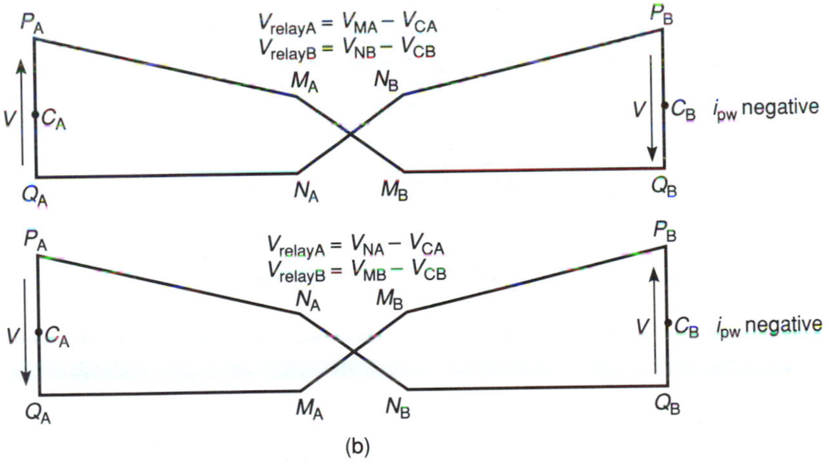

(b)

Fig. 9. 11 (b)

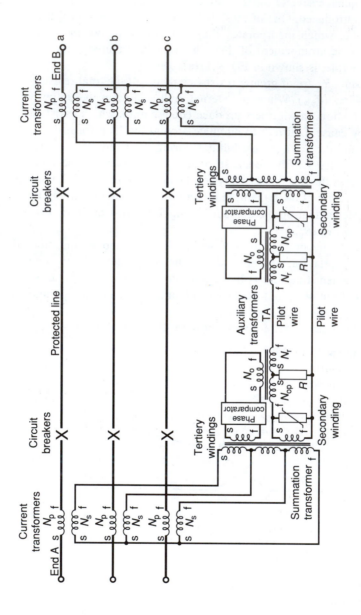

Fig. 9.12 Translay 'S' protective scheme. Note: Turns N_{op} on the auxiliary transformers provide operating m.m.f. Turns N_r on the auxiliary transformers provide restraining m.m.f.

Phase a to b	125%
Phase b to c	125%
Phase c to a	62%

Further information may be obtained from reference [6].

Translay S protective scheme (Relay type MBC1)

The original Solkor and Translay schemes described above did not operate particularly rapidly because of the inertias and travels of their relay movements and therefore schemes incorporating electronic equipment were introduced when it was felt that power-system engineers would consider them to be sufficiently reliable.

The Translay S protective scheme, which is produced by GEC Measurements, operates on the circulating-current principle. Comparisons are made over pilot wires and discrimination is achieved using phase comparators of the type described earlier in section 4.3.3 (page 119).

The basic circuit of the scheme is shown in Fig.9.12, from which it will be seen that summation transformers are fed by the current transformers at each end of a protected line. When faults are present at points external to the line, the secondary windings of the summation transformers provide equal e.m.f.s which cause current to circulate around the pilot wires and through the operating and restraining windings of the auxiliary transformers (TA). The phases of the output voltages from the auxiliary transformers are compared with those of the voltages produced by the tertiary windings of the summation transformers and the phase comparators are so set that operation will not occur during such conditions.

When faults are present on a protected line, the currents which circulate around the pilot-wire loop tend to be lower and the phases of the voltage outputs from the auxiliary transformers with respect to those provided by the tertiary windings of the summation transformers are such that tripping of the circuit-breakers will be initiated by the phase comparators.

The operating times of this scheme are dependent on the fault-current levels and other factors, the minimum value being about 25 ms.

Further details of this scheme, including the current transformer requirements and allowable pilot-wire resistances are provided in reference [7].

9.3.4 Pilot wires

It will be clear from the foregoing that the pilot wires form crucial parts of current-differential protective schemes.

In many cases special cables are provided for this duty and they may include additional conductors for other purposes. The individual conductors must be insulated to withstand the voltages which may be applied between them or to earth. As shown earlier, the r.m.s. voltages between pilot wires in balanced-voltage schemes may be limited to relatively low levels because of saturation occurring in the cores of summation transformers but nevertheless high peak levels may be reached when high currents are flowing in a protected circuit. It is therefore recognized that the insulation on the conductors and any equipment directly connected to them, such as summation transformers or relay windings, should be capable of withstanding voltages of standard levels such as 5 kV or 15 kV. It will be appreciated that devices such as voltage limiters or diverters may not be connected between the pilot wires used with some protective schemes because they would cause high currents to flow between the conductors and could thus affect the operation of the protective scheme when a high current was flowing to a fault outside the protected zone.

Whilst overhead air-insulated conductors could be used, they are particularly vulnerable to damage because of their exposure to atmospheric conditions. They are therefore not normally used for this purpose, buried cables being preferred in spite of their extra cost.

As an alternative, it is sometimes more economical to rent circuits from a telephone company. When this is done the circuits must be dedicated, i.e. connected continuously and solely to the protective scheme, because of the element of unreliability which would be introduced if circuit switching were permitted. Special contracts are necessary with clauses guaranteeing that notice will be given of any work that is to be done on the circuits by the telephone company. The maximum voltages which may be applied to the circuits and the currents which may flow in them are normally specified by the telephone company, typical values being 130 V (peak) and 60 mA (r.m.s.) and these may only be present during fault conditions.

Circuits available for rental have conductors of relatively small cross-sectional areas, their weights typically being 20 lb per mile and 40 lb per mile. Their resistances are therefore higher than those of the conductors which are specially provided for current-differential schemes.

Whilst the schemes described earlier may be capable of operating satisfactorily with relatively high resistances in their comparison circuits, they may not able to comply with the limiting voltage and current conditions set by telephone authorities. Such protective schemes are not generally suitable for operation with rented circuits and therefore special or modified schemes which may use them have been developed, one being described in detail in the next section.

9.3.5 Protective schemes which use rented telephone circuits

In addition to the limitations imposed by telephone companies on the voltages which may be applied between the conductors of rented circuits and the currents which may flow in them, the following conditions may also have to be complied with:

1. The conductors of the rented conductors may have to be isolated from most parts of a protective scheme and other circuits and earth by insulation able to withstand test voltages, typically of 15 kV.
2. Spark-gaps may have to be provided between the secondary circuits of current transformers and earth to ensure that high voltages, which could cause breakdowns or that could lead to excessive voltages being present on the rented conductors, will not be produced.

The Solkor Rf and Translay S schemes both satisfy the various conditions outlined above. Special transformers are available to isolate the telephone-circuit conductors and adequate high voltage insulation is included in the relay circuits and the summation transformers.

Equipment is always provided with the schemes to monitor the telephone-circuit conductors.

9.3.6 Monitoring of pilot wires

In spite of the precautions which are taken, faults may nevertheless occur on pilot wires. As examples, underground cables may occasionally be damaged by subsidence or when excavation work is being done.

The effects of faults on pilot wires depend on their nature and the type of protective scheme with which they are associated.

An extreme condition which could arise is an open-circuiting of a comparison loop because of a break in one of the pilot wires. If the associated protective scheme operated on the voltage-balance principle such a break would not cause relay operation should a fault occur outside the protected zone but it would prevent internal faults from being detected because current would not be able to flow in the operating windings of the relays. On the other hand, a break in a comparison circuit associated with a protective scheme operating on the circulating-current principle would interrupt one of the balancing currents and therefore relay operation could occur in the event of a fault outside the protected zone or because of the load current in the protected circuit if the fault settings were below the full-load level.

A second extreme condition would be a short-circuit between a pair of pilot wires. In this event and should the associated protective scheme employ the voltage-balance principle, then a current would flow from each end of the link and both relays would operate if a fault was present outside the protected zone. Relay operation could again be caused by the flow of load current in the protected circuit if low fault settings were being used.

Should the pilot wires of a circulating-current scheme be short circuited, then current could be diverted away from the operating windings of the relays when a fault was present within the protected zone and, as a result, opening of the associated circuit-breakers might not be initiated.

In addition to the extreme conditions considered above, other faults may arise on pilot wires. In the case of circuits rented from telephone companies, occasions have arisen where circuits have been replaced by others with different resistances and, as shown earlier, this could affect factors such as fault settings and the maximum current levels at which correct discrimination will be achieved when faults occur outside protected zones.

Experience has shown that specially installed cables, buried in the ground, are very reliable and therefore the likelihood of the above conditions arising are so small that monitoring of the cables may not be considered justified or necessary. Some authorities who are not willing to accept the relatively small risks however, do, arrange for supervisory equipment to be provided. Whilst the performance record of circuits rented from telephone companies is good because of the conditions specified in the contracts, such circuits are nevertheless considered to be somewhat weak links in protective schemes and therefore it is the usual practice to include supervisory equipment when they are used.

It will be clear from these considerations that supervisory equipment should be able to detect the following conditions:

(a) an open-circuit in the comparison loop of a protective scheme;
(b) a short-circuit between the pilot wires;
(c) a change in the resistances of the pilot wires.

For many years, monitoring has been achieved by injecting small direct currents into comparison circuits, as shown in Fig. 9.13. In this example, a balanced-voltage protective scheme is being supervised. It is common practice to include a relay (R_1) in the circuit at the opposite end of the protected zone to that at which the injected current is supplied from an auxiliary supply. The relay R_1 is continuously energized by the injected direct current and in addition

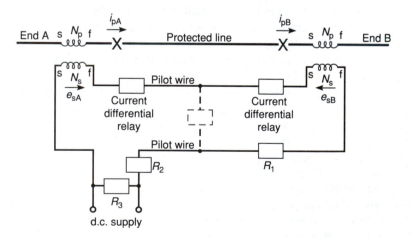

Fig. 9.13 Pilot wire supervision scheme.

a small power-frequency current may flow in it because of mismatching in the protective circuit and capacitance currents associated with the pilot wires.

In the event of an open circuit in the comparison circuit the direct current in the relay R_1 would fall to zero, causing it to reset and thereby initiate an alarm.

If a short-circuit occurred between the pilot wires, the direct current in the relay R_1 would again fall to zero and an alarm could be provided. If, however, significant resistance was present in a fault between the conductors, i.e. the R shown dotted in Fig. 9.13, the current in relay R_1 would reduce because of the resistance present in the pilot wires, but it would not fall to zero and therefore only faults in which the resistance R was below a certain value would be detected. Better performance is obtained by including an extra relay R_2, shown in Fig. 9.13, in the supply connection. The direct current in the operating winding of this relay would always rise if a fault occurred between the conductors and the condition would be detected by relay R_2.

Should the resistance of the comparison circuit rise the direct currents in both relays would fall. This condition, which could arise if a rented telephone circuit was replaced by another with smaller conductors and/or a greater route length, could be detected by the resetting of relay R_1. Should the reverse situation arise, namely the resistance of the comparison circuit fell significantly, the injected direct current would thus increase causing relay R_2 to operate.

It will be appreciated that the inclusion of the two relays would enable the presence of unacceptable conditions in a comparison circuit to be detected. It must be recognized, however, that the currents flowing in the relays would be affected if a fault occurred within a protected zone because alternating currents would then circulate in the comparison circuit. To ensure that alarm signals will not be provided in these circumstances, time delays are normally included so that the supervision relays must indicate that a comparison circuit is unhealthy for a period considerably longer than the times for which internal faults may exist before an alarm signal is initiated.

An additional relay (R_3 in Fig. 9.13) is often provided to detect the loss of the injection voltage. Supervisory arrangements of the above types have been applied successfully to many current-differential protective schemes. In recent years, schemes, based on the same principles, using different techniques have been developed and applied [8, 9].

9.3.7 Current-differential schemes incorporating optical-fibre links

It will be clear from the preceding sections that the limitations of current-differential schemes do not arise from the basic principle of comparison of the currents entering and leaving a power circuit but because of factors such as the resistances and capacitances of the interconnecting conductors over which comparisons are effected.

The introduction of equipment to enable communications to be effected over optical fibres caused groups of engineers concerned with protective systems to investigate the possibilities of developing current-differential schemes in which comparisons would be effected over optical-fibre links. Such schemes have now been produced and applied to feeders. Two schemes are described below.

Scheme produced by Asea Brown Boveri

This scheme, which is shown in block form in Fig. 9.14, has two main units at each end of a protected line or cable, namely a differential relay type DL91F and a fibre-optic system FOX6. In three-phase applications, three conventional iron-cored current transformers are needed at each end of the protected zone. The manufacturers recommend that these transformers should be designed for the maximum symmetrical short-circuit current expected in the protected circuit, without regard for the time constant of the network. They state that current transformers of class 5P20, $P_N = 30$ VA are usually adequate. Each of the relays (DL91-F) contains a summation transformer so that single phase comparisons may be made. An output (e_{AT}) provided by a summation transformer at one end of a protected zone is fed as an input T_x to an interface (N3AL) within the fibre-optic system unit (FOX6) and the digital signal produced is then transmitted over the optical fibre link to an interface (N3AL) in the unit (FOX6) at the other end of the zone. An analogue output (e_{BR}) is produced for comparison with the signal e_{BT}, provided by the local DL91-F relay. A similar comparison between the e.m.f.s e_{AR}, and e_{AT} is effected in the

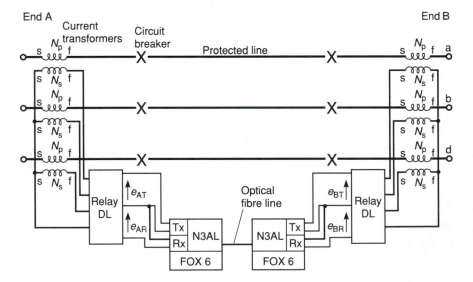

Fig. 9.14 Scheme incorporating relay DL91F. (Reproduced from *ABB Buyer's Guide 1989–90*, with the permission of ABB Relays AG, Baden, Switzerland.)

opposite direction. As with protective schemes using interconnecting conductors, the opening of the appropriate circuit-breakers is initiated by the relays when the e.m.f.s being compared are sufficiently different.

The fibre-optic system units (FOX6) contain monitoring circuits to ensure that the digital signals are transmitted without error and to provide blocking signals to the relays in the event of false operation. In addition, both the relays and the fibre-optic system units are able to detect an interruption of the optical-fibre link and they will initiate the operation of alarms about 3 seconds after such an occurrence.

It is recommended that multi-mode graded optical fibres 50/125 μm standardized by IEC for the wavelength 900 nm be used with this scheme and these should allow satisfactory performance to be obtained on circuits up to 8 km long. By using a variant of the fibre-optic system unit, which is designated FOX6L operating at 1300 nm, circuits up to 16 km long may be protected satisfactorily.

The above scheme, which may be applied to circuits operating at frequencies of 50 Hz or 60 Hz and which may be supplied from current transformers with secondary winding ratings of 1 A or 5 A, operates in 25 ms or less for faults of twice the setting values.

Further detailed information about this scheme is provided in reference [10]. Other schemes which are now available are described in references [11], [12] and [13].

LFCB scheme produced by GEC Measurements

Digital processing is used in this scheme in which the three-phase currents at each end of a protected line are sampled. Internal free running clocks at the ends control the sampling but they are not directly synchronized and this leads to phase differences of up to one half of the sampling period.

The data samples represent the instantaneous values of the phase currents which may contain d.c. offset, harmonic and high-frequency components in addition to the fundamental. The data are filtered and pre-processed to a form which enables the magnitudes of differential and bias currents to be determined. The one-cycle window Fourier method, which is used because of its stable transient characteristics, can be expressed as:

$$I_s = \frac{2}{N} \left\{ \sum_{n=1}^{N-1} \sin n\omega \Delta t \, i_n \right\}$$

$$I_c = \frac{2}{N} \left\{ \frac{i_0}{2} + \frac{i_n}{2} + \sum_{n=1}^{N-1} \cos n\omega \Delta t \, i_n \right\}$$

in which

N = number of samples per cycle

ω = fundamental angular frequence (rad/s)

Δt = sampling time (s)

i_n = instantaneous value of signal at time of sample n

I_s = Fourier sine integral of signal i

I_c = Fourier cosine integral of signal i

If the fundamental component of signal i is given by $I \sin(\omega t + \theta)$ then

$$I_s = I \cos \theta \quad \text{and} \quad I_c = I \sin \theta$$

From this information the instantaneous value of each phase current can be determined at any instant.

Data messages are sent between the ends of a protected line at regular intervals. As an example, a message will be sent about the phase a current (I_A) at end A of a line at a particular instant to end B of the line. The propagation time is known for the particular communication channel and therefore the current at end B (I_B) at the instant when the message was sent from end A can be determined. Differential and bias currents (I_{diff} and I_{bias}) can then be determined as:

$$|I_{diff}| = |I_A + I_B|$$

$$|I_{bias}| = \tfrac{1}{2}\{|I_A| + |I_B|\}$$

A per-unit biased characteristic is available, for which the tripping criterion is:

$$|I_{diff}| > k_1 |I_{bias}| + I_{s1}$$

in which k_1 is the pu bias setting and I_{s1} is the minimum differential-current setting.

An alternative bias characteristic is available to improve the stability of the scheme when high currents are flowing to external faults. In this case the tripping criterion is as before at low current levels but at the higher currents it is:

$$|I_{bias}| > I_{s2}$$

$$|I_{diff}| > k_2 |I_{bias}| - (k_2 - k_1) I_{s2} + I_{s1}$$

in which I_{s2} is the threshold at which the increased bias becomes effective and k_2 is the increased pu bias setting. This characteristic is shown in Fig. 9.15.

Typical settings are:

$I_{s1} = 0.2 \, \text{pu}$

$k_1 = 0.3 \, \text{pu}$

$I_{s2} = 2.0 \, \text{pu}$

$k_2 = 1.0 \, \text{pu}$

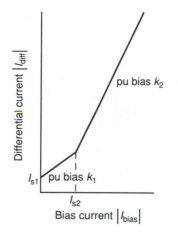

Fig. 9.15 Bias characteristic.

This scheme can readily incorporate extra features, such as monitoring, event recording, fault recording and inter-tripping. It may be operated over a dedicated link, fibre optics being embedded within the power system earth conductors run along the tops of the transmission-line towers or it may be multiplexed in a communications system link. It may also be operated over other communication links such as microwave or over shielded twisted pair cables.

The scheme does not incorporate summation transformers and direct comparisons are made of each of the phase currents. As a result a low and constant minimum fault setting is provided for each type of fault.

Further information about this scheme may be obtained from reference [14] and details of other schemes are provided in references [15] and [16].

9.4 THE APPLICATION OF CURRENT-DIFFERENTIAL SCHEMES TO MULTI-ENDED CIRCUITS

There are situations where it is economically advantageous to feed two or more loads from one source or set of busbars and certainly teed feeders are quite common in distribution networks. The protection of such circuits presents special problems [17].

Current-differential schemes are suitable for application to multi-ended circuits as was made clear in the previous chapter, in which the protection of busbars was considered. They have also been applied to teed feeders for many years and schemes such as Solkor and Translay have been widely used. In such applications, three or more currents must be summed or compared per phase to determine whether a protected unit is healthy or not. As an example, when the teed feeder shown in Fig. 9.16 is healthy, the currents fed into

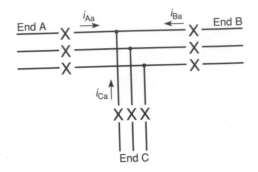

Fig. 9.16 A multi-ended (teed) circuit.

each phase must sum to zero at each instant, i.e. $i_{Aa} + i_{Ba} + i_{Ca} = 0$ or $i_{Aa} = -(i_{Ba} + i_{Ca})$. The magnitude of one of these currents must clearly be different to those of the others and indeed one current could be twice as great as each of the others. Because of this, linearity must be maintained in the components of a protective scheme up to the highest current levels which may be encountered during faults external to its zone. Saturable summation transformers, such as those used in the schemes described earlier, may not be used and air-gapped cores must be used to provide linear voltage-current characteristics. Other factors for which provision must be made in protecting multi-ended circuits are referred to in the next section.

Basic scheme in which metallic pilot wires are used

A basic voltage-balance scheme suitable for application to teed circuits is shown in Fig. 9.17. To enable balance to be achieved unless a fault is present on the protected circuit, summation transformers with air-gapped magnetic cores must be used as stated above. Their secondary winding e.m.f.s (e_A, e_B and e_C) must be proportional to the input m.m.f.s produced by their primary windings. The secondary windings of the summation transformers must be connected in series by two pilot wires (1 and 2) between each of the ends of the protected zone in such a way that the e.m.f.s (e_A, e_B and e_C) will sum, ideally, to zero unless an internal fault is present. In these circumstances, zero current should flow in the relay transformers but in practice small currents will flow because of mismatching. In the event of a fault occurring within the protected zone the e.m.f.s do not sum to zero and a current proportional to that in the fault circulates in the pilot wires and relay transformers. The same current will flow in the operating windings of all three relays, tripping of the three circuit-breakers being initiated if the fault current is above the setting level.

As stated above, mismatching causes currents to circulate and flow in the relay transformers when the protected circuit is healthy, and these currents might reach levels which could cause relay operation when large currents are flowing to faults external to the protected zone. To ensure that such maloperation will not occur, the relays have windings which provide an adequate percentage bias. It will be appreciated, however, that satisfactory performance would not necessarily be obtained if the restraining winding of each relay was connected only to a tertiary winding on the summation transformer at its own end of the circuit because little current might flow in that end of the circuit when high currents were flowing through the other two ends to an external fault. The technique, employed also in busbar protective schemes and described in section 8.7 (page 296), of biasing with a quantity derived from the currents at all the ends of the protected circuit must therefore be employed.

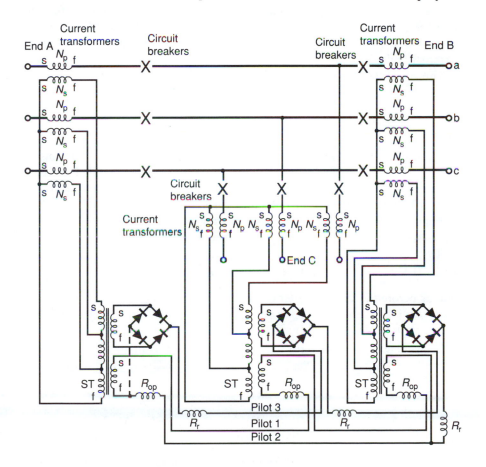

Fig. 9.17 Teed-feeder protective scheme. R_{op}, relay operating winding; R_r, relay restraining winding; ST, summation transformer (all have the same number of turns).

To achieve this the e.m.f.s provided by tertiary windings on the summation transformers may be rectified and connected in series via two interconnecting conductors (2 and 3). As a result a direct current will circulate and flow in the restraining windings of each of the relays. Clearly, the restraint will then be proportional to the sum of the magnitudes of the currents flowing at the ends of the protected zone and it will not be dependent on their directions.

Other devices such as diverter reactors may be included in schemes to assist further in obtaining the required discrimination between internal and external faults. Such reactors have two windings on a magnetic core, one of the windings being connected in the restraining circuit and the other across the operating winding of one of the relays. When external faults of high-current levels are present, the cores of the reactors saturate because of the high direct currents flowing in them and consequently the impedances of the windings connected across the relay operating windings decrease. As a result, increasing percentages of the imbalanced currents flowing in the relay operating windings tend to be shunted through the diverter reactors as the current levels of external faults increase.

The above techniques were used in the early Solkor and Translay schemes.

LFCB scheme produced by GEC Measurements

This scheme [14], which was described earlier in section 9.3.7, is suitable for application to teed feeders. In such applications the three-phase currents at each of the three ends of a line, i.e. nine currents, must be sampled and data must be sent between the ends via either metallic conductors or optical fibres.

A differential current (I_{diff}) and a bias current (I_{bias}) are produced at each of the ends (A, B, C), their values being given by:

$$|I_{\text{diff}}| = |I_A + I_B + I_C|$$

and
$$|I_{\text{bias}}| = 0.5\{|I_a| + |I_b| + |I_c|\}$$

and as before the tripping cirterion is

$$|I_{\text{diff}}| > k_1 |I_{\text{bias}}| + I_{s1}$$

or a characteristic of the form shown in Fig. 9.15 is available.

Further information on the protection of teed feeders is provided in references [18] and [19].

9.5 THE APPLICATION OF CURRENT-DIFFERENTIAL SCHEMES TO LINES AND CABLES TERMINATED WITH TRANSFORMERS

These circuits, which are usually referred to as feeder transformers, and in which there is no circuit-breaker between the feeder and transformer, are commonly used in distribution networks. Whilst separate current-differential protective schemes could be applied to the feeder and the transformer in such a circuit, it is clearly economically advantageous to employ a single overall scheme, thus reducing the number of current transformers and the amount of relaying equipment needed. It might appear that differential schemes suitable for transformers would be suitable for those applications and that it would only be necessary to position the current transformers on the primary side of the transformer at the input end of the feeder. Overall schemes must, however, be capable of coping not only with the special factors associated with transformers, such as their exciting currents and the need to use current transformers of different ratios, but also with the inclusion of long pilot wires or communication links in the comparison circuits.

A further variant of the basic voltage-balance protective scheme has been widely applied to feeder transformers. Figure 9.18 shows the arrangement needed to protect a circuit in which the power-transformer windings are delta-connected on the side connected to the feeder and star-connected on the other side, the neutral point of the latter winding being earthed. To achieve balance during healthy conditions, the current transformers on the star-connected side of the power transformer must be connected in delta and those at the other end of the feeder must be connected in star, as was explained earlier in section 6.3.2 (page 198). In addition, the secondary current ratings of the delta-connected current transformers must be $1/\sqrt{3}$ times that of corresponding rating of the star-connected transformers, i.e. either 2.89 A and 5 A or 0.578 A and 1 A respectively. Whilst a summation transformer with a primary winding having three sections of turns and four terminals could be connected to the star-connected current transformers, such a transformer could not be energized from a delta-connected set of current transformers from which only three connections may be made and therefore a summation transformer with only two sections and three terminals, as shown in Fig. 9.18, must be used.

As explained earlier in chapter 6, zero-sequence current components cannot be fed into a delta-connected winding and therefore, if an earth fault occurs on the star-connected output winding of a transformer, currents of equal magnitude and opposite phase will flow in two of the conductors feeding the delta-connected winding. Should such a condition occur with the arrangement shown in Fig. 9.17, the fault could appear to the protection to be an interphase fault for which the fault setting could be unacceptably high, say 240% or more. For this reason, the summation transformers at both ends have only two sections with three terminals as shown, and earth fault relays are included at each end of the protected zone. The earth fault relay in the neutral connection of the current transformers at the input end of the feeder will operate for earth faults on the feeder or the delta-connected windings of the transformer. The second earth fault relay, which must detect faults on the star-connected winding, is fed from an output winding on a core-balance transformer which has four input windings supplied by the current transformers in the phases and neutral of the star-connected side of the power transformer. A factor which

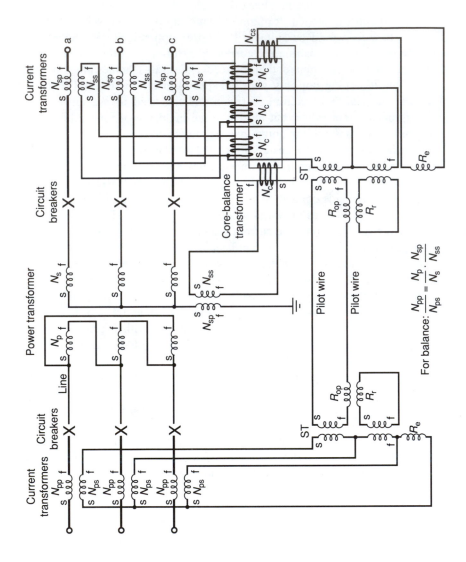

Fig. 9.18 Feeder-transformer protective scheme. R_{op}, differential relay operating winding; R_r, differential relay restraining winding; R_e, earth fault relay winding; ST, summation transformer (both have the same number of turns).

must be recognized with this arrangement is that any earth fault which occurs within the protected zone will only operate one of the earth fault relays. It is therefore necessary that intertripping arrangements be provided so that the operation of either of the relays will cause the circuit-breakers at both ends of the protected zone to be opened. This may be achieved independently of the protective scheme or the comparison circuit may be imbalanced by injecting a signal into it.

Other forms of current-differential schemes operating on current-balance and voltage-balance principles are available for application to feeder transformers, examples being Translay 'S' type MBCI [18] and the LFCB scheme, both of which are produced by GEC Measurements.

9.6 THE FUTURE

Because of the simplicity of the principles involved in current-differential schemes and the ability of such schemes to discriminate between faults within and external to their protected zones and also to provide sensitive fault settings and operate rapidly, it is probable that they will continue to be applied to short and medium length overhead lines and cables and feeder transformer circuits for many years.

Undoubtedly relay equipment will increasingly incorporate electronic circuitry and digital, rather than analogue, techniques. Whilst direct comparisons will continue to be effected using pilot wires because of their simplicity and reliability, increasing use of optical-fibre links is to be expected. Comparisons are already being effected over long distances by high-capacity, digital communication systems [17] owned by private and public companies and it is to be expected that increasing use of these techniques will occur. These modern methods of implementing current and voltage comparisons have the advantage that errors, and therefore imbalances, are not produced when faults are present external to protected zones because of the capacitances between metallic interconnecting conductors. In addition, non-linearities caused by the use of conventional current and summation transformers may be eliminated. As a result, it is probable that the lengths of circuits protected by current-differential schemes will increase from the present levels and in addition they are likely to be used in countries where supply authorities have been reluctant to apply schemes dependent on metallic conductor links because of the possibility of faults occurring on them.

REFERENCES

1. The 'Merz–Price' system of automatic protection for high-tension circuits, *Electrical Review*, August 28, 1908.
2. Leeson, B. H. and Leyburn, H. (1931) The principles of feeder protection and their application to three modern systems, *CIGRE Conference*, Paris 18–27, June paper 106, pp. 6–33.

3. Adamson, C. and Talkan, E. A. (1960) Selection of relaying quantities for differential feeder protection, *Proc. IEE*, **107A**, 37–47.
4. Solkor Protective Systems. Pamphlet 806, A Reyrolle and Co Ltd, 1954.
5. *Protective Relays – Application Guide* (1987) (3rd edn) GEC Measurements, Chapter 10.
6. High speed pilot-wire feeder protection Solkor-R and Solkor-Rf, Pamphlet published by NEI Reyrolle Ltd, 1990. Also *'Digital feeder protection – Solkor M'*, Data Sheet Solkor – M, 6/91.
7. Translay 'S' differential feeder and transformer feeder protection – Type MCBI, GEC Measurements, *publication R6011*.
8. Pilot supervision equipment for Solkor R and Solkor Rf pilot-wire feeder protection, *Data Sheet R/Rf SUP* (1985), NEI Reyrolle Ltd.
9. Supervision of AC pilot circuits – Relay type MRTP, *R-6026*, GEC Measurements.
10. Differential line protection with fibre optics-type LD91-F (1987), Asea Brown Boveri, *Publication CH-ES 63-54.10*.
11. Sun, S. C., Ray, R. E. (1978) A current differential relay system using fibre optic communications, *IEEE Transactions on Power Apparatus and Systems*, **PAS-102**, 410–419, February.
12. Takagi, T., Yamakoshi, Y., Kudo, H., Miki, Y., Tanaka, M., and Mikoshiba K. (1980) Development of an intrastation optical-fibre data transmission system for electric power systems, *Trans. IEEE*, **PAS-99**, 318–327, Jan.–/Feb.
13. Sugiyama, T., Kano, T., Hatata, M. and Azuma, S. (1984) Development of a PCM current differential relaying system using fibre-optic data transmission, *IEEE Transactions on Power Apparatus and Systems*, **PAS-103**, 152–159, January.
14. LFCB digital current differential relay, *Publications R-4054B and R-4028A*, GEC Measurements, England.
15. Kitagawa, M., Andow, F., Yamaura, M. and Okita, Y. (1978) Newly developed FM current-differential carrier relaying system and its field experience, *Trans. IEE*, **PAS-97**, 2272–2281.
16. Akimoto, Y., Matsuda, T., Matsuzawa, K., Yamaura, M., Kondow, R. and Matsushima, T. (1981) Microprocessor based digital relay application in TEPCO, *Trans. IEEE*, **PAS-100**, 2390–2398.
17. AIEE Committee Report (1961) Protection of multi-terminal and tapped lines, *Trans. AIEE*, **PAS-80**, pp 55–66.
18. Aggarwal, R. K. and Johns, A. J. (1986) The development of a new high speed 3-terminal line protection scheme, *Trans. IEEE*, **PWRD-1**, 125–134.
19. Aggarwal, R. K., Hussein, A. H. and Redfern, M. A. Design and testing of a new microprocessor-based current differential relay for EHV Teed feeders, 91 *WM 165-1 PWRD*.

10

Interlock and phase-comparison schemes for the protection of overhead transmission lines

INTRODUCTION

Protective schemes based on Kirchhoff's first law, namely those which determine the location of faults by comparing the instantaneous currents entering and leaving protected zones, were considered in Chapters 5–9. Their application to overhead lines and cables was described in Chapter 9, in which it was shown that there are factors which limit the lengths of circuits which may be protected when the comparisons are effected over interconnecting conductors (pilot wires). These limitations arise because of the need to convey continuous information with fairly high accuracy between the ends of protected circuits. To eliminate this need, alternative schemes were introduced in which only relatively simple signals need to be sent between the ends of protected zones, and because significant attenuation of these signals may be allowed it is possible to apply such schemes to very long circuits.

Interlock schemes which employ directional relays sited at each end of a protected zone initiate the opening of the circuit-breakers at each end of a line if both sets of relays indicate that currents are flowing into both its ends. Clearly, information must be conveyed between the ends of protected circuits but it is only of a yes–no nature, e.g. are the directional relays detecting current flow into the circuit or not? The magnitudes of the received signals in such schemes do not matter provided they are above particular threshold levels.

Phase-comparison schemes are somewhat related to Merz–Price schemes in that they compare the currents at the two ends of a protected circuit, but they do so by taking into account only the phases of the currents. Basically a signal is sent whilst the current at an end has a particular polarity, say positive, and this is compared with a similar signal at the other end. When the circuit is healthy the two signals will coincide because the currents at the two ends will be of similar phases whereas the phases will be considerably different if a fault is present within the protected circuit and in this event the signals will be displaced from each other. Clearly in these circumstances the opening of the circuit-breakers at the ends of the line must be initiated.

Whilst the principles involved in both the above forms of protection are simple, features have to be incorporated in practical schemes to both ensure that they will not operate incorrectly under certain system conditions and also minimize the amount of equipment required. As an example, it is desirable that only a single signal, rather than three, be transmitted between the ends of a three-phase line and therefore the most suitable signal derived from the three-phase currents should be used.

Practical schemes are considered in detail later in this chapter, after the following sections in which historical information is provided and the behaviour of long lines is examined.

10.1 HISTORICAL BACKGROUND

During the first decades of this century, the transmission of electricity was effected at relatively low voltages over fairly short distances and the amount of interconnection of networks was small. At that time most of the overhead lines and cables were protected by current-differential schemes based on the Merz–Price principles considered in the previous chapter. Various detailed improvements enabled these schemes to provide the fault settings and degrees of discrimination needed for the applications which arose.

The situation changed quite significantly in Britain, however, around 1930, with the advent of the 132 kV interconnection and transmission (grid) network. Much longer lines than those which had been used before were constructed and the fault-current levels were significantly greater than those previously encountered. Current-differential schemes, because of their limitations, which were examined in the previous chapter, were not suitable for some of the new circuits. Other schemes were therefore developed at that time, including a group which employed simple interlocking signals to obtain the necessary discrimination between internal faults on protected lines and other faults or healthy conditions. These schemes were applied not only to long lines but to shorter lines where simple signals could be transmitted over telephone circuits or specially provided pilot wires. In addition, schemes were produced in which high-frequency signals were transmitted via the conductors of the lines being protected and, of course, such schemes contained electronic equipment.

Interlock schemes discriminate by determining the directions of current flows in lines, i.e. by determining the phase relationships of the line currents and voltages. They therefore require inputs derived from both current and voltage transformers. These latter items are costly and therefore, when developments in electronic equipment made it possible, schemes were developed in which the phases of the currents at the two ends of a protected line were compared. Two such phase-comparison schemes, namely Telephase and Contraphase, were introduced around 1950 by A. Reyrolle & Co Ltd and Metropolitan-Vickers Electrical Co Ltd respectively. In both these schemes, which are considered in detail later in this chapter, the necessary comparisons were effected by transmitting high-frequency signals over the lines being protected.

10.2 THE CONSTRUCTION AND BEHAVIOUR OF TRANSMISSION LINES

Because of the great cost of high voltage underground cables, the relatively long-distance transmission of electricity is effected over bare conductors supported by string insulations which are suspended from towers. In many cases, two three-phase circuits are carried on the same set of towers, one circuit being supported on each side of a tower, as shown in Fig. 10.1(a). In other cases towers may only support a single three-phase circuit, in which case the conductors may be mounted in several configurations, the two most commonly used arrangements being shown in Fig. 10.1(b) and (c).

In all cases, conductors are run along the tops of the towers to which they are directly connected. In many installations, only a single earthed conductor is provided except near the ends of lines where it is common to run two, spaced apart in a horizontal configuration. In some cases, however, two conductors are used throughout the length of a line. These conductors are primarily to screen the phase conductors in the event of lightning storms in the vicinity of a line and it is well known that earthed conductors shield a zone between the vertical beneath them and a certain angle each side of that vertical. Two horizontally spaced earth conductors shield a larger zone than a single conductor and, as stated above, they are usually installed at the ends of lines to reduce the likelihood of direct lightning strikes close to vulnerable terminal equipment.

It must be recognized that such shielding may not be fully effective and that a strike may nevertheless occur to a phase conductor, thus raising its potential and then causing flashovers across supporting insulators as the surge propagates in each direction. Such events do not occur often and breakdowns result

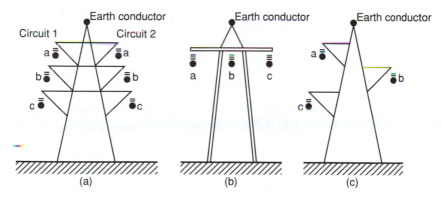

Fig. 10.1 Overhead line arrangements.

more frequently from lightning strikes to the overhead earth conductors. When such a strike occurs, the potential of the earth conductor is raised and surges propagate along it in both directions. When each of these surges reaches a tower a large current flows down the tower to earth and the voltage at the top of the tower is raised above earth because of the impedance of the tower to the surge. This can cause the potentials between the phase conductors and the top ends of their supporting insulators to be so high that flashovers occur across one or more of the insulators. Thereafter an arc or arcs may be maintained by the normal power-frequency voltages.

It will be clear from the above that both phase to earth and interphase faults can occur as a result of lightning. In addition, faults can occur due to mechanical failures in spite of the high standards to which overhead lines are designed and produced. Conductors do occasionally break due to overstressing, a condition which may arise when thick coatings of ice are present on them. Severe arcing may occur at a break in a conductor and the associated ionization could cause flashover to the other phase conductors. In addition, one or both parts of the broken conductor may fall onto the ground or on to other conductors, causing earth and/or interphase faults.

It will be clear from the above that faults must be expected on long transmission lines and indeed the numbers of faults are likely to increase with line length because of the increased amount of exposed conductor.

As was stated in the previous chapter, there is capacitance between the phases and to earth of three-phase overhead lines and there is also capacitance between the conductors of parallel circuits which are carried on the same towers. As a consequence, quite significant capacity currents flow between the phases and to earth on high voltage transmission lines and these increase with line length. Quite clearly the instantaneous currents entering individual phase conductors are not equal to those which leave at the other ends, and in steady state both the magnitudes and phases of the currents at the two ends of a line are different. Because of the series impedances associated with the conductors, the instantaneous voltages at the two ends of each phase conductor must also be different and, again, in steady state the pairs of voltages must be different in phase and/or magnitude.

The well-known steady state relationships between the quantities at the ends A and B of a long, healthy single-phase line are:

$$\bar{V}_A = \bar{V}_B \cosh \gamma l + \bar{I}_B \bar{Z}_0 \sinh \gamma l \tag{10.1}$$

$$\bar{I}_A = \bar{I}_B \cosh \gamma l + \frac{\bar{V}_B}{\bar{Z}_0} \sinh \gamma l \tag{10.2}$$

in which
$\bar{V}_A$ and $\bar{V}_B$ are the r.m.s. voltages at ends A and B respectively
$\bar{I}_A$ is the current flowing into end A
$\bar{I}_B$ is the current leaving end B

l	is the line length
γ	the line propagation constant
and $\overline{Z}_0$	the line surge (or characteristic) impedance.

If an interlock protective scheme incorporating directional relays were to be applied to the above single-phase line the magnitudes of the voltages and currents would not be important. The angular zones over which the relays were set to operate would, however, have to be such that both relays would never indicate that currents were entering both ends of the line unless there was a fault on it. To ensure that this condition would be satisfied, the displacements between the voltages and currents at each end of the line, i.e. between $\overline{V}_A$ and $\overline{I}_A$ and $\overline{V}_B$ and $\overline{I}_B$, would have to be examined for healthy and external fault conditions.

Should a phase-comparison scheme be applied to a single-phase line then the only quantities of significance would be the phases of the currents $\overline{I}_A$ and $\overline{I}_B$ and clearly their phase displacements, when the line was healthy, should never be in the range in which tripping of the circuit-breakers would be initiated. In addition, the phase displacements possible during all internal faults should ideally cause operation.

In practice, long overhead lines are invariably three-phase and their protective schemes must be set so that the principles outlined above will obtain under all conditions. The behaviour of overhead lines is considered in more detail in sections 10.3.1 and 10.5.1, in which details of particular protective schemes are provided.

10.3 FEATURES OF INTERLOCK PROTECTIVE SCHEMES

As indicated earlier, relays or detectors are installed at each end of a protected line to determine the directions of current flow in the conductors and, when necessary, signals are sent between the line ends to ensure that the appropriate actions are taken.

In the following sections the relaying and signalling arrangements are considered.

10.3.1 Relaying arrangements

The construction and behaviour of electro-mechanical directional relays was considered in some detail in section 4.2.4 (page 113). These relays must be energized by both voltages and currents and basically they operate when the phase displacements between the voltages and currents supplied to them are within certain ranges. As an example, a relay may operate when the voltage and current supplied to it are displaced from each other by $\pi/2$ rad or less and it will then restrain for other phase displacements, i.e. those greater than $\pi/2$ rad.

Induction-type electro-mechanical relays tend to operate over an angular range of about π rad, the actual values for particular relays being somewhat affected by the magnitudes of the voltages and currents supplied to them. Relays which use more complex circuitry and electronic processing can provide operating ranges of any desired value and these may not be affected by the magnitudes of the voltages and currents.

All types of relays may have their operating ranges positioned asymmetrically; for example, a relay could be set to operate for currents with displacements between $\pi/3$ rad leading and $2\pi/3$ rad lagging the voltages applied to it.

The directions of currents associated with three-phase overhead lines may be detected in several ways and these are considered below.

Basic detection arrangements

To detect each of the types of fault which could occur on or external to a three-phase line, six directional relays could be provided at each of its ends, i.e. three to detect phase to earth faults and three to detect interphase faults.

It was shown earlier in section 4.5 (page 135) that earth fault directional relays supplied with voltages and currents from the same phase, e.g. V_a and I_a, are unable to discriminate correctly when faults occur near the end of a line at which they are sited, because the voltages supplied to them may then be too low. Clearly phase-fault relays are also unable to discriminate correctly for faults near to them if they are supplied with voltages from the phases they are monitoring, i.e. if a voltage proportional to that between phases a and b of a line was supplied to the relay required to detect faults between these phases it might operate incorrectly for close-up faults.

In practice, therefore, the voltage supplied to each relay is obtained from a phase or phases not associated with the faults which it is to detect. As an example, the 'a' phase earth fault relay could be energized by a voltage proportional to that between the phases b and c of the line being protected.

Whilst the cost and complexity associated with the use of six directional relays at each end of a long overhead line will not usually be prohibitive, a reduction in the number of relays required in a scheme is clearly desirable provided that performance is not impaired. Such a reduction could be achieved by using a single relay to detect the directions of all earth faults. The relay would be energized by the sum of the outputs of the current transformers in the three phases and the sum of the phase voltages, this latter quantity being obtained from either open-delta-connected tertiary windings on the main voltage transformers or from an extra auxiliary transformer. Such an arrangement is shown in Fig. 10.2(a) and the relay quantities obtained thereby during internal and external fault conditions are shown in Fig 10.2(b). It will be clear that a directional relay could be set to discriminate between the two conditions.

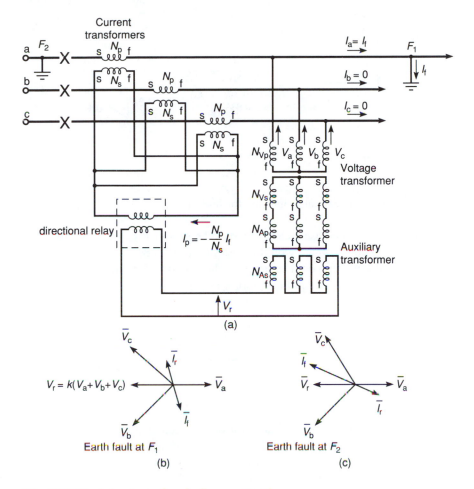

Fig. 10.2 The behaviour when faults are present.

As an alternative a single relay could be used to determine the directions of both a phase to earth fault and an interphase fault. As an example, a relay could be supplied with inputs proportional to the 'c' phase voltage and the 'a' phase current of a line. By setting the relay to operate in a suitable angular zone it could be made, as shown later, to detect the directions of faults to earth on phase a and faults between phases a and b of a protected line. It will be clear that this practice would enable schemes to be produced with only three relays at each end.

Several other methods could be used to reduce the number of relays required in a scheme. A single relay could be used at each end of a protected circuit if current transformers of different ratios were employed or, alternatively, summation transformers connected to current transformers, each of the same ratio,

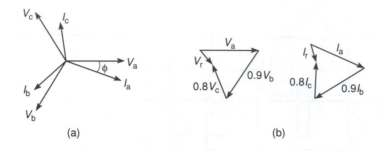

Fig. 10.3 Voltage and current summation for internal fault.

could be used. In each case a current would be provided to energize the relay and a corresponding voltage would have to be provided.

The relay current (i_r) would be equal to the sum of different fractions of the currents in the phases of a protected line, i.e.

$$i_r = k_1 i_a + k_2 i_b + k_3 i_c \qquad (10.3)$$

The outputs of the voltage transformers would also be summated to provide a voltage (v_r) to energize the relay. The same proportions of the phase voltages would be summed, i.e.

$$v_r = k_1 v_{an} + k_2 v_{bn} + k_3 v_{cn} \qquad (10.4)$$

In the event of a three-phase fault on a protected line the voltages at the ends of the line would be depressed and the currents could be well above the rated value for the line. The phases of the currents would lag those of the respective voltages by angles of φ, as shown in Fig. 10.3(a) and the outputs from the summation processes, based on equations (10.3) and (10.4), using values of $k_1 = 1$, $k_2 = 0.9$ and $k_3 = 0.8$ would be as shown in Fig. 10.3(b). It will be seen that the output current (I_r) would lag the relay voltage (V_r) by the angle φ, i.e. by the displacement between the phase voltages and their associated currents.

In the event of a three-phase fault external to a line, at end A as shown in Fig. 10.4(a), the phase voltages could be depressed and the currents I_a, I_b and I_c would be of the opposite polarities to those above, as can be seen from Fig. 10.4(b). The summated current output (I_r) would then lag the output voltage (V_r) by ($\pi + \varphi$) rad. As a result a single directional relay could be set to restrain for this condition but operate for the internal fault condition considered above.

Similar behaviour would be obtained for other types of faults. As an example, the conditions which would obtain during internal and external faults to earth on the 'a' phase are shown in Fig. 10.5. In both cases the voltage of

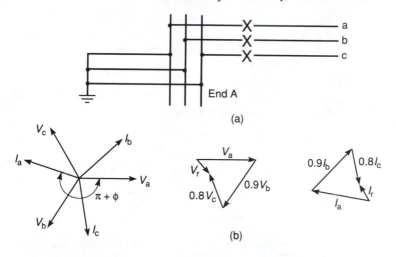

(a)

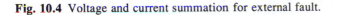

(b)

Fig. 10.4 Voltage and current summation for external fault.

the 'a' phase at end A could be depressed and the currents for the two faults would be approximately in antiphase with each other. A directional relay could clearly be set to differentiate between the two conditions.

Summations of the above type produce outputs from the positive-, negative- and zero-sequence components of the voltages and currents present during fault conditions. The outputs caused by zero-sequence components of a given magnitude are much greater than those produced by the other sequences, as shown in Fig. 10.6. The actual outputs are dependent on the ratios k_1, k_2 and k_3 and thus the turns in the primary winding sections when summation trans-

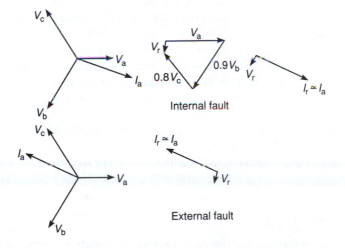

Fig. 10.5 Conditions when single-phase to earth faults are present.

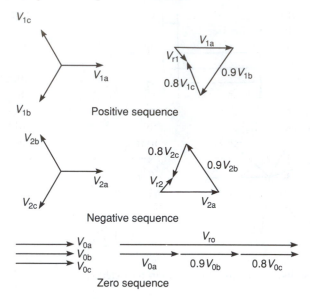

Fig. 10.6 Sequence outputs of a summation transformer.

formers are used. Whatever ratios are used, however, the sensitivities to positive- and negative-sequence inputs are equal to each other.

Clearly, the same performance could be obtained from sequence networks connected to current and voltage transformers of the same ratio in each phase. In addition, however, the outputs obtained from quantities of a given magnitude of each of the sequences could be different from each other and, of course, zero output could be obtained from one or more of the sequences if desired. Outputs could be obtained for all types of faults from positive-sequence networks but the outputs provided by such networks during phase to earth faults are only one third of those obtained for three-phase faults of the same current or voltage magnitudes. The resulting low sensitivity to earth faults is normally unacceptable and therefore it is usual to employ networks which provide relatively low outputs when their inputs are of positive-sequence but higher outputs when the inputs are of zero sequence, e.g. $I_r = k_1 I_1 + k_0 I_0$, k_0 being larger than k_1.

Whatever methods are used and however many directional relays are used, the angular operating and restraining zones of the relays must be such that they will discriminate between faults within protected zones and those external to them.

Relay settings

On very high voltage transmission networks, the neutral points are usually earthed solidly and in these cases earth fault currents lag their phase voltages by angles (φ) approaching $\pi/2$ rad. A typical condition during a short circuit to earth on phase a of a line is shown in Fig. 10.7. The voltage of the faulted phase (V_a) could be low but the voltage between phases b and c (V_{bc}) would be unaffected.

In a protective scheme with three earth fault and three phase-fault induction-type relays at each end, the earth fault relays could be set to provide maximum torque for short-circuits to earth, e.g. maximum torque would be provided when I_a lagged V_a by φ rad. As these relays would have angular operating zones approaching π rad, boundary conditions would exist near the line PQ in Fig. 10.7. Should the 'a' phase relay be supplied with a voltage proportional to that between the healthy phases (V_{bc}), it would have to be set to operate whenever the fault current (I_a) lead the voltage V_{bc} by an angle less then ($\pi - \varphi$) rad or lagged by an angle less than φ rad.

The phases of the currents relative to the input voltages of a line on which an interphase short-circuit is present are dependent on the line impedance between its input and the fault. As a result, fault currents lag the corresponding voltages by angles (φ) approaching $\pi/2$ rad, i.e. for a fault between phases a and b, the current (I_a) will lag the voltage V_{ab} by almost $\pi/2$ rad and the current I_b will be in antiphase with the current I_a as shown in Fig. 10.8(a) and (b). Again the voltages between faulted phases could be very depressed and therefore the relays are usually energized with the voltage of the unaffected phase. As an example, the relay which is to detect faults between phases a and b could be supplied with a current proportional to that in the 'a' phase of the line (I_a) and a voltage proportional to that of the 'c' phase of the line (V_c). In these circumstances the operating angular zone of the relay would be positioned as shown in Fig. 10.8(c).

It will be appreciated that all the voltages may be depressed in the event of a symmetrical three-phase short-circuit near the end of a line and, in these circumstances, the above cross-polarizing measures will clearly not be effec-

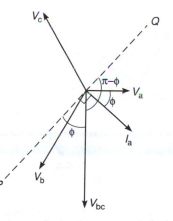

Fig. 10.7 Conditions when a single-phase to earth fault is present.

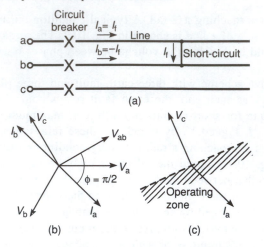

Fig. 10.8 Conditions when an interphase fault is present.

tive and relays might not operate. This situation might be thought acceptable in some applications because of the unlikelihood of such faults. If this is not so, however, some form of memory action, such as tuned circuits, could be included to ensure that the voltages supplied to the relays will not collapse suddenly when faults occur, but decay slowly, thus enabling the relays to operate if necessary.

It was stated in the previous section that the number of directional relays needed at each end of a scheme could be reduced from six to three by so setting each of the relays that it will determine the directions of both a phase to earth fault and an interphase fault. As an example, a relay could be supplied with inputs proportional to the 'c' phase voltage and the 'a' phase current of a line. In the event of a short-circuit to earth on the 'a' phase of the line, the fault

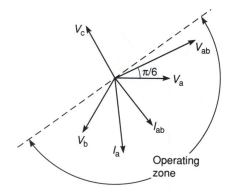

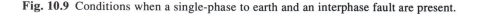

Fig. 10.9 Conditions when a single-phase to earth and an interphase fault are present.

current (I_a) would lag the 'a' phase voltage by an angle approaching $\pi/2$ rad, as shown in Fig. 10.9. Should a fault occur between phases a and b, the fault current (I_{ab}) would lag the line voltage (V_{ab}) by a similar angle. The angle between the 'a' phase currents for the two faults would be of the order of $\pi/6$ rad and the operating zone of the relay could be set to operate for both of them, as shown in Fig. 10.9.

When a single relay fed with summated quantities is used at each end of a scheme, suitable angular operating zones can be readily provided.

Relay sensitivity

In the previous section, the phase displacements between the currents and voltages associated with transmission lines during short-circuit conditions were considered and it was clear that the displacements when faults occur within a protected zone are very different to those encountered when faults are present on or beyond the busbar near the relaying position. As a result, the angular operating zones of directional relays can be set so that correct discrimination will always be achieved for such conditions.

It must be recognized, however, that directional relays may not operate in ways which will enable internal faults of limited current magnitudes to be detected. To illustrate this in a simple manner, the behaviour obtained in the event of a resistive fault to earth on a single-phase line is examined below.

To further simplify the treatment it is assumed that the fault is at the mid-point of the line, as shown in Fig. 10.10(a) and that the line conductor has no resistance. For the short-circuit condition ($R_f = 0$), the currents I_A and I_B would flow. The current I_A would lag the voltage V_A by $\pi/2$ rad, whereas the current I_B would lead the voltage V_B by $\pi/2$ rad, as shown in Fig. 10.10(b) and the currents would be given by:

$$I_A = -j\,V_A/x \quad \text{and} \quad I_B = j\,V_B/x$$

The directional relays would be set so that the presence of an internal fault would be detected.

In the extreme case of a fault with a resistance approaching infinity ($R_f \to \infty$), the currents I_A and I_B would be almost the same, i.e.

$$I_A \simeq I_B \simeq -j\,\frac{V_A - V_B}{2x}$$

This condition is shown in Fig. 10.10(c). It is very different from that shown in Fig. 10.8(b) and the relays would not indicate an internal fault.

Conditions as the fault resistances increase from zero are shown in Fig. 10.10(d), from which it can be seen that the phase displacement at end B would increase from $\pi/2$ rad as the fault resistance rose from zero to a peak value of about 2.8 rad. Thereafter the phase displacement would fall with increase of

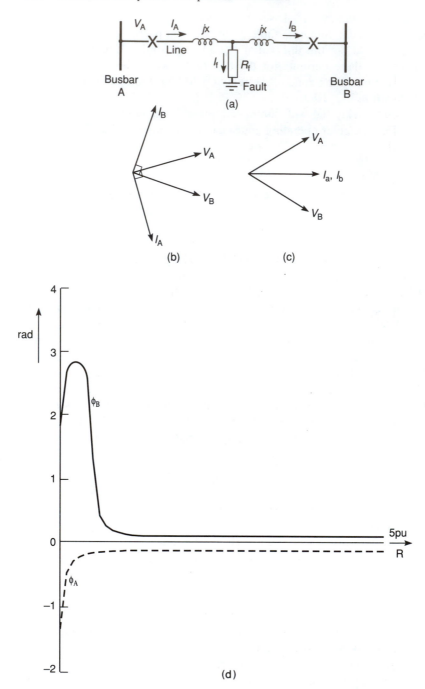

Fig. 10.10 Conditions when resistive faults are present on a single-phase line. **(a), (b), (c), (d)**

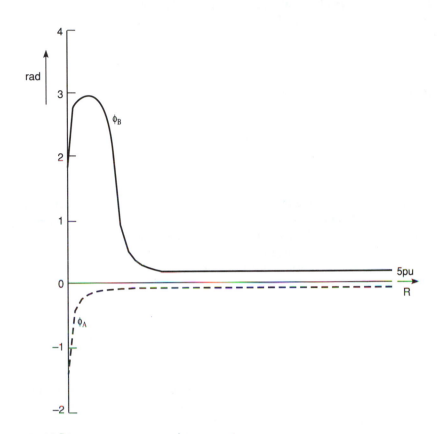

Fig. 10.10 (e)

fault resistance, the change for resistances between 1 pu and infinity being small.

At end A, however, the displacement decreases continuously from a value of about 1.3 rad. when the fault resistance is zero.

Similar results would occur for faults at a point quarter the way along a line, as is shown in Fig. 10.10(e) and it is therefore clear that directional relays cannot be set to detect internal faults with resistances above certain levels, i.e. internal faults of current levels below particular values would not be detected.

It will be appreciated from the preceding sections that several factors have to be considered when the operating zones of directional relays are being selected for particular applications. Because of their experience, manufacturers should be able to supply interlock schemes which incorporate relays with settings which will enable given lines to be adequately protected. If desired, however, the suitability of the proposed operating zones of relays could be determined by using either the three-phase long-line equations or approximate methods to calculate the phase relationships between the various voltages and currents which could be present during healthy conditions and also those possible during both internal and external fault conditions.

10.3.2 Interlocking signals

If an interlock scheme associated with a transmission line has only one directional relay at each of its ends, then each relay should detect that current is flowing into its end when a fault is present within the protected zone. In these circumstances, opening of the circuit-breakers should be initiated.

When the line and the network of which it forms a part is healthy or when faults are present outside the zone covered by the line protective scheme, one relay should detect current flowing into the line, whereas the other relay should detect current leaving the line. For these conditions the line circuit-breakers should not be opened.

The above performance could be obtained by using one of the following signalling arrangements.

Permissive signalling

In this method, also referred to as the transfer trip method, a signal is sent when a directional relay detects current flowing into its end of a line. On receipt of this signal at the other end of the circuit, tripping of the circuit-breaker at that end is initiated if its directional relay also detects current flowing into the line.

With this method, signals have to be sent from each end of a protected line when a fault occurs on it so that tripping of the circuit-breakers at each of the ends is initiated.

A disadvantage of this method is that the non-arrival of the signals would prevent an interlock scheme from initiating the opening of the circuit-breakers when internal faults were present. In these circumstances, reliance would then have to be placed on the operation of other protective equipment or an alternative method of permitting tripping, such as that described later in section 10.4.2, must be used.

Because current normally flows into one end of a healthy line a signal would be transmitted continuously unless the directional relays were prevented from operating at currents up to levels somewhat above the rated values. It is clearly desirable, however, that signals should only be sent when faults are present and therefore current-sensitive starting relays must normally be provided. The settings of these relays are considered later in section 10.3.3 ('Permissive signalling').

Blocking signalling

In this method, signals are sent to prevent or block the tripping of circuit-breakers when necessary. These signals are transmitted when a directional relay detects that current is leaving its end of a protected circuit. Clearly this method has the advantage that the non-arrival of a signal cannot prevent circuit-breakers being opened when internal faults are present, but it could cause a circuit-breaker at the end of a healthy line into which current is flowing, to be allowed to open incorrectly.

Under healthy conditions, current must be flowing out of one end of a protected line and a blocking signal could be sent out continuously, and indeed such a signal would be required to prevent the directional relay at the other end of the line from initiating the opening of its associated circuit-breaker. This is clearly undesirable because any interruption of a blocking signal could cause the opening of a circuit-breaker. For this and other reasons which are considered later, current-operated starting relays must be included in interlock schemes which use this type of signalling.

These considerations have been based on schemes with a single directional relay at each end of a protected zone. Most schemes do, however, have several relays at each end. In these cases permissive signals must be sent when any directional relay detects current flowing into its end, whereas, when blocking signals are employed, a signal must only be sent when all the directional relays at an end detect currents leaving the protected zone.

10.3.3 Starting relays

It was made clear in the previous section that the signals sent between the ends of lines to either permit or block the tripping of circuit-breakers should not be sent continuously but only when faults are present. Extra relays, which are usually referred to as starting relays, must therefore be provided to detect abnormal conditions, namely the presence of both phase to earth and inter-phase faults on or external to a protected line. Interphase faults can be detected by including simple overcurrent relay with settings above the full load level of the line. These need only be provided in two of the phases, say 'a' and 'c', at each end of a line. Because the current flowing in the event of an earth fault may be restricted, it is necessary that earth fault relays should have sensitive settings, say 0.2 pu, and a single relay carrying the sum of the outputs of the current transformers in the three phases of the line should be included at each end of a scheme.

Should a scheme incorporate summation transformers then they could feed starting relays which would operate above certain output levels.

Because the arrangements required with permissive and blocking signalling are different, they are considered separately below.

Permissive signalling

The operation of one or more of the starting relays at an end of an interlock scheme should permit an interlocking signal to be sent from that end if a fault is detected in the appropriate direction.

With this arrangement, the opening of the circuit-breakers at both ends of a line will be initiated when currents above certain levels are fed into both its ends. Should the current infeed to one end of a line be below the level needed to operate one or more of the starting relays, then neither of the circuit-breakers would be opened even though the current at one end of the line could be very high. Because of this limitation, this type of signalling should only be applied to lines in networks which are so interconnected that large currents will flow into both ends of a faulted line or some alternative provision should be made for such situations.

Blocking signalling

When this method of signalling is used it is imperative that the directional relays at the end of a line to which a signal is being sent should not be allowed to initiate the tripping of their associated circuit-breaker at a current level

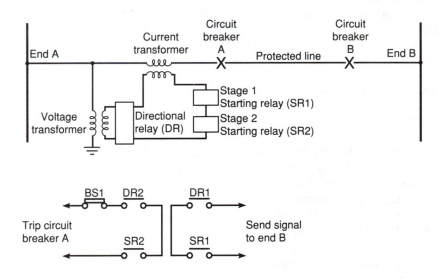

Fig. 10.11 Blocking signalling. Contact BS1 closes when a blocking signal is received from end B. Contact DR1 closes when current flows from the line to the busbar at end A. Contact DR2 closes when current flows into the line from the busbar at end A. Contact SR1 closes when fault currents are above the Stage 1 level. Contact SR2 closes when fault currents are above the Stage 2 level.

below that at which the signal may be sent. To meet this requirement and ensure that a margin of safety will exist, relays, termed Stage 1, should be provided to allow signals to be sent and a second set of starting relays, with higher current settings than the Stage 1 relays, must be provided at each end of a line. One or more of these latter relays, which are termed Stage 2 starting relays, must operate at an end of a line before the circuit-breaker at that end may be tripped.

A typical arrangement of the circuitry needed at each end of a line to ensure correct operation is shown in Fig. 10.11. It will be seen that the receipt of a blocking signal must cause the trip circuit to be opened, i.e. contact BS1 must open. It is essential that the trip circuit should not be completed before the blocking signal has had time to arrive and to achieve this an adequate time delay must occur before the Stage 2 relay allows tripping to be initiated. This requirement prevents such schemes from operating very rapidly in the event of internal faults including those of high current magnitudes.

This arrangement does, however, have the advantage that should a fault on a line be fed from only one end, then it will be cleared by opening the circuit-breaker at that end.

10.3.4 Signalling channels

Because of the simple nature of the information which must be transferred between the ends of interlock schemes, a wide range of signalling channels may be employed. These are briefly considered below.

Interconnecting conductors (pilot wires)

Signals may clearly be sent over pilot wires specially provided for the purpose. This practice may be thought desirable when major lines are to be protected and in such cases a relay or electronic circuitry may be used to detect the arrival of signals and then tripping of the circuit-breaker at the receiving end will be either allowed or prevented depending on the type of signalling being employed.

When less important lines are to be protected, signalling may be effected over conductors used for telephony. In such cases, operation of starting relays must cause the conductors to be available for the transmission of interlocking signals if necessary. This requires special arrangements with the telephone company so that transmission is always available within an acceptable time delay. An analysis of pilot-relaying performance requirements can be found in reference [1].

The conductors of protected lines

To avoid the expense of providing special interconnecting conductors for major long lines, signals may be transmitted over the conductors of the lines

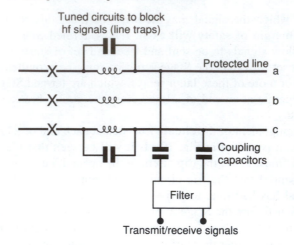

Fig. 10.12 Coupling equipment for PLC signalling.

being protected by interlock schemes. This method of transmission is referred to as Power Line Carrier, often abbreviated to PLC.

Relatively high-frequency (typically 30–300 kHz) signals are transmitted. These must be produced by electronic circuits including oscillators and amplifiers and they must be fed to and from two of the phase conductors of a line via coupling equipments turned to the signal frequencies. A typical arrangement is illustrated in Fig. 10.12.

The power supplies to the transmitting and receiving equipment used in interlock schemes must be secure and must not be affected by faults on or near the lines being protected.

When a scheme depends on blocking signals to obtain stability, correct operation will be obtained when faults occur on circuits connected to the line it protects, i.e. external faults, because the signals will be transmitted satisfactorily over the line. In the event of faults occurring on the protected line, blocking signals are not required and therefore the presence of the faults in the signalling channel is of no consequence.

When a scheme employs permissive signalling, however, signals must be received satisfactorily when faults are present on the line being protected. This requirement may not be met under all conditions and therefore permissive signalling over protected lines is not usually acceptable.

A discussion of the transmission behaviour of power lines at PLC frequencies can be found in reference [2].

Optical-fibre links

Optical-fibre links are now employed for the transmission of interlocking signals, and because such links are separate from the conductors of lines being protected, they are not directly affected by system faults. Both permissive and blocking signals can therefore be sent over these links. Again, reliable power supplies must be provided for the equipment associated with this method of signalling. Information about an interlock scheme employing optical links is provided later in section 10.4.2. A review of protective relaying using fibre-optic communications may be found in reference [3].

Microwave links

Line-of-sight communication links may be established using microwave signals at frequencies ranging from 1 to 10 GHz. Such links can rarely be justified for protective purposes alone and thus when they are used they perform a number of protection, monitoring and control functions.

A summary of the various protection signalling methods is given in reference [4].

10.4 INTERLOCK PROTECTIVE SCHEMES

A variety of interlock protective schemes have been applied to overhead lines over the years. One of the first, which was produced by A. Reyrolle and Co Ltd, was referred to earlier in section 10.1. It was designated as 'interlock protection' and details of it are provided below and then information about a recently developed scheme is provided.

10.4.1 Reyrolle interlock protective scheme

This scheme was described in papers [5, 6] presented at the International Conference on Large Electric High-Tension Systems held in Paris in June 1931. The diagram which was included to illustrate the scheme in the second of these papers is reproduced as Fig. 10.13.

It will be seen that the scheme, which is shown in single-phase form for simplicity, incorporated directional relay elements (1), starting relays (2) and interlocking relays (4). The interlocking relays (4) opened the tripping circuits when they operated and therefore the scheme employed blocking signals which were sent over pilot wires normally used for telephony. To meet the requirement that time must be allowed for blocking signals to arrive before tripping can be initiated, which was referred to in section 10.3.3 (page 361), the tripping elements (7) operated after a time delay of 0.3 s.

10.4.2 Modern schemes

Modern interlock schemes incorporating directional relays are based on relaying quantities derived from incremental voltage and current signals. Incremental signals Δv and Δi are obtained after removal of the prefault

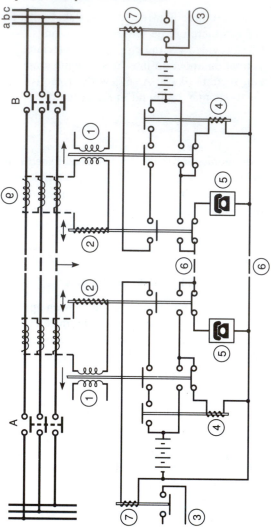

Fig. 10.13 Diagram illustrating the 'Interlock' protective scheme. (1) Three-phase directional stabilizing element. (2) Three-phase over-current element. (3) Trip circuit. (4) Interlock relay. (5) Telephone. (6) Pilot. (7) Tripping element with time lag. (Reproduced from Leeson and Leyburn, 1931, *CIGRE Conference Paper 106* with the permission of CIGRE.)

components and their relative polarity indicates the direction of propagation of the disturbance caused by the fault. Comparison of this information obtained at the two ends of the protected line can be used to discriminate between internal and external faults. Practical implementations of these ideas are described in references [7, 8, 9] and the schemes are examined in more detail in Chapter 12.

A directional earth fault comparison relay 75N21, produced by Siemens, is supplied with the sum of the outputs of the three line current transformers and either a voltage proportional to the sum of the three line voltages or a polarizing current obtained from the station transformer neutral. The relay has a two-stage (1 and 2) zero-sequence current detector. The first, more sensitive stage initiates the directional measurement whilst the second stage allows circuit-breaker tripping to be initiated. Current transformers with secondary windings rated at 1 A or 5 A may be used and Stage 2 settings of 0.1 to 0.85 pu in 0.05 pu steps are available, i.e 0.1–0.85 A or 0.5–4.25 A. Angular operating zones of π rad, i.e. $\pm \pi/2$ rad, are provided, and operating times are 25 ms or more depending on the fault conditions.

Permissive signals are employed with this scheme and a special feature is provided so that should a low current be fed into one end of a line to an internal fault then tripping of the other end nevertheless will be initiated, if the fault current is high enough, after the receipt of an echo of the signal sent out by that end.

This scheme, which provides very sensitive earth fault settings, has been developed to detect high-resistance faults on transmission lines and it is intended to be used with other schemes which will detect the high-current interphase faults.

Clearly, both the above schemes must be fed from current transformers capable of faithfully reproducing the waveforms of their primary currents.

10.5 FEATURES OF PHASE-COMPARISON PROTECTIVE SCHEMES

The principle involved in comparing the phases of the currents at the two ends of a line is related to that of comparing the instantaneous values of the current, as in Merz–Price type protective schemes. It is also related to the directional-interlock principle because the phase of the current at an end of a line changes significantly when the direction of the current flow reverses to feed a fault on a line.

Phase-comparison schemes have an advantage over current-differential schemes in that only information about the phases of the currents must be transmitted between the ends of lines and therefore attenuation of signals is acceptable. They also have an advantage over directional-interlock signals in that they do not require information about the line voltages and therefore voltage transformers are not needed to supply them.

As stated earlier in section 10.1, phase-comparison schemes in which high-frequency signals were transmitted along the conductors of protected lines were introduced about 1950, by which time suitable electronic equipment was available.

Several factors which affect phase-comparison schemes are considered in the following sections.

10.5.1 The phase displacements of line currents

If there was no capacitance or leakage resistance between the conductors of a healthy transmission line, then the currents at the two ends of each of its phases would be the same at every instant and therefore they would be in phase with each other.

If either a phase to earth or an interphase short circuit was present on such a line this condition would not hold and the phase currents would be displaced from each other by considerable angles. As an example, the currents for a short circuit to earth at the mid-point of one phase of a line would be almost in antiphase with each other.

The phase-displacements between the currents in the event of highly-resistive faults could however be quite small as was shown earlier in section 10.3.1 and Fig. 10.10.

In practice, of course, significant capacitance is present between the phase conductors of long lines and therefore, although the leakage resistances are very high, the instantaneous currents in each phase at the two ends of lines are not the same. As an example, the capacitive current per phase of a British 400 kV line, 100 miles long, is about 140 A and the resultant phase displacement between the currents at the two ends when operating at rated current is approximately 0.036 rad.

It will be clear from the above that phase comparison schemes must be so arranged that they will not initiate tripping of the circuit-breakers when the phase displacements are less than a certain set value but that circuit-breaker opening will be initiated when displacements exceed the set value.

The phase-displacement settings provided by particular manufacturers are quoted later in section 10.6 and experience has shown that these provide satisfactory discrimination and also enable internal faults above certain current levels to be detected. If it is felt to be necessary the phase displacements which may occur in particular applications may be calculated using either the full three-phase long-line equations [(10.1) and (10.2)] or simpler approximate equations and in this way the suitability of proposed angular settings may be determined.

10.5.2 The production of comparison signals

Comparison of the phases of each of the currents at the ends of a line would require three separate signals to be sent in each direction, and whilst this could be done it is clearly preferable that only a single signal should be transmitted, when necessary, from each end.

Summation processes using current transformers of different ratios or summation transformers fed from current transformers of the same ratios, as described earlier in section 10.3.1, could be used to provide a single output at each end of a line. It has, however, been the usual practice to use symmetrical-component sequence networks for this purpose. The sensitivities to the separate components of the networks used in practical schemes are quoted in section 10.6.

10.5.3 The comparison process

During healthy conditions the currents at the opposite ends of each of the phases of a line are almost in phase with each other, whereas they are displaced by large angles when internal faults are present. These conditions are illustrated in Fig. 10.14(a). If the current transformers are connected in the reverse direction at one end relative to that at the other or the outputs from the sequence networks are taken in reverse directions, then these outputs will be almost in antiphase with each other when a line is healthy, as shown in Fig. 10.14(b), and, of course, the outputs will be almost in phase with each other when internal faults are present.

As a result, if a signal is produced at end A of a line whenever the sequence-network output at that end is positive and a signal is generated at the other end B when the output there is positive then the receipt at end A of this latter signal from end B together with the locally derived signal at end A will provide a continuous signal when the line is healthy and this can be used to prevent tripping of the circuit-breaker at end A. Clearly, a similar process would occur at the second end of the line (B). These conditions are illustrated in Fig. 10.14(c).

During internal faults, the two signals overlap, as shown in Fig. 10.14(d), and there are significant periods when there is no resultant signal and this condition can be detected and used to initiate the opening of the circuit-breakers.

In practice, the phase currents at the two ends of long healthy lines are not exactly in phase and therefore there will be a short break each cycle in the signals at each end with the arrangement described.

In addition, if the outputs of the sequence networks have to reach a particular positive level before signals are generated and transmitted, then the durations of each of the high frequency signals will be less than the half period of the power system quantities, as shown in Fig. 10.15. As a result there will be short breaks each half cycle in the signals fed to the receivers at each end of a scheme and these breaks will be longer the lower the outputs of the sequence networks, i.e. the effect will be more pronounced the lower the current flowing in the line.

To allow for the above effects, the detection process must be so arranged that circuit-breaker tripping is only initiated when the breaks exceed a certain period, say a period corresponding to $\pi/12$ rad of the power-frequency cycle.

It will be clear that the above results may be produced in other ways. As an example, the current transformers and sequence network output-windings could be connected in the same directions and signals could be derived at one

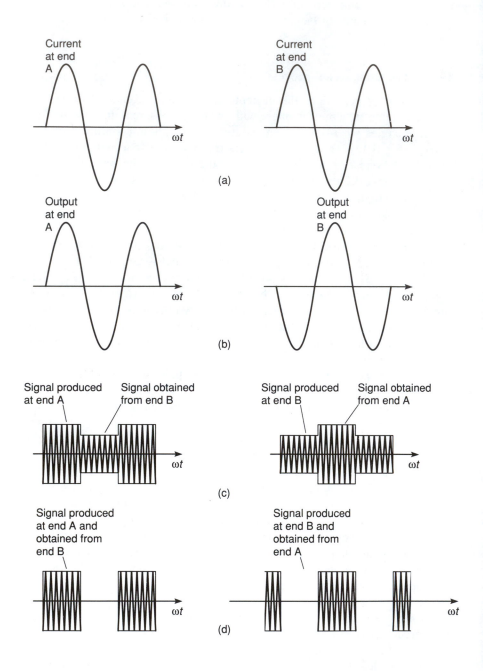

Fig. 10.14 Conditions when faults are present on a line and external to it.

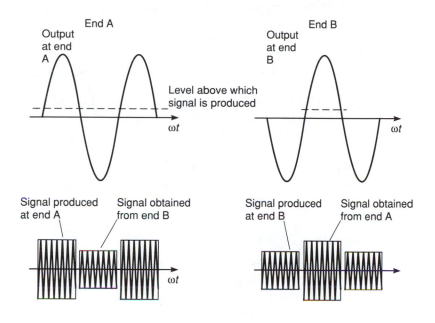

Fig. 10.15 Conditions obtained on a healthy line.

end when the network output was positive whilst signals would be sent from the other end when the corresponding output was negative.

10.5.4 Comparison signals

These schemes were developed for application to long lines and they were designed so that the signals could be transmitted along the conductors of two of the phases. Initially high frequency signals in the frequency range 80–500 kHz were used.

A signal must be sent from each end of a line to the other end and two frequencies sufficiently far apart must be used in a scheme so that filters of reasonable bandwidths may be provided to pass the signals being transmitted and received at each end without interference occurring. As an example, the signals could be at frequencies of 140 kHz and 148 kHz.

During period of transmission, i.e. when the output of a sequence network is positive, the transmitted signal is of constant amplitude.

10.5.5 Signalling equipment

It will be clear that an oscillator and amplifier must be provided at each end of a line to produce the outgoing signals and a receiver is needed to feed incoming signals to the comparison circuits. Coupling equipments to enable signals to be both injected into and received from a protected line are required as also are the filter circuits referred to above.

These items have varied in form over the years and details of the equipments provided by some manufacturers are provided in the later sections of this chapter.

It will be appreciated that alternative methods of signalling may be used and certainly phase-comparison schemes could operate using pilot wires or optical fibres to link the ends of lines although such links would often be very costly. VHF radio links could also be used.

Whatever equipment is employed, reliable power supplies must be provided to ensure that signals will be transmitted whenever they are required to ensure that correct operation of the protective scheme will be obtained.

10.5.6 Starting relays

As with interlock protective schemes, it is undesirable that signals should be sent continuously and therefore starting relays are necessary.

To ensure that the opening of a circuit-breaker will not be initiated because a signal has not been received from the other end of a healthy protected line, Stage 1 and Stage 2 relays must be provided at each end of a scheme. These relays, which are fed from the sequence networks, are current-operated, the Stage 2 relays having the higher settings.

Operation of the Stage 1 relays permits the comparison signals to be sent and the operation of the Stage 2 relays allows the circuit-breakers to be tripped if necessary. It is also desirable that the Stage 1 relays should operate slightly faster than the Stage 2 relays.

10.5.7 Current transformers

Because phase-comparison schemes operate when there are significant breaks in the signals sent to their receivers it is essential that such breaks should not be produced as a result of the distortion of the output waveforms of current transformers, because of saturation of their cores when large currents are flowing due to faults on circuits connected to them.

Current transformers must therefore be so designed that they will not saturate during the transient conditions which may occur on the circuits in which they are to be used. Protective equipment manufacturers specify the minimum knee-point voltages of the current transformers which may be used with their schemes and examples are quoted in section 10.6.

10.6 PHASE COMPARISON SCHEMES

Details of a number of schemes are provided in the following sections.

10.6.1 Telephase protective schemes

This scheme was developed jointly by A. Reyrolle and Co Ltd, and GEC (Telephones) Ltd, prototype equipments [10] being produced in 1945. After these were operated on an experimental basis, a final production version was applied to one of the transmission lines between Loch Sloy Hydro-Electric Station and Windyhill Switching Station, north of Glasgow in 1950. This line was about 40 miles in length.

The equipment, which was supplied by current transformers with secondary windings rated at 1 A, incorporated sequence networks. The connections were such that their outputs were in antiphase with each other when the phase currents at the ends of the protected line were in phase with each other.

The sequence networks produced voltage outputs given by:

$$V_{out} = k \left(I_{1a} + 8 I_{2a} \right)$$

in which I_{1a} and I_{2a} are the 'a' phase positive- and negative-sequence currents respectively and k is a constant.

The general arrangement of the equipment was as shown in Fig. 10.16(a). When the output of a network exceeded the level needed to operate its associated Stage 1 relay, the high frequency signal produced in the oscillator was modulated by the output signal of the sequence network, and a high frequency signal of constant amplitude was injected into the line conductors, via coupling equipment, during those periods when the output of the sequence network was positive. Because of the tuned-line traps, inserted in series at each end of the line, the injected signal travelled to the other end of the line, where after passing through the line-coupling equipment it was fed to a receiver. Each receiver not only received signals from the other end of the line but also the high frequency signals produced at its own end.

As a result, signals would have been received continuously by each receiver, as shown in Fig. 10.16(b) if the currents in the two ends of the line were in phase. When the line currents were not in phase, however, the signals overlapped as shown in Fig. 10.16(c) and there were periods during each power-frequency cycle when the receivers did not receive signals. When the duration of a period in which no signal was received exceeded $1/12^{th}$ of a cycle of power frequency, i.e. $\pi/6$ rad, an output relay operated. This relay then initiated the opening of the associated circuit-breaker provided that the local Stage 2 relay supplied by the sequence network had operated.

The phase-displacements between the currents at the ends of a healthy line tend to be considerably less than $\pi/6$ rad. As an example, the phase displacement on a three-phase, 400 kV 3800 A line of 100 miles in length is 0.036 rad. Should a fault occur on a circuit connected to such a line, the voltage at the end near the fault would be depressed below normal, causing reductions in the capacitive current and phase displacement.

Experience has shown that the above angular range of $\pm \pi/6$ rad in which

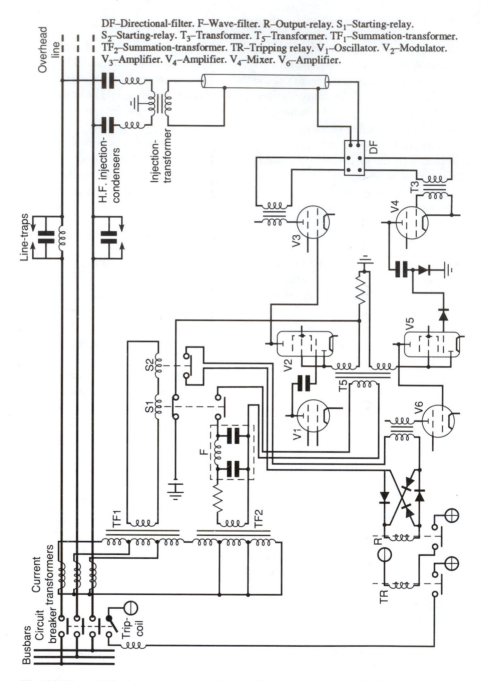

Fig. 10.16 (a) Telephase protective scheme, (b) basic operating principle when currents are in phase, (c) and out of phase. (Reproduced from *The Reyrolle Review, No 142, Jan–Mar 1950*, with the permission of Reyrolle Protection.)

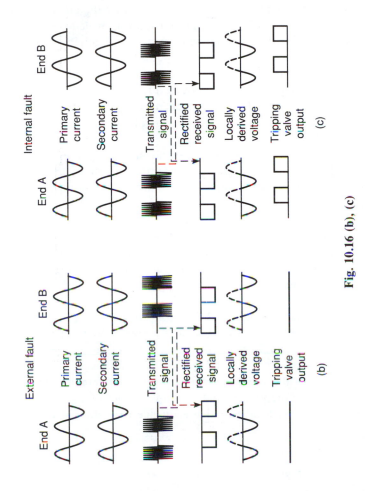

Fig. 10.16 (b), (c)

operation does not occur, does usually provide an adequate factor of safety and assures correct discrimination. On very long lines, however, a larger range may be needed and schemes are arranged so that ranges up to $\pm \pi/3$ rad may be selected.

The various electronic units in the early Telephase schemes, such as the oscillators and amplifiers, which were mounted on racks, employed thermionic valves. These had limited lives and therefore the circuits were monitored to detect failures and in addition provision was made to enable operators to check the condition of the equipments regularly. Equipment could be provided to initiate checks automatically at regular intervals.

It will be appreciated that the schemes needed secure power supplies and standby generating equipment was provided for this purpose.

The fault settings available on these schemes, which were controlled by the sequence networks and the Stage 2 starting relays, were as shown in Table 10.1.

It will be seen that the settings were dependent on the phases on which faults occurred and the setting for three-phase faults was above the rated current level so that the comparison process was not initiated by normal load currents. This high three-phase setting was considered unacceptable for some applications and later versions of the scheme incorporated sequence networks which provided separate voltage outputs (v_1 and v_2) proportional to the positive- and negative-sequence inputs supplied to them. The separate outputs were fed to modulation amplifiers which produced outputs proportional to (kv_2 and v_1) and these outputs were used to modulate the high frequency carrier signals which were injected into the protected lines.

With this arrangement four level detectors were provided at each end of a scheme, i.e. a pair of low-set devices and a pair of high-set devices. These replaced the Stage 1 and Stage 2 relays used in the original Telephase scheme. The low-set detectors again allowed the comparison process to be initiated and the high-set devices allowed tripping to take place when faults occurred within the protected zone. One low-set and one high-set detector at an end were fed with the output of the local negative-sequence network and because this is

Table 10.1

Type of fault	Setting (percentage of current transformer secondary rating)
Phase 'a' to earth	75
Phase 'b' to earth	90
Phase 'c' to earth	90
Phase 'a' and 'b'	45
Phase 'b' and 'c'	60
Phases 'c' and 'a'	45
Three-phase	225

zero under normal healthy conditions, the detectors could be given settings well below the value corresponding to the rated current of the protected line. The other low-set and high-set detectors were fed with the output of the positive-sequence network via impulse networks which only provided outputs when the line currents changed suddenly. This second set of detectors was not therefore sensitive to the passage of normal load currents and could thus be set to detect relatively small three-phase fault currents, unlike the earlier scheme.

Various forms of Telephase have been applied to many transmission lines over the years and they have performed well, operating for internal faults in times of about 20 ms.

The latest form of Telephase protection which is designated T3 [11], is produced by Reyrolle Protection, NEI Electronics Ltd. It incorporates the above impulse-detection arrangements and is arranged as shown in Fig. 10.17. As a result it can provide the range of fault settings shown in Table 10.2.

Because the scheme utilizes modern electronic equipment, several facilities such as automatic checking procedures are incorporated. A further feature of this scheme is that the angular zone in which operation will not occur, i.e. the stabilizing zone is not fixed as in the early schemes, but increases as the line current decreases. It varies between limits of about $\pm \pi/6$ rad to $\pm \pi/3$ rad. It thus takes account of the fact that the phase displacement caused by the capacitive current of a line decreases as the current fed through it increases.

The operating time of this scheme is somewhat dependent on the levels of fault currents and also on the transient components present and as can be seen from Fig. 10.18 it may vary between about 18 ms and 40 ms.

To ensure that incorrect operation will not occur because of saturation of the cores of current transformers feeding Telephase T3 schemes, it is recommended that the secondary knee-point voltage of the transformers should satisfy the following conditions:

$$V_k \geqslant 1.5 \frac{X}{R} I_f (A + B + C)$$

in which
1.5 is a stability factor,

Table 10.2

Type of fault		Fault setting (percentage of current transformer secondary rating)				
Balanced	Three-phase (impulse)	30	40	50	60	70
Unbalanced	Phase to earth	36	48	60	72	84
	Phase to phase	20	27	34	41	48

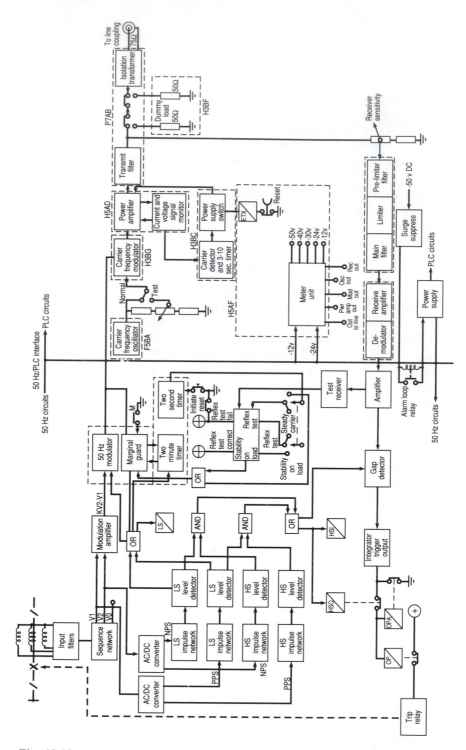

Fig. 10.17 Block diagram of Telephase Protection T3. (Reproduced from *Telephase T3, PLC Phase-Comparison Protection, T3.4/86*, with the permission of NEI Reyrolle.)

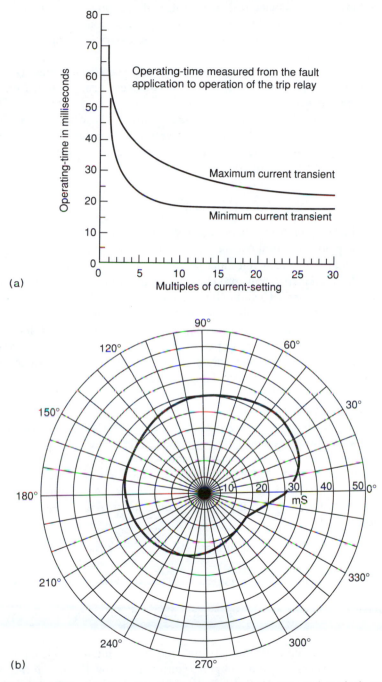

Fig. 10.18 Overall operating time/current characteristic (a), and typical operate time/point-on-wave initiation b–e fault at three times fault-setting (b).

$X/R =$ Inductive reactance/resistance ratio of the power-system impedance corresponding to the maximum through-fault conditions,

I_f is the maximum through-fault in the secondary winding of the current transformers (A),

A is the resistance of the current-transformer secondary windings (Ω),

B is the resistance of the connection between each current transformer and the sequence network (Ω),

C is the impedance of the sequence network (Ω), which has maximum values of $0.1\ \Omega$ and $0.2\ \Omega$ respectively when current transformers with 1 A and 5 A secondary current ratings are used. The value of C is normally very small compared with $A + B$.

10.6.2 Contraphase protective schemes

The original Contraphase scheme was produced by Metropolitan Vickers Electrical Company Ltd, at about the same time as Telephase and both schemes worked on the same principles and performed in similar ways.

GEC Measurements now produce two modern versions of Contraphase which are designated P10 and P40 [12]. The scheme P10, which has an output signal power of 10 W and can operate over links with attenuations up to 36 dB, is suitable for use on transmission lines up to 200 miles in length. The scheme P40 has an output power of 40 W and it can operate over links with 40 dB attenuation. It may therefore be applied to lines longer than 200 miles.

These schemes are so connected that the transmitted signals are in antiphase with each other when the currents at the two ends of a protected line are of the same phase.

As in the Telephase scheme, low- and high-set starters are provided which operate when the negative-sequence outputs of the sequence networks exceed pre-set levels and low- and high-set impulse starters are also provided which operate when there are sudden increases above set levels in the positive- and negative-sequence outputs of the sequence networks. As a result the schemes have a high sensitivity to all types of faults, the settings available being as follows:

Positive-sequence impulse currents	30–60% of rated current in 5% steps
Negative-sequence impulse currents	5–20% of rated current in 2.5% steps
Negative-sequence non-impulse currents	eight times the impulse value selected.

To ensure that current comparisons may be maintained when faults are present, it is arranged that the impulse starters have dwell times during which they remain operated after being initiated, provided that fault current is still flowing. The high-set dwell time is 0.5 s and the low-set dwell time is 0.6 s.

In addition, the low-set starters have time delays which maintain the transmission of carrier signals for 100 ms after faults are cleared.

The carrier signals which are transmitted are modulated by an output derived from the positive- and negative-sequence networks, i.e. $k_1 I_{1a} + K_2 I_{2a}$ in which $k_1 = -1$ and $k_2 = 16$.

The stability angle, i.e. the phase displacement over which operation will not occur, is nominally about $\pm \pi/6$ rad but, as with Telephase T3, this angle increases at low-current levels for which the effects of the capacitive currents in a line are greater.

This scheme, which has an operating time between 10 and 40 ms for fault currents of five times the setting levels, incorporates in-built testing features and secure power supplies.

To ensure that Contraphase schemes will operate satisfactorily, the manufacturers recommend that current transformers with knee-point voltages (V_k) of the following levels be used:

$$V_k \geqslant 1.3 \frac{X}{R} I_f (R_{CT} + 2R_w)$$

in which
X/R is the primary system reactance/resistance ratio,
I_f is the maximum through-fault current (secondary A),
R_{CT} is the resistance of the secondary winding (Ω),
R_w is the resistance of leads between the current transformers and the protective equipment (Ω).

10.6.3 Other schemes

Schemes with similar characteristics to those described above are produced by other manufacturers, an example being protection relay 7SD32 [13] which is available from Siemens AG.

10.7 AUTO-RECLOSING

Many of the faults which occur on overhead lines are caused by flashovers between phase conductors or between one or more of the phase conductors and earthed metal or the ground. After the initial breakdown quite large currents flow in the arcs which are formed and considerable ionization is produced.

Such conditions cannot be allowed to persist and they must therefore be detected by protective equipment which will initiate the opening of the circuit-breakers associated with the faulted line.

Clearly the arc or arcs will extinguish after the line is de-energized and the ionized products will then disperse. As a result, the line could be re-energized after a certain period, which is termed the dead time.

It will be appreciated, however, that the disconnection of a line connected between sources affects power transfers and it could lead to asynchronous operation, i.e. instability. It is therefore necessary that dead times be so chosen that the line circuit-breakers will be reclosed before instability will occur. In most cases a dead time of about 0.2 s proves sufficiently long to allow arc products to disperse and yet enables stability to be maintained.

The above process, which is referred to as auto-reclosing, is now widely used on both transmission and distribution lines. It is implemented in several different ways. In each case, operation of the protective equipment associated with a faulted line initiates the opening of the circuit-breakers which then automatically reclose after the pre-set dead time. Should the isolation of the line clear the fault then no further action is initiated and the line is left in service. If, however, the fault is not cleared at the end of the dead time, either because it is a solid fault, i.e. direct contact between conductors or to ground, or because the products of arcing have not dispersed sufficiently, then the protective equipment detects the presence of the fault when the line is reclosed and initiates the opening of the circuit-breakers again. Because it is clearly undesirable to repeatedly reclose a major circuit on to a fault, schemes are arranged so that either only one reclosure is possible, i.e. if a fault has not been cleared when the circuit-breakers reclose initially, then when they are opened for the second time a further reclosure will not occur. Two or more reclosures are allowed, however, on less important circuits such as distribution lines. To differentiate between the arrangements described above, they are referred to as single shot reclosing and multi-shot reclosing.

It will be appreciated that the disconnection of a major transmission line significantly affects other parts of the network of which it forms a part and, as stated above, it can lead to asynchronous operation after a relatively short period of time. Clearly the effect of opening only one phase of a line is less than that caused by opening all three phases because power can be transferred on the two unopened phases and therefore instability might not be caused or it will not occur as quickly as it would with three-phase opening. It is therefore advantageous, when single-phase to earth faults occur on a line, to open only the faulted phase and then to use one of the reclosure methods described above and this is the usual practice when suitable circuit-breakers are provided, i.e. those in which the phases may be operated independently.

To enable this arrangement to be implemented, it is clearly necessary that the protective schemes should be able to determine the types of faults which may occur on a transmission line, e.g. interphase and single-phase to earth, and the phases on which they are present so that the appropriate circuit-breaker operation and reclosure procedures may be initiated.

Faults involving all three phases are relatively rare and when they do occur it is usually because of errors, such as the failure to remove earths, applied while maintenance work has been done, before a line is re-energized. Reclosure in such circumstances is undesirable and, as a result, some users prefer to inhibit automatic reclosing when three-phase faults are detected by the protective equipment.

In some cases, when voltage transformers are present on a network in positions which enable the voltages on both sides of line circuit-breakers to be measured, check-synchronizing equipment is provided to ensure that automatic reclosures do not take place unless the phase displacements across circuit-breakers are within acceptable limits. It will be appreciated that when a line is disconnected at both ends to clear a fault, the voltages on the line sides of both its circuit-breakers will be zero and therefore the phase displacements at neither end can be determined. It is thus necessary that the circuit-breaker at one end be allowed to close after the necessary dead time and that check-synchronizing equipment at the other end should determine whether conditions are suitable for the second circuit-breaker to be reclosed. Further details of problems associated with auto-reclosing may be found in reference [14].

10.8 THE FUTURE

It will be evident from the preceding sections that both interlock and phase-comparison schemes have been used successfully to protect transmission lines for many years and modern versions of these schemes are being implemented using the latest electronic circuitry and equipment. It therefore seems likely that they will continue to be applied in the foreseeable future and although their hardware and that of the communication links will continue to be up-dated, the basic principles and features will not be changed.

REFERENCES

1. Pilot relaying performance analysis – IEEE Committee Report, (1990) *IEEE Trans*, **PWRD-5**, 85–102.
2. Eggimann, F., Senn, W. and Morf, K. (1977) The transmission characteristics of high voltage lines at carrier frequencies, *Brown Boveri Review*, **8**, 449–459.
3. Fibre optic channels for protective relaying, report prepared by the Fibre Optics Relay Channels Working Group for the IEEE Power System Relaying Committee (1989): *IEE Trans, on Power Delivery*, **4**, (1), 165–176.
4. *Power System Protection*. Vol 1 *Principles and Components* (1981), (ed. Electricity Council), Peter Peregrinus 2nd edn, Chapter 7.
5. Clothier, H. W. (1931) Overhead line and feeder protection, *CIGRE Paris*, 1931, Paper No. 88.
6. Leeson, B. H. and Leyburn, H. (1931) The principles of feeder protection and their application to three modern systems, *CIGRE Paris*, 1931, Paper No. 106.
7. Chamia, M. and Liberman, S. (1978) Ultra high speed relay for EHV/UHV transmission lines – development design and application, *Trans. IEE*, **PAS-97**, 2104–2112.
8. Johns, A. T. (1980) New Ultra-high-speed directional comparison technique for the protection of ehv transmission lines, *Proc. C, IEE*, **127**, 228–229.

9. Vitins, M. (1981) A fundamental concept for high speed relaying, *Trans. IEE*, **PAS-100**, 163–168.

10. Lackey, C. H. (1950) A review of British practice in the protection of electric power systems, *The Reyrolle Review*, No 142, pp 1–19.

11. Telephase T3, PLC Phase-Comparison Protection, *Pamphlet T3.4/86*, (12 pages), NEI Reyrolle Ltd.

12. Type Contraphase P10/P40, Phase Comparison Carrier Protection, *Publication R-5247B* (11 pages), GEC Measurements.

13. Phase Comparison Protection Relay 7SD32, Protective Relays, Siemens *Catalog R.1989*, pp 6/25–6/33.

14. Automatic reclosing of transmission lines – an IEEE Power Systems Relaying Committee Report (1984), *Trans. IEE*, **PAS-103**, 234–245.

11

Distance-type protective schemes for overhead lines and cables

In Chapters 7–10 various schemes which are used to protect overload lines and cables have been considered. All of these schemes are of the unit type in that they should operate whenever faults above certain levels of current occur on the protected units, i.e. between the current transformers mounted at each of the ends of the units, and they should not operate when external faults are present or during any healthy conditions. In each case this desired performance is obtained by making use of information about the conditions at both ends of a protected unit at any time. As examples, the instantaneous input and output currents of lines are compared in current-comparison (Merz–Price) schemes, the phases of currents are compared in 'phase-comparison' schemes and the directions of currents are taken into account in interlock schemes. To effect the necessary comparisons, signals must be sent between the ends of protected units which may, of course, be quite long distances apart, and either pilot wires or optical fibres or other high frequency electronic communication equipment may be required for this purpose.

An alternative method of protecting overhead lines and cables, without the need for comparisons between the quantities at both ends of the protected units, is based on measuring the input impedances of lines and cables and to do so clearly requires that information about both input currents and voltages be available. It will be clear, therefore, that such schemes must be fed from both current and voltage transformers mounted near the ends of protected circuits.

It is shown later that it is not possible to measure impedance with such accuracy that discrimination can be achieved between faults a short distance from the end of a long line and those just beyond its end, and therefore these schemes are not of the unit type and hence features, such as time delays, must be incorporated to enable correct discrimination to be obtained between internal and external faults.

The input impedance of a short-circuited line or cable varies from zero for a fault at its input end to a finite value for a fault at its remote end, the actual impedance value increasing with the distance to the fault; therefore schemes based on such measurements are referred to either as 'distance- or distance-measuring protection' or alternatively as 'impedance-measuring protection'.

These schemes have been applied successfully to many circuits over the years and they are still being produced today. Detailed information of various schemes and their behaviour is given later in this chapter. In the following section, however, details of the original impedance-measuring schemes and other historical information are provided.

11.1 HISTORICAL BACKGROUND

About 1920 some difficulties were being experienced in obtaining correct discrimination when existing relays were applied to power lines. This caused P. Ackerman to examine the possibility of using the input impedance to a line to obtain discrimination. He subsequently produced a relay with a beam, pivoted at its centre, which had armatures at each of its ends, these being attracted to electromagnets. One of the electromagnets carried a current proportional to the input voltage of the protected circuit and the other carried a current proportional to the input current to the circuit. Under normal conditions the net torque on the beam tilted it towards the voltage-energized electromagnet but when the ratio of the input voltage to current of the circuit fell below a particular value, the beam tilted in the opposite direction, causing the operation of contacts. This relay, which was described in a paper [1] published in the *Journal of the Engineering Institute of Canada* in December 1922, was the forerunner of the beam relays which were produced in large numbers in the following decade.

Shortly afterwards Crichton described a different design of impedance-measuring relay in a paper [2] published in the *Transactions* of the IEE in April 1923. This relay had an induction-type element in which the disc was driven by an electromagnet energized by a current proportional to the current in the protected circuit. When the disc rotated it wound up a helical spring, one end of which was connected to an armature attracted by an electromagnet energized with a current proportional to the input voltage of the protected circuit. This relay, which was similar to that described later in section 11.3.1, operated in a time which was approximately proportional to the input impedance of the protected circuit.

Development work continued and it was recognized that impedance relays could not discriminate between faults on a protected circuit and those on other circuits connected to the same busbars, i.e. between forward and reverse faults. The need to use impedance relays in conjunction with directional relays was therefore appreciated and certainly suitable directional elements with watt-metric-type movements were available when McLaughlin and Erickson of Westinghouse Electrical and Manufacturing Co had a paper published in the *Transactions* of the IEE in 1928 [3].

At this time it was realized that the behaviour of impedance relays was affected by the resistances of fault arcs and ground-return paths and relays which were only affected by the input reactances of protected circuits had been produced.

The publication of a paper [4] by George in March 1931 in the *Transactions* of the IEE indicates that the use of distance protection was increasing. It was stated in the paper that both impedance- and reactance-measuring relays produced by the Brown Boveri, General Electric and Westinghouse companies were in use on circuits of the Tennessee Electric Power Company in the USA.

11.2 THE BEHAVIOUR OF OVERHEAD LINES

The conductors of overhead lines possess resistance and inductance and in addition there are capacitance and leakage conductance between them. These parameters are distributed throughout the lengths of the conductors, R, L, C and G usually being used to represent the values per metre length.

The effects of capacitance and leakage conductance on the behaviour of relatively short lines operating at power frequencies are very small and they will be neglected in the following section in which the conditions which may be encountered on such lines are examined. To further simplify the treatment a single-phase line will be considered.

11.2.1 The input impedance of a short single-phase line

It will be evident from Fig. 11.1(a) that the input impedance (Z_{in}) to end A of the line P during steady state conditions if a short-circuit was present at a point a distance x along it would be given by:

$$\bar{Z}_{in} = \frac{\bar{V}_A}{\bar{I}_A} = (R + j\omega L) x \ \Omega$$

This impedance clearly increases linearly with the distance to the fault (x) until it reaches its maximum value for a fault at the remote end of the line, i.e. $\bar{Z}_{in} = (R + j\omega L) \, l\Omega$. This variation is shown in Fig. 11.1(b). On most lines, the ratio of the inductive reactance to resistance, ($\omega L/R$), is high and therefore the phase angle φ of the input impedance is quite high. As an example the $\omega L/R$ ratio of the 400 kV transmission lines in Britain is 16 and the corresponding value of φ is 1.508 rad.

If the line P supplied a similar line Q on which a short-circuit occurred at a point a distance x_Q along it, then the input impedance to line P would be given by:

$$\bar{Z}_{in} = (R + j\omega L) (l + x_Q) \ \Omega$$

Clearly therefore, an input impedance (Z_{in}) in the range between the points O and M on Fig. 11.1(b) would indicate a fault on line P and a value above the

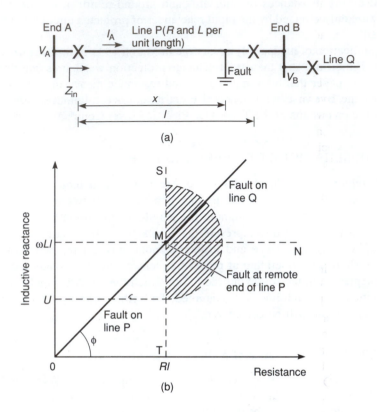

Fig. 11.1 Input impedance of a faulted line.

value at point M would indicate that line P was healthy. This is one of the simple principles on which distance- or impedance-measuring schemes are based.

It will be appreciated that discrimination must also be achieved between short-circuits on a protected line and normal healthy operating conditions. During the latter conditions a protected line (P) may be supplying a range of loads connected to the busbar at its remote end (B). Should these loads be resistive then the input impedance (Z_{in}) to line P would be given by:

$$\overline{Z}_{in} = (R + j\omega L)\,L + R_L \quad \Omega$$

in which R_L is the effective resistance of the loads.

The locus of the input impedance would thus be on the line MN in Fig. 11.1(b). Considering also the limiting cases of highly-inductive and capacitive loads, the measured impedances would lie along the lines MS and MT respectively.

Impedances with values in the zone to the right of the line ST in Fig. 11.1(b) could therefore be encountered during normal healthy conditions. In practice, however, the load impedance has a minimum magnitude related to the total rated current (I_r) of the loads, i.e. V_B/I_r, and therefore the input impedances (Z_{in}) to line P would not have values within a semicircle such as that shown shaded in Fig. 11.1(b).

From the above considerations it is clear that the input impedances which may be present when short circuits occur on a protected line (P) are different to those which may be present at other times and therefore satisfactory discrimination is achievable. A further factor which must be taken into account, however, is that significant resistance may be present in a fault path. The input-impedance (Z_{in}) to a protected line when such a fault occurs at a point a distance x along it will be given by:

$$\bar{Z}_{in} = (R + j\omega L)\, x + R_f$$

in which R_f is the fault resistance.

The range of input impedances which may occur when faults with resistances up to a certain value (R_{fmax}) occur on a protected line will thus be within the shaded parallelogram shown on Fig. 11.2(a).

It will be seen from Figs. 11.1(b) and 11.2(a) that correct discrimination could be achieved for faults with resistances up to a particular value at which the part of the parallelogram to the right of the vertical through the point RL in Fig. 11.2(a) lies in the shaded semicircle in Fig. 11.1(b), i.e. when the vertical height OA in Fig. 11.2(a) is greater than the height OU in Fig. 11.1(b). This condition is shown in Fig. 11.2(b).

Should a protected line form part of a highly interconnected network, then current may be fed through it in the reverse direction to normal when faults occur on the circuits connected to its input end. This situation is illustrated in

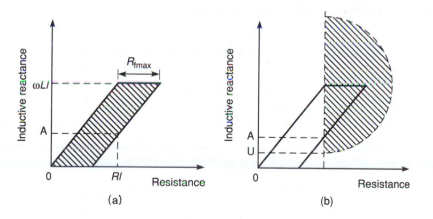

Fig. 11.2 Range of input impedances for resistive faults.

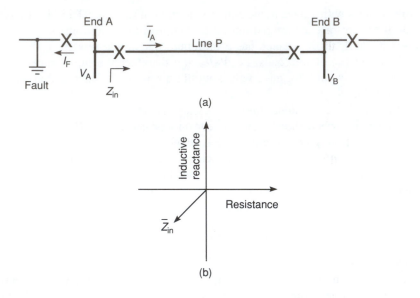

Fig. 11.3 Apparent input impedance of a line in a highly-interconnected network.

Fig. 11.3. In these circumstances the fault current $(\overline{I}_F)$ would usually lag the busbar voltage $(\overline{V}_A)$ and the apparent input impedance of line P $(\overline{Z}_{in})$ would then have both a resistance and inductive resistance which are negative, an example being shown in Fig. 11.3(b). Should the fault be on busbar A then the apparent impedance would be zero.

It will be clear from the above that faults on a protected line could be detected and that correct discrimination could be achieved in this simple case by providing an impedance-measuring relay which would operate for impedances in a section of the Z plane, such as that shown in Fig. 11.2(b).

11.2.2 The input impedance of a long single-phase line

When long lines are being considered it is necessary to take into account the distributed nature of the impedance parameters R, L, C and G. The last component (G) usually has a very low value which is difficult to assign. It has little effect on the line behaviour and is therefore usually neglected.

In the general case, however, the voltage ~ current relationships are expressed by the well-known long-line equations below.

$$\overline{V}_x = \overline{V}_A \cosh \gamma x - \overline{I}_A \overline{Z}_0 \sinh \gamma x \tag{11.1}$$

$$\overline{I}_x = \overline{I}_A \cosh \gamma x - \frac{\overline{V}_A}{\overline{Z}_0} \sinh \gamma x \tag{11.2}$$

in which $\overline{V}_x$ and $\overline{I}_x$ are the quantities at a point a distance x from the input end of a line and $\overline{V}_A$ and $\overline{I}_A$ are the input-end voltage and current. The quantities γ, Z_0 are given by the expressions

$$\gamma = \sqrt{(R + j\omega L)\,(G + j\omega C)}$$

and

$$Z_0 = \sqrt{\frac{R + j\omega L}{G + j\omega C}}$$

where R, L are the series resistance and inductance per unit length and G, C the shunt admittance and capacitance per unit length.

It can be seen from equation (11.1) that the input impedance to a line with a short circuit at a distance x along it, when V_x would be zero, would be:

$$\overline{Z}_{in} = \frac{\overline{V}_A}{\overline{I}_A} = \overline{Z}_0 \tanh \gamma x \ \ \Omega$$

The form of the variation of $\overline{Z}_{in}$ with the distance x is shown in Fig. 11.4(a), from which it can be seen that the input impedance becomes very large at a

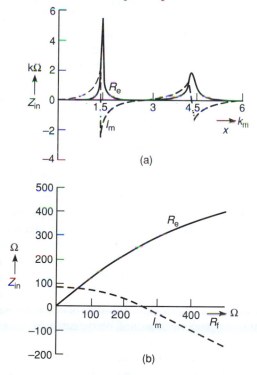

(a)

(b)

Fig. 11.4 (a) Typical input impedance of a line with a short circuit fault at a distance x. (b) Variation of input impedance with fault resistance.

particular distance which is dependent on frequency. For the frequencies used in power systems, however, these distances are great, for examples 1500 km for 50 Hz operation and 1250 km for 60 Hz operation.

In practice, therefore, the positions of short circuits on all the lines in use today could be determined by impedance measurement and the non-linearity of the input impedance variation with fault position does not pose problems.

Should resistance (R_f) be present in the path of a fault on a long line then the value of $\overline{V}_x/\overline{I}_x$ would be R_f and it can be shown from equations (11.1) and (11.2) that the input impedance ($\overline{Z}_{in}$) would then be given by:

$$\overline{Z}_{in} = \frac{\overline{V}_A}{\overline{I}_A} = \frac{R_f \cosh \gamma x + \overline{Z}_0 \sinh \gamma x}{\cosh \gamma x + \dfrac{R_f}{\overline{Z}_0} \sinh \gamma x}$$

The variation of the value of the input impedance with fault resistance is not as simple as that which was shown in the previous section, in which relatively short lines were considered. The effect is similar, however, as can be seen from Fig. 11.4(b), which shows the variation which would be obtained for a fault at a point 300 km along a single-phase line with parameters related to those of the 400 kV, three-phase lines in Britain. The same variation would clearly be obtained if current was supplied to a resistive-load connected to the end of a 300 km-long line with the same parameters but in this case the ohmic value of the load resistance would be relatively high.

Clearly the effect produced by the presence of resistance in faults close to the input end of a line is the same as that obtained on short lines.

The above considerations demonstrate that correct discrimination could be achieved between faults on protected lines and other conditions by detecting input impedances within certain areas in the Z plane.

11.2.3 The input impedances of a three-phase line

All major transmission and distribution lines are three-phase and therefore several different types of faults may occur either on them or on other associated circuits, namely single-phase to earth, phase-to-phase, phase-to-phase to earth and, more rarely, three-phase. Clearly the positions of all these types of faults cannot be determined by a single impedance measurement and therefore a protective scheme must be energized with voltages and currents proportional to those associated with the phases of the line being protected. As examples, faults between phase a and earth would be detected by monitoring a current and voltage proportional to the 'a' phase current and the phase a to earth voltage whilst faults between phases a and b would be detected by monitoring a current proportional to either the current in the 'a' phase or the 'b' phase and a voltage proportional to the voltage between phases a and b.

It will be evident from the above that distance-protection schemes to be used with three-phase lines require that a large number of impedance measurements be performed.

11.3 IMPEDANCE MEASUREMENT

It must be accepted that impedance-measuring relays or other devices cannot be produced with the accuracy needed to enable them to discriminate between faults on an overhead line and all faults elsewhere. As an example, a relay could not be expected to differentiate between a fault one metre from the remote end of a line 100 km long and a fault on the busbars to which the remote end is connected as this would represent a difference of only 0.001% of the setting.

As a consequence, impedance-measuring schemes do not inherently provide unit protection and extra features must be incorporated in them to enable correct discrimination to be achieved in service. Before considering these features, however, methods available for measuring impedance are examined below.

11.3.1 Relays dependent on the magnitude of impedance

Two different types of electro-mechanical relays were used in early protective schemes. One of them operated in a time related to the magnitude of the ratio of the voltage to the current applied to it whilst the other operated rapidly whenever the magnitude of the ratio fell below a set level. These two basic forms are considered below.

Relays which operated in times related to impedance

The earliest impedance-measuring relays were intended to employ time grading similar to that of the IDMT relays in use at the time and they were based on the induction elements then available. A particular design is illustrated in Fig. 11.5(a), from which it can be seen that a current proportional to that in the circuit being protected was fed to the winding on the driving magnet of the induction element. When the current exceeded a set value the disc rotated at a speed dependent on the current and wound up the spiral spring attached to it. The other end of the spiral spring was attached to the armature of an electromagnet which was energized with a voltage proportional to that of the protected line. The greater the level of this voltage, the greater was the spiral spring torque needed to move the armature and operate the contacts. This required the disc, which was braked by a permanent magnet, to travel a greater distance. As a result, the operating time increased with increase of voltage and decreased with increase of current and therefore it increased with the ratio of voltage to current, i.e. impedance. Linearity was not achieved however, the actual characteristics being of the form shown in Fig. 11.5(b), which also shows how discrimination was obtained between circuits in series.

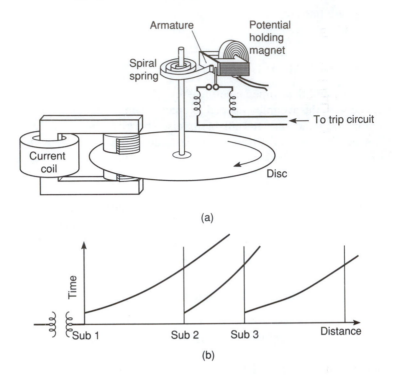

Fig. 11.5 Principle and application of time distance relay.

Relays which operated rapidly when impedances were below set levels

The above relays did not provide rapid clearance of faults on protected lines and the operating times clearly increased with the distance of a fault from the relaying position. This was not always acceptable because high fault currents could persist for quite significant periods and therefore relays were produced which operated rapidly for all faults which caused the magnitudes of the input impedances to protected circuits to be below appropriate set levels. In early distance-protection schemes, relays with a pair of electromagnets above which was a balanced beam were used. The basic form of such relays is shown in Fig. 11.6(a).

The beam, because of its inertia, could not respond to the power-frequency variations of the fields set up by the two electromagnets. Under normal conditions the voltage applied to the winding on the left electromagnet was high and a current flowed in the winding producing a large m.m.f. (I_vN_1) and therefore a large flux φ_1. The current in the protected line at such times would be relatively low and therefore both the m.m.f. (IN_2) in the right electromagnet and the flux φ_2 would be low. During these conditions the beam was restrained by the backstop.

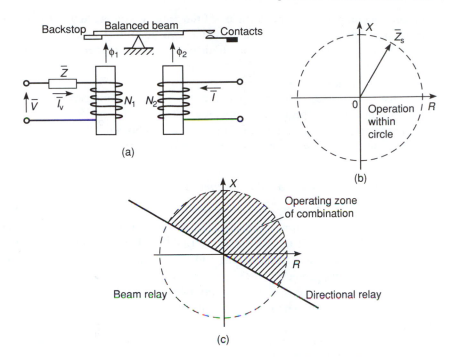

Fig. 11.6 Balanced-beam relay.

During periods when faults were present on the line, the voltage (V) would be depressed and the current (I) would be above normal and as a result the magnitude of the flux φ_2 would exceed that of the flux φ_1, causing the beam to tilt and close the contacts. These relays were therefore amplitude comparators.

Because of the air gaps in the magnetic circuits, almost linear behaviour was obtained and operation occurred when the ratio of the magnitude of the m.m.f. IN_2 to that of m.m.f. I_vN_1 exceeded a set value. As a result operation occurred when the magnitude of $\bar{I}/\bar{V}$ exceeded a set value. This provided an impedance setting, independent of phase, below which operation occurred.

These relays therefore had a circular impedance characteristic centred on the origin of the Z plane, as shown in Fig. 11.6(b) and not offset characteristics of the forms shown earlier in Fig. 11.2. They did not therefore have directional properties and had to be used in conjunction with separate directional elements so that the combination operated only in semicircular zones of the form shown in Fig. 11.6(c).

11.3.2 Relay characteristics

Relays with circular operating zones centred on the origin of the Z plane are known as plain-impedance relays. As was shown above, this characteristic is obtained by comparing the magnitudes of the applied voltage and current and operation is obtained when the ratio of the two quantities falls below a set value. Over the years a number of methods have been developed to effect such comparisons and the devices used for this purpose have been designated 'amplitude-comparators'.

Because relays with plain-impedance characteristics could not be used on their own in distance protection schemes, relays with other characteristics were produced. One of the first of these was the 'mho' characteristic, which was first obtained using beam relays as described below.

Mho characteristic obtained using an amplitude comparator

As shown in Fig. 11.7(a), the restraining electromagnet had two windings. One of them, which had N_1 turns, was supplied via an impedor to make the total impedance of the circuit have a set value ($Z_1 \underline{|\beta}$). This is referred to as the replica impedance. The voltage applied to the circuit ($k_1 V \underline{|\alpha}$) was proportional to that of the protected circuit, i.e. $V \underline{|\alpha}$. The second winding on this electromagnet had N_2 turns which carried a current ($k_2 I$), I being the current in the protected circuit. The current $k_2 I$ also flowed in the single winding of N_3 turns on the operating electromagnet.

For a relay in which operation would occur when the m.m.f. applied to the operating electromagnet exceeded that applied to the restraining electromagnet the following relationship would exist:

$$k_2 \left| IN_3 \right| \geqslant \left| \frac{k_1 N_1 V \underline{|\alpha}}{Z_1 \underline{|\beta}} - k_2 IN_2 \right| \tag{11.3}$$

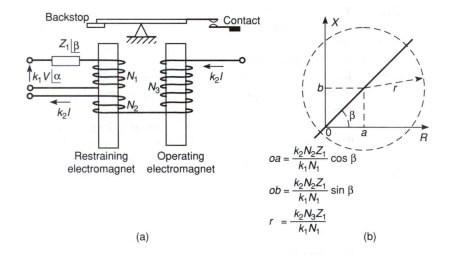

$$oa = \frac{k_2 N_2 Z_1}{k_1 N_1} \cos \beta$$

$$ob = \frac{k_2 N_2 Z_1}{k_1 N_1} \sin \beta$$

$$r = \frac{k_2 N_3 Z_1}{k_1 N_1}$$

(a)   (b)

Fig. 11.7 Mho characteristic.

The limiting condition, expressed in complex form, is thus:

$$|k_2 IN_3| = \left| \frac{k_1 VN_1}{Z_1} \cos(\alpha - \beta) - k_2 IN_2 + j \frac{k_1 VN_1}{Z_1} \sin(\alpha - \beta) \right|$$

Substituting R and X, the input resistance and input inductive-reactance of the protected circuit respectively, for the quantities $V \cos \alpha / I$ and $V \sin \alpha / I$ in the above expression leads to the following equation:

$$\left(R - \frac{k_2 N_2 Z_1 \cos \beta}{k_1 N_1} \right)^2 + \left(X - \frac{k_2 N_2 Z_1 \sin \beta}{k_1 N_1} \right)^2 = \left(\frac{k_2 Z_1 N_3}{k_1 N_1} \right)^2 \tag{11.4}$$

This is the circle of radius

$$\frac{k_2 N_3 Z_1}{k_1 N_1}$$

and centre

$$\frac{k_2 N_2 Z_1}{k_1 N_1} \cos \beta, \quad \frac{k_2 N_2 Z_1}{k_1 N_1} \sin \beta$$

shown in Fig. 11.7(b).

It can be seen from the above expressions that the circle can be made to pass through the origin by making the turns of the two current windings (N_2 and N_3) the same. Relays with this characteristic are described as 'mho' relays whereas the term 'offset mho' is used for other relays with circular characteristics in the Z plane. It is worthy of note that the term 'mho' was adopted because the circular characteristic in the Z plane becomes linear when it is represented in the Y (admittance, i.e. mho) plane.

It can also be seen from Fig. 11.7(b) that the position of the circle with respect to the axes is dependent on the phase angle of the impedance of the voltage circuit, i.e. $\underline{\beta}$, and this can be adjusted to provide the characteristics needed by the circuits being protected.

Although this characteristic was produced using beam relays it can clearly be produced by other forms of relays including those which employ electronic circuitry, in which the amplitudes of the quantities on the two sides of equation (11.3) are compared. Mho characteristics can also be produced by relays incorporating digital circuitry and in such cases direct amplitude comparison is not necessary.

Ohm characteristic obtained using an amplitude comparator

Beam relays were also used to produce the so-called 'ohm' characteristic, which is linear in the Z (impedance, i.e. ohm) plane, by having a winding of N_1 turns on each electromagnet, these being connected in series and fed through an impedor to make their total impedance have a value of $Z_1 \underline{|\beta}$ as before. This circuit was energized by a voltage $(k_1 V \underline{|\alpha})$ proportional to that of the protected circuit. Each electromagnet also had a winding which carried a current $(k_2 I)$, I being the current in the protected circuit. The windings on the restraining and operating electromagnets had N_2 and N_3 turns respectively and the current flowed through them in opposite directions.

Operation thus occurred when the following condition applied:

$$\left| \frac{k_1 N_1 V \underline{|\alpha}}{Z_1 \underline{|\beta}} + k_2 I N_3 \right| \geqslant \left| \frac{k_1 N_1 V \underline{|\alpha}}{Z_1 \underline{|\beta}} - k_2 I N_2 \right| \tag{11.5}$$

By again substituting $R = V \cos \alpha / I$ and $X = V \sin \alpha / I$, the following expression is obtained:

$$X = -R \cot \beta + \frac{Z_1 k_2 (N_2 - N_3)}{2 k_1 N_1 \sin \beta} \tag{11.6}$$

The resulting operating characteristic of these relays is therefore of the general form shown in Fig. 11.8(a).

It will be clear that the characteristic would pass through the origin, as shown in Fig. 11.8(b), if the current windings had the same number of turns, i.e. $N_2 = N_3$, and such a relay would operate for currents varying in phase over π rad relative to the applied voltage, i.e. it would behave as a directional relay.

It will also be clear that a horizontal ohm characteristic could be produced by making the phase angle β equal to $\pi/2$ rad. A relay with such a characteristic could be set to operate when the inductive reactance component of the input impedance of a line fell below a particular value and it would therefore not be affected by the presence of resistance in a fault path. Relays with this characteristic, which were described as reactance relays, were produced using induction type elements, and were employed to detect earth faults in early protective schemes.

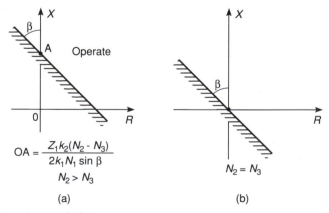

$$OA = \frac{Z_1 k_2 (N_2 - N_3)}{2 k_1 N_1 \sin \beta}$$

$N_2 > N_3$

(a)   (b)

Fig. 11.8 Ohm characteristic.

The above characteristics can be produced by other forms of relays in which the amplitudes of the quantities on the two sides of equation (11.5) are compared.

Characteristics obtained using phase comparators

Phase comparators, as their name implies, are relays or devices which are supplied with two quantities, for example a voltage and a current, and they operate whenever the phase displacement between the quantities is in a particular range. As an example, a device could be set to operate when the two quantities are within $\pi/2$ rad of each other. A directional relay is thus one form of phase comparator.

The characteristics considered on page 396 may be obtained using a phase comparator supplied with two quantities produced by summing and taking the differences of voltages and currents proportional to those associated with a protected line.

As an example, if the following quantities are produced:

$$\bar{S}_1 = \bar{k}_v \bar{V} + k_i \bar{I}$$

and
$$\bar{S}_2 = \bar{k}_v \bar{V} - k_i \bar{I}$$

and operation occurs when the phase angle between the quantities is $\pi/2$ rad or less, either lagging or leading, then the limiting condition would occur when:

$$\frac{|V|}{|I|} = \frac{k_i}{|k_v|} = \text{constant}$$

i.e. such a relay would have a plain-impedance characteristic.

As a further example, a 'mho' characteristic can be produced by a relay which operates when the following two quantities are displaced in phase by $\pi/2$ rad or less:

$$\bar{S}_1 = k_r \bar{I} + \bar{k}_v \bar{V} \quad \text{and} \quad \bar{S}_2 = k_0 \bar{I} - \bar{k}_v \bar{V}$$

The resulting circle has a radius of

$$\frac{k_0 + k_r}{2|k_v|}$$

and a centre at

$$\frac{k_0 - k_r}{2|k_v|} \cos\theta, \quad \frac{k_0 - k_r}{2|k_v|} \sin\theta$$

in which θ is the phase angle of k_v. This is an offset-mho characteristic unless k_r is zero, in which case the circle passes through the origin of the Z plane.

Ohm relays and directional relays can also be produced using phase comparators.

Quadrilateral and other characteristics

To reduce the possibility of incorrect operation of impedance-measuring relays it is desirable that the areas of the zones in the Z plane in which they will operate are limited to those needed to detect faults on the circuits being protected. It was shown earlier in section 11.2.1 that faults up to a certain resistance can be detected using the characteristic shown in Fig. 11.2, i.e. a parallelogram.

Such characteristics cannot be produced simply by a relay employing either an amplitude or phase comparator. As these relays can, however, have linear characteristics, i.e. an ohm characteristic, then four of them with suitable characteristics could be used to produce an overall quadrilateral characteristic. This would, however, be expensive.

Over the years combinations of relays have been used to produce desired characteristics. As stated earlier, plain-impedance relays have been used with directional (ohm) relays to obtain the semicircular characteristic shown in Fig. 11.6(c). Similarly, an ohm relay could be used with a mho relay to produce the operating characteristic shown in Fig. 11.9.

Impedance relays have been produced in several different forms, e.g. beam relays, induction relays and rectifier comparators [5]. The range of characteristics which could be produced by such relays, was limited but the introduction of electronic digital processing in recent times has made it possible for any desired characteristic to be provided including quadrilaterals and those with lenticular shapes with a range of aspect ratios.

11.3.3 The effects of system transients

The operating characteristics considered in section 11.3.2 were all presented in the complex Z plane which is related to sinusoidal steady state conditions. In practice, however, relays must operate relatively quickly after the incidence

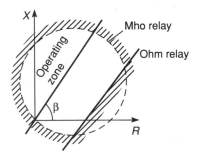

Fig. 11.9 A combined mho and ohm characteristic.

of faults, say in 20 ms or less, and steady state conditions will not usually have been obtained in such times and indeed r.m.s. values determined over fractions of a power frequency cycle are not those used in phasor diagrams. In addition, quite significant unidirectional transient components may be present in the voltages and currents at the ends of faulted lines.

To ensure that large errors will not occur in the assessments of the distances to faults, relays must be so designed that the effects of the above factors will be either eliminated or minimized.

As an illustration of the way in which this can be done, the behaviour of a 'mho' element incorporating a beam relay as an amplitude comparator is examined below.

Behaviour obtained during transient conditions resulting from faults on short lines

To simplify the treatment, the application of the relay to a relatively short, single-phase line is considered. The capacitive current between the conductors of such a line would be very low and therefore to detect short circuits up to a distance ℓ along a line, which has a resistance R per unit length and an inductive reactance (ωL) per unit length, the diameter of the mho characteristic would be made equal to $\ell \, |(R + j\omega L)|$ in steady state.

It was shown on page 397 that the diameter of the mho circle is given by:

$$\text{Diameter} = \frac{2k_2 N_3 Z_1}{k_1 N_1}$$

and therefore to provide the necessary setting the magnitude of the impedance of the voltage circuit of the relay must be:

$$|Z_1| = \frac{k_1 N_1 \ell}{2k_2 N_3} |R + j\,\omega L| \tag{11.7}$$

If the impedance Z_1 is made up of a resistance

$$R_1 = \frac{k_1 N_1 \ell R}{2k_2 N_3}$$

in series with an inductance

$$L_1 = \frac{k_1 N_1 \ell L}{2k_2 N_3}$$

then the transient and sinusoidal components of the current in the voltage winding of the relay will match those in the current windings at all instants. As an example, should the voltage at the input to a line remain sinusoidal during a short circuit condition, i.e. $v = V_{\text{pk}} \sin(\omega t + \alpha)$ in which t is measured from the time of fault occurrence, then the instantaneous line current would be given by:

$$i = \frac{V_{pk}}{\ell \, (R^2 + \omega^2 L^2)^{\frac{1}{2}}} \left\{ \sin(\omega t + \alpha - \beta) - \sin(\alpha - \beta) \exp\left(-\frac{R}{L} t\right) \right\}$$

in which $\tan \beta = \omega L / R$

The current (i_v) in the relay voltage winding would be given by:

$$i_v = \frac{2 k_2 N_3 V_{pk}}{N_1 \ell \, (R^2 + \omega^2 L^2)^{\frac{1}{2}}} \left\{ \sin(\omega t + \alpha - \beta) - \sin(\alpha - \beta) \exp\left(-\frac{R}{L} t\right) \right\}$$

and the quantities on both sides of equation (11.3) would be equal at all instants.

As a result the m.m.f.s applied to the two electromagnets would be the same. The impedance Z_1 given in equation (11.7) is thus a replica of that in the line and its use would ensure that in practice only the small errors caused by neglecting the capacitive current of the line would occur during transient conditions. In addition the behaviour would not be affected by the relay operating time and operation would not therefore need to be delayed to allow r.m.s. values to be determined.

It will be clear that errors will occur, however, in the assessments of the distances to short circuits if the phase angle of the impedance Z_1 is not the same as that of the line impedance.

It can be shown that accurate assessments of the distances to short circuits can be obtained with relays having the other characteristics considered earlier and this is so when either amplitude or phase comparators are used.

It must be recognized that significant errors may be made in determining the input impedances of lines on which resistive faults are present because under such conditions the ratio of the inductance to resistance in the replica impedor (Z_1) will not be the same as that in the line input impedance. In these circumstances, the waveforms of the currents in the voltage circuits of the comparators will not be the same as those in the current circuits when transient components are present and certainly those in the current circuits will decay more rapidly than those in the voltage circuits. As a result, the values being compared will not be in a constant ratio and the phase differences will not be constant.

Behaviour during transient conditions which result from faults on long lines

The capacitive current between the conductors of long lines tends to be small relative to the normal or fault currents in lines, but nevertheless the input impedances when short circuits occur on lines do not increase linearly with the distances to faults as shown in section 11.2.2. In addition the phase angles of the impedances are affected by the distances to faults.

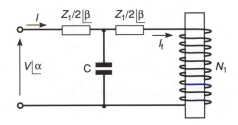

Fig. 11.10 T-section line model connected to a relay.

Clearly, a relay could be set to detect when the input impedance of a line fell below a certain value during steady state conditions, but the technique of matching the impedance in the voltage circuit of a relay to that of the protected line does not enable completely accurate assessment of the positions of faults to be achieved. This is basically because the distributed nature of the impedors in a long line cannot be matched by a single impedor.

The behaviour of a line would be approximately mirrored by making allowance for its capacitance by connecting a replica capacitor in the voltage circuit of a relay as shown in Fig. 11.10. In this way a T-section modelling of the line up to the distance at which short circuits should be detected would be produced. This would provide improved accuracy of measurement but it would not be completely accurate because the current (I_f) in the relay winding would not contain the components caused by the travelling waves which are present under fault conditions.

In practice a degree of error in the assessment of the distances to short circuits or long lines must be accepted [6] and errors will again be present when resistive faults occur.

11.4 BASIC SCHEMES

It is desirable that faults on major transmission and distribution lines should be cleared rapidly and therefore relays which operate in times proportional to the distances of faults along lines are not acceptable and relays which operate in short times independent of the fault positions are always used. Such relays cannot be set so that they will detect faults anywhere on a protected line and yet not operate for faults elsewhere on a network. They must therefore be set to operate when faults occur on most of the length of a line, a percentage of 80% being usual. Such a setting provides a factor of safety to ensure that operation for faults beyond the end of a line will not be caused by measuring errors. In addition to this setting, which is designated Stage 1 or Zone 1 reach, another setting which can detect faults both on the protected line and on circuits beyond the remote end of the line must be provided and operations within this setting, which is designated Stage 2 or Zone 2 reach, must initiate circuit-breaker tripping after a short time delay (t_2).

To ensure the isolation of faults on a protected line, similar sets of relays must be provided at each of its ends, the arrangements being as shown in single-phase form in Fig. 11.11. It will be clear that such a scheme would ensure that faults in the centre 60% of the line, i.e. $0.2\ell \rightarrow 0.8\ell$ would cause the opening of the circuit-breakers at both ends to be initiated after a time t_1 say 20 ms. Faults in the 20% end sections would cause the opening of the nearer and more remote circuit-breakers to be initiated after times of t_1 and t_2 respectively. To ensure that circuit-breaker opening will not be initiated in the event of faults on the other circuits connected to the busbars at the line ends, the operating time t_2 must be greater than that of the protective equipments associated with the other circuits.

In some applications, a further setting (Stage 3 or Zone 3) is included, this being greater than the Stage 2 setting, and in this way faults which are more remote than those in the zones covered by the Stage 2 setting can be detected. Stage 3 operation occurs after a longer time delay than that associated with the Stage 2 setting and therefore it only initiates the opening of the circuit-breakers if faults on connected circuits are not cleared by their own protective schemes. This setting thus provides a form of back-up protection.

As stated above, a fault near one end of a protected line would not be completely cleared until after the operation of the Stage 2 relays at the remote

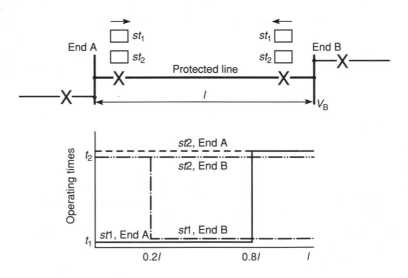

Fig. 11.11 Application of relays with Zone 1 and Zone 2 settings.

end, i.e. after a time t_2. In some applications, such relatively slow clearance of faults might be considered unacceptable. In these circumstances, the tripping of the remote circuit-breaker can be accelerated by sending inter-tripping signals between the line ends as described later in section 11.12. The resultant scheme is then clearly a form of interlock protection incorporating impedance-measuring relays.

11.4.1 Application to three-phase lines

To ensure that the various types of faults which may occur on three-phase networks will be correctly detected, separate relays could be used on each of the phases to determine whether faults to earth are within the protected zones, i.e. Stage 1, 2 and possibly 3. Separate relays could also be provided to detect interphase faults. Clearly one or more of the above relays would operate in the unlikely event of a three-phase fault.

It will be seen from the above that 18 impedance-measuring elements would be needed at each line end in such an arrangement to provide three impedance zones and certainly when electro-mechanical relays were used the costs were significant and the burdens placed on the current and voltage transformers supplying the relays were high. This latter factor in turn made for greater expense because of the need to use larger cores and/or windings, particularly in the current transformers, to ensure that they would perform satisfactorily under all the conditions which could be encountered.

At the other extreme, schemes having only a single impedance-measuring element may be produced by including simple elements to sense the types of fault present on lines and using these to provide the necessary quantities to the impedance-measuring element. As an example, simple overcurrent starting elements may be used to detect interphase faults and in the event of a fault between, say, phases a and b, the appropriate overcurrent elements would operate and connect an input voltage proportional to that between the 'a' and 'b' phases of the line and an input current proportional to either the 'a' or 'b' phase line current to the impedance-measuring element. Should this element not operate within a certain time (t_1) because the fault was not within its Stage 1 reach, then its setting would be increased to that of Stage 2 and this procedure could, if desired, be repeated to extend the reach to that of Stage 3 after a further period.

Over the years a variety of schemes ranging between these two extremes have been produced. Clearly the advantage obtained by including starting elements is the reduction in the number of impedance-measuring elements needed. The disadvantage of these schemes is the element of unreliability introduced by the inclusion of switches in the circuits supplying the measuring relays and certainly some authorities have considered the risk to be unacceptable. Nevertheless such schemes, which are designated as switched-distance schemes, have been used in many applications and have given good service.

11.4.2 The detection of faults close to the input ends of lines

Various relay characteristics which may be produced in the Z plane were considered in section 11.3 and it was shown in some cases that the characteristics could pass through the origin. Such behaviour implies a limiting condition at zero impedance and thus zero voltage.

In the case of a semi-circular characteristic, shown in Fig. 11.6(c), being produced by combining a directional element with a plain-impedance element, it will be evident that the direction of a current cannot be determined when the reference voltage being used is zero. Such a relay could not therefore be expected to discriminate between faults close to the current and voltage transformers on a protected line and those on the adjacent busbars or near the ends of other circuits connected to the busbars. This situation was considered earlier in section 4.5 (page 135) when the behaviour of directional relays was examined, and it was then shown that satisfactory operation could be obtained by energizing relays with voltages derived from phases other than those from which the currents were obtained, e.g. a directional relay associated with phase a to earth faults could be supplied with a current proportional to that flowing in phase a of the protected circuit and a voltage proportional to that between phases b and c. This technique could also be used successfully with the relay considered above. To correctly assess the input impedance during an earth fault on phase a of a line, the plain-impedance element would be supplied with a voltage and a current proportional to those of phase a whilst the directional element would be supplied with quantities proportional to the phase a current and the 'b' to 'c' phase voltage of the line.

The technique cannot, however, be applied directly to relays with 'mho' characteristics which have only two inputs. In such applications if a current from one phase and the voltage between the other two phases were used then the limiting conditions would not be those of the required circle. To produce an acceptable compromise, three inputs must be provided as in the previous example, e.g., the 'a' phase voltage (v_a) and current (i_a) and the voltage between the other phases (v_{bc}).

The relay would combine the 'mho' and 'ohm' characteristics and to do so, in the case of an amplitude comparator, an equation containing the terms of both equations (11.3) and (11.5) would define the conditions needed for operation, i.e.

$$\left| \frac{k_3 N_3 V_{bc} \underline{|\gamma}}{Z_3 \underline{|\delta}} + k_2 I_a N_2 \right| \geqslant \left| \frac{k_3 N_3 V_{bc} \underline{|\gamma}}{Z_3 \underline{|\delta}} + \frac{k_1 N_1 V_a \underline{|\alpha}}{Z_1 \underline{|\beta}} - k_2 I_a N_2 \right| \quad (11.8)$$

Such a relay does not have a single operating characteristic in the Z-plane because it is controlled by four variable terms, i.e. V_a/I_a, V_{ab}/I_a, γ and α, unlike a mho relay which is only controlled by two of these terms, namely V_a/I_a and γ.

It can be seen, however, that in the event of a short circuit on the 'a' phase near the current and voltage transformers, for which the voltage V_a would be zero the equation (11.8) would reduce to that of an ohm relay, i.e. equation (11.5), and it would pass through the origin of the Z plane. It would thus act as a directional relay and the angle of the linear characteristic could be set at the required angle in the Z plane by choosing an impedor Z_3 with the necessary phase angle.

For short circuits on phase a of a line near the point beyond which operation is not to occur, the voltage V_a would be quite high and provided that the magnitude of the term $k_3 N_3 / Z_3$ is very much less than that of the term $k_1 N_1 / Z_1$, the effect produced by the polarizing voltage V_{bc} would be small compared with that produced by the phase a voltage (V_a). As a result the behaviour would approach that of an mho relay. Such relays are referred to as being cross-polarized and clearly the same performance can be obtained by providing the necessary inputs to other forms of amplitude or phase comparators.

It will be evident that neither self- or cross-polarized relays would operate correctly in the unlikely, but nevertheless possible, event of a three-phase short circuit on a line at a point near the current and voltage transformers because under such a condition all the voltages would collapse to near zero levels.

In the past, so-called memory circuits were connected at the outputs of the voltage transformers. These were oscillatory circuits which would continue to provide power-frequency voltages that would decay in amplitude slowly should the power-system voltages collapse. These voltages which were related to the pre-fault system voltages were then used to provide polarization to the various relays.

More recently, with the introduction of modern electronic equipment, voltages have been continuously sampled and stored for short periods and these have been used to provide a measure of polarization in the event of system faults.

It will be appreciated that the basic circular mho characteristic is distorted by the use of cross-polarization and clearly the degree of distortion increases as the amount of cross-polarization is increased. This feature is utilized on occasions and the amount of cross-polarizing is so set that the circle is distorted to increase the value of the intercept on the R axis of the Z plane diagram whilst retaining the cut-off impedance value at the characteristic angle β. In this way, relay operation for relatively high-resistance faults at points along a protected line can be obtained.

To differentiate between the types of mho characteristics, the term 'self-polarized' mho is now used for relays with a single voltage input, the term 'cross-polarized' being used for those with two voltage inputs.

11.4.3 The settings of impedance-measuring relays

The setting of relays used in three-phase applications must take account of the current distributions which can exist during the various types of faults which may occur on a line. Those which arise during phase to earth faults and interphase faults are considered separately below.

Phase to earth faults

If a line had a completely symmetrical conductor configuration and it was supplied from an earthed source, as shown in Fig. 11.12(a), and there were no other earths on the system, then the conditions during a short circuit between phase a and earth could be determined from the sequence network shown in Fig. 11.12(b). This is demonstrated in Appendix 3. The fault current would be given by:

$$\bar{I}_a = \bar{I}_{1a} + \bar{I}_{2a} + \bar{I}_{0a} = \frac{3\,\bar{V}_{1a}}{x\,(\bar{Z}_1 + \bar{Z}_2 + \bar{Z}_0)}$$

and therefore the apparent input impedance to phase a (Z_e) would be given by:

$$\bar{Z}_e = \frac{\bar{V}_{1a}}{\bar{I}_a} = x\,\frac{\bar{Z}_1 + \bar{Z}_2 + \bar{Z}_0}{3} \tag{11.9}$$

in which $\bar{Z}_0$, $\bar{Z}_1$ and $\bar{Z}_2$ are the sequence impedances up to the fault position.

Each of the earth fault impedance-measuring relays could be set to operate at impedances up to the value given by the above expression for a fault at the desired cut-off or reach, say 80% of the line length.

In practice, lines are not symmetrically spaced in triangular configurations and therefore the impedances per unit length of the phases differ and measuring errors would be present if the above setting was used on all phases. The

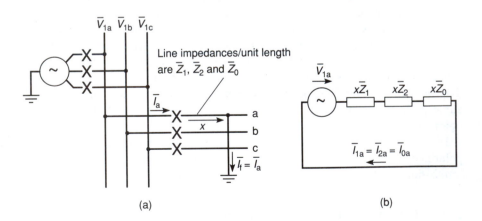

Fig. 11.12 A phase-to-earth fault and the associated sequence network.

imbalance is usually small, however, and the resulting errors tend to be acceptable.

It will be realized that the current $(\overline{I}_a)$ would flow in the faulted phase during the above condition and it would all return through the earth. In multiply-earthed systems, however, different current distributions can occur and currents can flow in the healthy phases. This can be seen by considering the circuit shown in Fig. 11.13(a) in which a line is connected between two earthed sources. The sequence network would then be connected as shown in Fig. 11.13(b) to determine the conditions which would obtain during a short circuit to earth on phase a of the line. It will be clear that the positive-sequence current $\overline{I}_{A1a}$ would not be the same as the zero-sequence current $\overline{I}_{A0a}$ if the source voltages $\overline{V}_{A1a}$ and $\overline{V}_{B1a}$ were not of the same magnitudes and phases, and under these conditions currents would flow in phases b and c of the line. As a result, the apparent input impedance $(\overline{Z}_{Ae})$ to phase a at end A of the line, given by $\overline{Z}_{Ae} = \overline{V}_{Aa}/\overline{I}_{Aa}$, would not be equal to the value given by equation (11.9), i.e.

$$\overline{Z}_{Ae} \neq x \frac{\overline{Z}_1 + \overline{Z}_2 + \overline{Z}_0}{3}$$

A relay set to operate up to this latter value would thus assess the distance to the fault incorrectly and it could either under-reach or over-reach.

Sound-phase compensation To avoid the above situation, allowance for the currents in the healthy phases can be made as explained below.

It can be seen from Fig. 11.13(b) that the input voltage to phase a and end A of the line is given by:

$$\overline{V}_{Aa} = \overline{V}_{Aa1} = x \{\overline{I}_{A1a} \overline{Z}_1 + \overline{I}_{A2a} \overline{Z}_2 + \overline{I}_{A0a} \overline{Z}_0\}$$

For a line $\overline{Z}_1 = \overline{Z}_2$, therefore,

$$\overline{V}_{Aa} = x \{\overline{Z}_1 (\overline{I}_{A1a} + \overline{I}_{A2a}) + \overline{Z}_0 \overline{I}_{A0a}\}$$

$$= x \{\overline{Z}_1(\overline{I}_{A1a} + \overline{I}_{A2a} + \overline{I}_{A0a}) + \overline{I}_{A0a} (\overline{Z}_0 - \overline{Z}_1)\}$$

$$= x \{\overline{I}_{Aa} \overline{Z}_1 + 1/3 (\overline{I}_{Aa} + \overline{I}_{Ab} + \overline{I}_{Ac}) (\overline{Z}_0 - \overline{Z}_1)\}$$

$$= x \left\{ \overline{I}_{Aa} \left(\frac{2\overline{Z}_1 + \overline{Z}_0}{3} \right) + (\overline{I}_{Ab} + \overline{I}_{Ac}) \frac{(\overline{Z}_0 - \overline{Z}_1)}{3} \right\}$$

The distance to the fault (x) on phase a is thus given by:

$$x = \frac{\overline{V}_{Aa}}{\overline{I}_{Aa} \left(\dfrac{2\overline{Z}_1 + \overline{Z}_0}{3} \right) + (\overline{I}_{Ab} + \overline{I}_{Ac})(\overline{Z}_0 - \overline{Z}_1)}$$

A method based on the above equation was employed in Ratio-Balance protective schemes [6] produced by A. Reyrolle and Co Ltd, a sound-phase compensation transformer of the form shown in Fig. 11.13(c) being included for the purpose. It will be seen that the phase output currents of the transformer were of the form:

$$\bar{I}_{\text{outa}} = \bar{I}_{\text{Aa}} \frac{N_1}{N_3} + \frac{N_2}{N_3}(\bar{I}_{\text{Ab}} + \bar{I}_{\text{Ac}})$$

in which

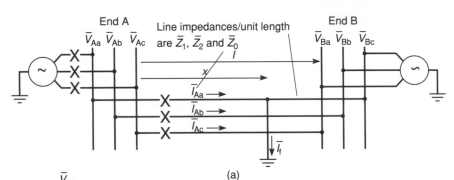

(a)

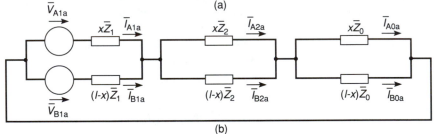

(b)

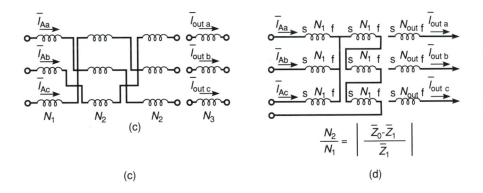

$$\frac{N_2}{N_1} = \left| \frac{\bar{Z}_0 - \bar{Z}_1}{\bar{Z}_1} \right|$$

(c)

(d)

Fig. 11.13 Current flow during earth faults in multiply-earthed systems.

$$\frac{N_2}{N_1} = \frac{|\overline{Z}_0 - \overline{Z}_1|}{|2\overline{Z}_1 + \overline{Z}_0|}$$

Whilst account should strictly be taken of the phase angles of the impedances $\overline{Z}_1$ and $\overline{Z}_0$, this form of compensation does enable earth fault relays to perform with acceptable accuracy because the line resistances are much lower than the reactances and the assessments of the positions of faults are not therefore significantly affected by the distributions of the currents in the phases.

To give an output current of $I_{outa} = x I_{Aa}(\overline{Z}_1 + \overline{Z}_2 + \overline{Z}_0)$ when there are no currents in the healthy phases the turns ratio N_3/N_1 must be unity.

Residual compensation The equation above relating the voltage $\overline{V}_{Aa}$ to the three-phase currents $\overline{I}_a$, $\overline{I}_b$ and $\overline{I}_c$ may also be expressed in the form:

$$\overline{V}_{Aa} = x \left\{ \overline{I}_{Aa} Z_1 + (\overline{I}_{Aa} + \overline{I}_{Ab} + \overline{I}_{Ac}) \left(\frac{\overline{Z}_0 - \overline{Z}_1}{3} \right) \right\}$$

$$= x \{ \overline{I}_{Aa}\overline{Z}_1 + \overline{I}_{A0}(\overline{Z}_0 - \overline{Z}_1) \}$$

An alternative form of compensation, known as residual compensation, can thus be provided and if the phase displacements between the impedances $\overline{Z}_1$ and $\overline{Z}_0$ are neglected then the compensation can be obtained using a transformer connected as shown in Fig. 11.13(d). The turns ratio must be

$$N_2/N_1 = \left| \frac{\overline{Z}_0 - \overline{Z}_1}{\overline{Z}_1} \right|$$

Interphase faults

In the event of a short circuit between two phases of a line, as shown in Fig. 11.14(a), the sequence network would be connected as shown in Fig. 11.14(b).

The magnitude of the fault current would be given by:

$$|\overline{I}_f| = \frac{\sqrt{3} |\overline{V}_{1a}|}{x |\overline{Z}_1 + \overline{Z}_2|}$$

and the magnitude of the apparent input impedance $(\overline{Z}_{bc})$ given by

$$\frac{|\overline{V}_{bc}|}{|\overline{I}_b|}$$

would be:

$$|\overline{Z}_{bc}| = \frac{|\overline{V}_{bc}|}{|\overline{I}_b|} = \frac{\sqrt{3} |\overline{V}_{1a}|}{|\overline{I}_b|} = x |\overline{Z}_1 + \overline{Z}_2|$$

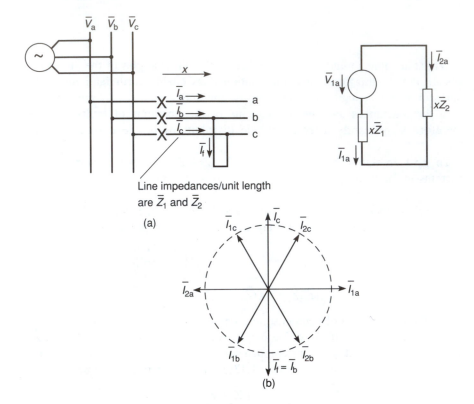

Line impedances/unit length
are $\bar{Z}_1$ and $\bar{Z}_2$

(a)

(b)

Fig. 11.14 An interphase fault and associated sequence network.

In practice, the positive and negative sequence impedances of lines are equal and therefore $|\bar{Z}_{bc}| = 2x\,|\bar{Z}_1|$.

During a three-phase fault the fault currents are given by:

$$I_f = \frac{\bar{V}_{1a}}{x\bar{Z}_1}$$

The magnitude of the apparent input impedance $\bar{Z}_{bc}$ in these circumstances would be:

$$|\bar{Z}_{bc}| = \frac{|\bar{V}_{bc}|}{|\bar{I}_b|} = \frac{\sqrt{3}\,|\bar{V}_{1a}|}{|\bar{I}_f|} = \sqrt{3}\,x\,|\bar{Z}_1|$$

It can be seen from this that phase-fault relays would operate for three-phase faults at distances up to $2/\sqrt{3}$ times those for which they would operate for phase to phase faults. As a result, if Stage 1 phase-fault relays were set to have a reach of 80% of the length of a protected line in the event of a three-phase fault, then they would only detect phase to phase faults on 69% of the line length.

This situation may be avoided by supplying each of the phase-fault relays with a line voltage and the difference of two phase currents. Considering again a short circuit between phases b and c, the magnitude of the apparent input impedance ($|\bar{Z}_{bc}|$) would be determined from the expression:

$$|\bar{Z}_{bc}| = \frac{|\bar{V}_{bc}|}{|\bar{I}_b - \bar{I}_c|}$$

For this fault

$$\bar{I}_b = \bar{I}_f = -\bar{I}_c$$

$$\therefore \; |\bar{Z}_{bc}| = \frac{\sqrt{3} \; |\bar{V}_{1a}|}{2 \; |\bar{I}_f|} = \frac{x \; |\bar{Z}_1 + \bar{Z}_2|}{2}$$

During a three-phase fault, the fault currents are given by:

$$\bar{I}_a = \frac{\bar{V}_{1a}}{\bar{Z}_1} \; ; \bar{I}_b = \frac{\bar{V}_{1a}}{\bar{Z}_1} \left\lfloor -\frac{2\pi}{3} \right. , \bar{I}_c = \frac{\bar{V}_{1a}}{\bar{Z}_1} \left\lfloor -\frac{4\pi}{3} \right.$$

and

$$\bar{I}_b - \bar{I}_c = \frac{\sqrt{3} \; \bar{V}_{1a}}{\bar{Z}_1} \left\lfloor -\frac{\pi}{2} \right.$$

For this condition the magnitude of the apparent input impedance ($|\bar{Z}_{bc}|$) would be:

$$|\bar{Z}_{bc}| = \frac{|\bar{V}_{bc}|}{|\bar{I}_b - \bar{I}_c|} = \frac{\sqrt{3} \; |\bar{V}_{1a}|}{\sqrt{3} \; |\bar{V}_{1a}| \, / x \, |\bar{Z}_1|} = x \, |\bar{Z}_1|$$

As $\bar{Z}_1 = (\bar{Z}_1 + \bar{Z}_2)/2$, the relays would operate for all interphase faults up to the same distance from the relays and therefore this method of measurement is now used in most schemes.

The effects of the physical asymmetry of lines

The settings considered in the previous sections have been based on complete line symmetry, whereas in practice conductors are run in asymmetrical configurations and therefore the self and mutual impedances associated with the phases are not the same. Whilst conductors could be transposed to reduce the effects of their physical asymmetry, this practice is now seldom adopted and certainly it would be difficult to implement on major lines because of the large spacings which would have to be maintained at the points where transpositions were made.

Clearly errors will be present in the assessments of fault positions if the impedance-measuring relays in each of the phases of a scheme applied to an untransposed line are given the same settings and, of course, these could be reduced by employing different settings.

In a study [6] on the behaviour of an untransposed line it was shown that the apparent impedances seen by relays during a three-phase fault at a point 500 miles from the relaying position were affected by the zero-sequence source impedance, variations up to 10% being possible. In addition, the impedances seen on the individual phases could differ by about 20%.

In general, the above effects are insignificant when schemes are being applied to relatively short lines but they should be taken into account when the protection of very long lines is being considered.

The effects of mutual coupling between lines

When overhead lines run close to each other, there is mutual inductance and capacitance between them and therefore they do not operate independently of each other. The effects of coupling can be quite significant on pairs of lines which are carried on double-circuit towers because of the close proximities of the various conductors.

The mutual inductances between pairs of conductors in the two lines are not all the same and therefore e.m.f.s are produced in the conductors of one line by the currents in the conductors of the other line. Although the effect is greatest when zero-sequence currents are flowing in a line there is nevertheless a small effect when the line currents are balanced. As a result, the apparent input impedances of a line on which a fault of any type is present are affected by load currents or currents being fed to the fault by a parallel healthy line and clearly the operation of impedance-measuring relays can be affected. This may be seen by considering the situation which would arise in the event of a

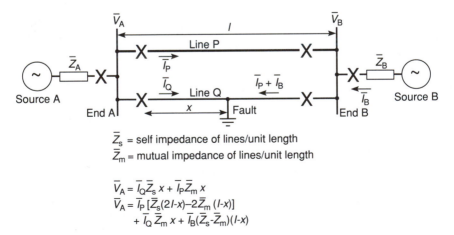

$\bar{Z}_s$ = self impedance of lines/unit length
$\bar{Z}_m$ = mutual impedance of lines/unit length

$$\bar{V}_A = \bar{I}_Q \bar{Z}_s x + \bar{I}_P \bar{Z}_m x$$
$$\bar{V}_A = \bar{I}_P [\bar{Z}_s(2l-x) - 2\bar{Z}_m (l-x)]$$
$$+ \bar{I}_Q \bar{Z}_m x + \bar{I}_B(\bar{Z}_s - \bar{Z}_m)(l-x)$$

Fig. 11.15 Effects of mutual coupling on line input impedance.

fault on one of a pair of single-phase lines connected in parallel, as shown in Fig. 11.15.

For this condition, the lines would feed the currents $\overline{I}_P \overline{I}_Q$ and $\overline{I}_B$ to the short circuit a distance x from end A along line Q and the apparent input impedances to the lines, $\overline{Z}_{inP}$ and $\overline{Z}_{inQ}$, at end A would be:

$$\overline{Z}_{inQ} = \frac{\overline{V}_A}{\overline{I}_Q} = \left(\overline{Z}_s + \frac{\overline{I}_P}{\overline{I}_Q} \overline{Z}_m \right) x \qquad (11.10)$$

and

$$\overline{Z}_{inP} = \frac{\overline{V}_A}{\overline{I}_P} = \overline{Z}_s (2\ell - x) - 2\overline{Z}_m (\ell - x) + \frac{\overline{I}_Q}{\overline{I}_P} \overline{Z}_m x + \frac{\overline{I}_B}{\overline{I}_P} (\overline{Z}_s - \overline{Z}_m)(\ell - x) \quad (11.11)$$

It will be seen that the input impedance of the faulted line ($\overline{Z}_{inQ}$) is not directly proportional to the distance to the fault (x) because the ratio $\overline{I}_P/\overline{I}_Q$ is affected by this distance. It is clear, however, that the correct distance to the fault could nevertheless be determined from equation (11.10) if the ratio of the two currents was known and in practice this could be achieved by feeding the relays with the currents from both lines.

The input impedance to the healthy line ($\overline{Z}_{inP}$) is dependent on the distance to the fault (x) and also the ratio $\overline{I}_Q/(\overline{I}_P + \overline{I}_B)$. This ratio varies in practice with the operating conditions of the sources and as the current input at end B ($\overline{I}_B$) cannot readily be provided as an extra input to the relays at end A, the distance to the fault (x) cannot be determined by the relay on the healthy line when it is only compensated with the current at end A of the other line ($\overline{I}_Q$).

When an uncompensated relay, i.e. one in which $\overline{I}_P$ is taken to be zero in equation (11.10), is present on a line on which a fault occurs, it will clearly not determine fault positions correctly and the errors will increase with increase in the value of the current (I_P) in the healthy line. These errors will cause the distance to the fault to appear greater than the actual value and therefore the reach at any given setting will be less than the nominal cut-off value unless the current in the healthy line (I_P) is zero. The relay would therefore never over-reach.

When a line is protected by a relay which is compensated only with the input current from the other line, i.e. I_Q in equation (11.11) then errors will be present in the determinations of the distances to faults as stated earlier, but in this case operation can occur during faults beyond the nominal cut-off position and indeed Stage 1 operation could occur for faults beyond the remote end of the protected line. Such over-reach is not acceptable and it is therefore much more serious than the under-reach, referred to above, which can occur with uncompensated relays.

The behaviour of both uncompensated and compensated relays applied to three-phase, parallel-connected lines was examined in detail by Davison and Wright [7] in 1963 and it was concluded that the normal arrangements used on single-circuit lines should be used on double-circuit lines and that compensation for the mutual effects should not be provided. The study showed that over-reach could occur with certain faults but that maloperation would not result whatever first-stage setting was adopted.

In the past, some schemes produced by manufactures in the USA did incorporate compensation dependent on the currents in parallel lines, but today uncompensated relays are usually employed.

The micromho relay [8] produced by GEC does, however, include provision for compensation derived from the zero-sequence current in the parallel circuit. It is occasionally applied to long double circuit lines but care must be taken when choosing the settings.

11.4.4 The operating times of relays

Electro-mechanical relays, because of the inertias of their movements and the displacements needed to close their contacts, have finite operating times and these vary with the operating forces applied to them.

If a beam relay is considered, it will be clear that the torques applied to it will be equal at the limits of its operating zone, i.e. the nett operating torque will be zero on the limiting circle of a mho characteristic. As a result, the operating times will be quite significant when faults occur on a line near the current and voltage transformers and also when faults occur near the point beyond which operation does not occur. When faults occur at intermediate points, however, the operating torques will exceed the restraining torques by significant amounts, and therefore more rapid operation will occur.

A further factor which must be taken into account when considering relay operating times is the effective source impedance ($\overline{Z}_s$) at the input end of a protected line. It can be seen from Fig. 11.16(a), which shows a single-phase line fed from another circuit of impedance $\overline{Z}_s$, that the current fed to a short circuit a distance x along the line would be:

$$\overline{I} = \frac{\overline{V}_s}{\overline{Z}_x + \overline{Z}_s}$$

The voltage ($\overline{V}_{in}$) at the input end of the line would be

$$\overline{V}_{in} = \overline{I}\,\overline{Z}_x = \frac{\overline{Z}_x}{\overline{Z}_x + \overline{Z}_s} \cdot \overline{V}_s \qquad (11.12)$$

The impedance seen by an impedance-measuring relay would be given by:

$$\frac{\overline{V}_{in}}{\overline{I}} = \overline{Z}_x$$

and it would thus assess the fault position correctly but both the voltage and current would decrease with increase in the effective source impedance. As a

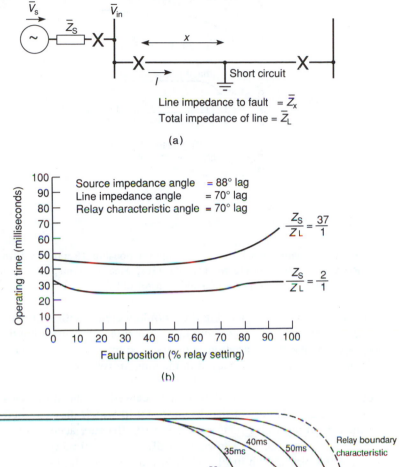

(a)

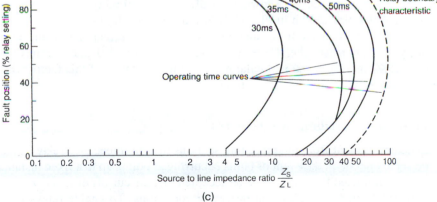

(c)

Fig. 11.16 Effects of source impedance. (a) Network configuration. (b) Typical operating characteristics for Zone 1 phase-fault mho relay. (c) Isochronic time characteristics for mho relay (three-phase faults).

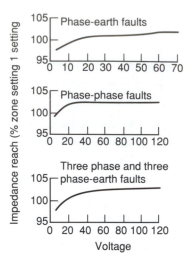

Fig. 11.16 (d) Typical reach to relay voltage relationship. (Reproduced from *Protective Relays – Application Guide, 3rd edn*, GEC Measurements, 1987 with the permission of GEC Alsthom Protection and Control Ltd.)

consequence, the relay operating time would increase with source impedance. It is therefore customary to present information on relay operating times either on graphs of the time variation with distance to a fault for several ratios of $\bar{Z}_s/Z_L$ as shown in Fig. 11.16(b) or in the contour form shown in Fig. 11.16(c).

In addition to affecting the relay operating times, the reductions of the voltages and currents applied to relays, caused by the presence of source impedance, can also affect the distances up to which they can detect faults, i.e. the cut-off or limiting distances may vary. Because account must be taken of such variations, curves showing cut-off distance to relay voltage relationships, of the type shown in Fig. 11.16(d), are produced.

It will be appreciated from equation (11.12) that the effect of a given source impedance increases the shorter the length of a protected line, because the ratio $\bar{Z}_s/Z_L$ increases and particular account must be taken of this factor when impedance-measuring relays are to be applied to very short lines.

11.4.5 Relay performance standards

It will be evident from the preceding sections that an impedance relay cannot be set to operate in the event of faults occurring at points up to a fixed distance along a protected line and it must be accepted that the cut-off distance or reach will be affected by the system operating conditions. To enable relays to be applied it is necessary that their accuracies be known and therefore British Standard BS 3950, 1965 was produced to define their performance requirements. In this Standard the following two quantities are defined:

1. System Impedance Ratio (SIR), which is the ratio of source impedance to relay setting impedance, expressed at the same impedance level, either primary or secondary.

   As an example, should the source impedance (Z_s) at primary level be 40Ω and a relay be set to detect faults on a line, with an impedance of 0.4Ω/km, at distances up to 200 km, i.e. a setting (Z_L) of 80Ω at primary level, then the system impedance ratio would be:

$$\text{SIR} = \frac{|Z_s|}{|Z_L|} = 0.5 \text{ or } 50\%$$

2. Characteristic Impedance Ratio (CIR) is the maximum value of system impedance ratio up to which a relay will operate at its prescribed accuracy. This ratio enables the voltage range over which a relay must be able to operate with the prescribed accuracy to be determined.

   It will be clear from equation (11.12) in the previous section that at the cut-off distance or reach,

$$V_{in} = \frac{1}{1 + Z_s/Z_x} \cdot V_s = \frac{1}{1 + \text{SIR}} \cdot V_s$$

If therefore, a relay were required to operate with a CIR value of 24, then it would have to maintain its prescribed accuracy of say ±5% at voltages (V_{in}) down to $V_s/25$, i.e. if the rated secondary voltage provided by the voltage transformers was 110 V then the relay would have to maintain its accuracy at input voltages down to 4.4 V.

Because of the significant changes in relaying in recent years, such as the use of digital processing, the above standard was thought to be inappropriate and it has therefore been withdrawn. If necessary, manufacturers should now be approached for details of the performances of their schemes to ensure that they will be suitable for application to particular circuits.

11.5 CONDITIONS WHEN A HEALTHY LINE IS AFFECTED BY ASYNCHRONOUS SYSTEM OPERATION

In the earlier sections, the input impedance values of a protected line which may be present when faults occur have been considered and various relay characteristics which are available have been outlined.

As with all protective schemes, it is necessary that incorrect operations should not occur and certainly the opening of line circuit-breakers should not be initiated when a protected line is healthy. By setting relays so that they will only operate when the input impedances of a protected line are below particular levels, the requirement that faults on circuits connected to it will not cause incorrect operation will be met.

A condition which could, however, cause incorrect operation is the asynchronous operation of the system of which a protected line forms a part, i.e. system instability. The behaviour during such a condition is considered below and a single-phase treatment is again employed in the interest of simplicity.

For the circuit shown in Fig. 11.17(a) in which the line is considered to be short and in which the voltages V_A and V_B have slightly different frequencies, the phase angle between them would vary slowly over 2π rad. At an instant (t) when the phase angle between the voltages was θ, as shown in Fig. 11.17(b), the line current I would be given by:

$$I = \frac{(V_A - V_B \cos\theta) + jV_B \sin\theta}{R + j\omega L}$$

The circular loci of the current I for several ratios of the magnitudes of V_A to V_B for values of θ over 2π rad are shown in Fig. 11.17(c) in which the circular loci of the apparent input impedances at end A, which is given by V_A/I, are also shown.

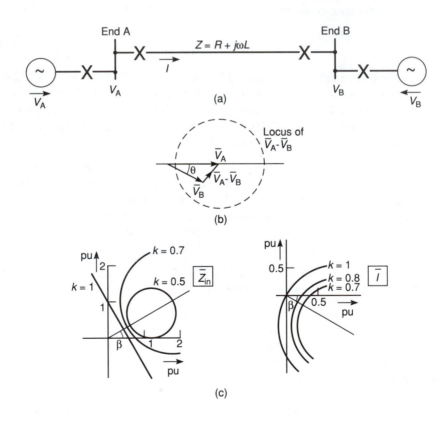

Fig. 11.17 Conditions during asynchronous operation. (a) Network configuration. (b) Locus of $V_A - V_B$. (c) Loci of I and Z_{in} for various values of $k = \left| \overline{V}_A / \overline{V}_B \right|$

It will be seen that the magnitudes of the impedances vary between very large and very small values when the voltages at the two ends are almost equal and when the frequency differences are small, the apparent input impedance changes slowly. In these circumstances, tripping of the line circuit-breakers would be initiated when the apparent impedance came within the Z plane operating zone of an impedance relay.

Now although two sources which are operating asynchronously should be disconnected from each other, it would not be satisfactory if the whole network was split into several sections because of the operation of several distance-protection schemes and ideally the separation should be effected by opening circuit-breakers near the mid-point between the sources. It is therefore necessary that the instability should be detected and the appropriate actions then be taken.

A method which has been used for many years is to include in a distance-protection scheme an extra relay element (A) with a characteristic within which the operating zones of the other measuring elements all lie. As an example an offset-mho characteristic could be used, as shown in Fig. 11.18. It will be appreciated that system instability affects all the phases in a similar manner and that its presence can be detected by a single element supplied for example by the 'a' phase voltage and current. In the event of instability the apparent input impedance to a line changes quite slowly initially around its particular circular locus, as shown in Fig. 11.18. As a result the instability detector element A would operate a significant time before any of the other impedance-measuring elements. Subsequently, the measuring element with the greatest reach, either Stage 3 in a three-zone scheme or Stage 2 in a two-zone scheme, would operate. This behaviour, which would clearly indicate the onset of instability, would be used to prevent the impedance-measuring

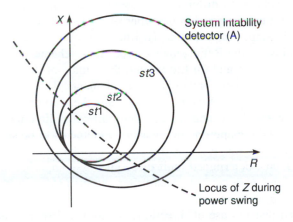

Fig. 11.18 Principle of the instability detector element A.

relays from initiating the opening of their associated circuit-breaker. A signal could also be provided to cause the opening of one or more circuit-breakers to split the network at the most suitable points.

When faults occur within the zones protected by distance-protection schemes, the input impedances of the lines change suddenly and in these circumstances the instability detector A could operate, but it would do so at the same time as the appropriate impedance-measuring relays; in these circumstances the latter relays would be allowed to initiate the opening of their associated circuit-breaker.

Should faults occur which would not cause the detector A to operate, e.g. a fault on a phase different from that to which the detector is connected, then operation of one or more of the impedance-measuring relays would again be allowed to initiate the opening of their associated circuit-breaker.

Summarizing, the opening of circuit-breakers as a result of the operation of impedance-measuring relays would only be inhibited if the instability detector relay A operated more than a certain time interval before the other relays.

This method of detecting instability is quite widely used. There are, however, some conditions when the impedance trajectory could occasionally enter the tripping zone of the impedance relays quite quickly, causing them to operate incorrectly.

An alternative method, based on the rate of change of voltage, is now used in the LU91 relay [9] produced by Asea Brown Boveri.

11.6 SCHEMES APPLIED IN THE PAST

As indicated in the previous sections, early distance-protection schemes employed quite large numbers of separate relay elements to provide two- or three-stage cover to overhead lines.

The schemes produced by different manufacturers operated on the principles outlined earlier but they differed in the types of relays used to determine fault directions and line-input impedances. To illustrate the methods used, details are provided below of the Ratio Balance type XZA scheme [10] which was produced by A. Reyrolle and Co Ltd around the middle of the century.

This scheme incorporated directional relay elements. Three were provided to detect phase to phase and three-phase faults and these controlled the phase-fault distance-measuring elements, i.e. when particular directional elements operated they caused the appropriate measuring elements to be supplied with input voltages and currents. A single directional element, which was included to detect earth faults, determined fault directions by comparing the phase of the sum of the three main current-transformer outputs ($3I_0$) with the phase of the current in the system neutral.

It was claimed that the use of directional control had several advantages, one being that the delay which occurred in energizing the measuring elements, after the incidence of a fault, typically 20–40 ms, greatly reduced the measuring errors which could be caused by the transient components in the system currents and voltages.

As a further feature, three earth fault phase selectors were included, each of which was fed with the sum of the output currents from the main current transformer and the output of one phase of a sound-phase compensation transformer, the latter being connected as described earlier in section 11.4.3 (page 407). Operation of both the earth fault directional relay and an earth fault phase selector caused the appropriate measuring element to be energized.

Three-beam type relays with plain-impedance characteristics were provided to deal with phase to phase and three-phase faults and three induction-type relays with reactance (ohm) characteristics were included to deal with earth faults. The latter elements were used because of their insensitivity to resistance in the fault path.

Two timing relays were included and these were energized when any of the directional relays operated. The measuring relays were set to operate for faults within the Stage 1 reach, but their reaches were extended when the first timing relay operated, i.e. Stage 2, and later, if the second timing relay operated, the reaches were further extended to Stage 3.

Several contactors were necessary to enable the various operations described above, for example the extension of the reaches, to be achieved.

The phase-fault directional relays were set so that they would not operate at current levels below 120% or 150% of the rated current of the line. The measuring elements were therefore not energized during normal healthy conditions. The ohmic settings of the measuring elements could be selected to enable different lengths of lines to be protected.

It will be appreciated that a complete scheme contained a considerable amount of equipment including at least eighteen electro-mechanical relays as shown in [10]. Nevertheless, this scheme and similar schemes produced by other manufacturers were applied to many lines and their performance records were very good.

11.7 PRESENT-DAY SCHEMES

Schemes based on modern electronic equipment which utilize digital-signal processing are now produced by the manufacturers of protective equipment. These schemes are extremely versatile and can include features which could not readily be provided in the earlier schemes which employed electro-mechanical elements.

Two schemes which are currently available are described in some detail below.

11.7.1 Optimho static-distance protection schemes

These schemes, designated LFZP, which are produced by GEC Measurements and described in detail in their literature, [11], are suitable for the whole range of overhead lines from those employed as main transmission links to those used in distribution circuits. To cover this range several different models are available.

All the models have a full set of measuring elements for each main zone of protection to avoid the need to rely on phase selection hardware or software to provide the appropriate voltage and current inputs to a single measuring element. As a result 18 measuring units are provided in schemes with three distance zones.

Each measuring element uses a micro-controller to produce a software equivalent of the hardware phase comparators which were used in the past. To enable schemes to be suitable for the various applications which may arise, the measuring elements may be arranged to have a wide range of characteristics in the Z plane, including mho, quadrilateral and lenticular of variable aspect ratio by selecting from the various forms of polarizing which are available, including self-polarizing and a range of degrees of cross-polarizing.

Partial synchronous polarizing, derived from the pre-fault system voltages, is also applied to the measuring elements associated with phase to phase faults to ensure that they will operate in the event of three-phase faults near relaying positions during which all the voltages provided to a scheme by the transducers will be very low. The pre-fault voltages, used to provide the necessary polarization, are stored for 16 cycles. Because of this feature, operation of the measuring elements can be obtained for all types of faults with applied voltages down to zero and with currents down to very low levels.

The characteristic angles of the measuring elements are adjustable and the impedance settings can be adjusted over a wide range enabling lines of all types and lengths to be protected. As an example, impedance settings, referred to the voltage- and current-transformer secondary rated values (110 V and 1 A) of 0.2 to 250Ω, are available with accuracies of $\pm 5\%$.

Because these schemes derive power from auxiliary supply units they impose very low burdens on the voltage and current transformers, typical values being 0.1 VA per phase on the voltage transformers at rated voltage and 0.08 VA per phase on current transformers with 1 A secondary windings.

The operating times for faults in Zone 1, which vary with the types of characteristics of the measuring elements and the power system frequency (50 or 60 Hz), are between 14 and 23 ms.

The schemes can detect power swings associated with system instability and tripping is inhibited under these conditions. Several other features can be included, such as, (i) single-pole tripping of the line circuit-breakers in the event of single phase to earth faults, (ii) fault location and (iii) the recording of fault data.

11.7.2 Distance protection relay 7SL32 [12]

This relay, which is produced by Siemens, incorporates six starting elements, which detect the various types of faults which can occur and a memory feature

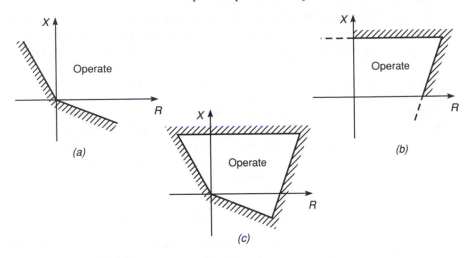

Fig. 11.19 Production of a polygonally shaped characteristic. (a) Starting element (directional). (b) Measuring element (distance). (c) Tripping characteristic.

is provided to store pre-fault data so that close-up faults can be correctly detected. The starting elements, which are cross-polarized, have directional characteristics of the form shown in Fig. 11.19(a) and those which detect phase to phase faults only operate at current levels above the rated values.

There are also six measuring elements which have offset characteristics of the form shown in Fig. 11.19(b) and these elements are energized when one or more of the starting elements operate. As a result, polygonally-shaped operating zones are produced in the Z plane, a typical zone being shown in Fig. 11.19(c). Should a fault be present in the Zone 1 reach, tripping will be initiated in a minimum period of 14 ms. If this does not occur then the reach is extended to cover Zone 2 after a fixed delay and then it can be further extended to produce a three zone scheme after a further interval.

This scheme can also detect power swings and can, if required, initiate single-pole opening of the circuit-breakers. It imposes the low burdens of 0.1 VA per phase and 0.5 VA per phase respectively on the current and voltage transformers.

11.8 THE DUPLICATE PROTECTION OF TRANSMISSION LINES

Transmission lines form vital parts of networks and therefore, as with busbar protection, it is common practice to apply two protective schemes to each major line.

In Britain many of the 400 kV transmission lines are provided with both phase-comparison and distance schemes. This practice is adopted because phase comparison schemes can detect faults with high resistances but they require signalling equipment which may at some time be defective. In such an event, the distance scheme would nevertheless operate to clear the fault. Clearly, in such applications, the two schemes must be able to initiate the tripping of the circuit-breakers independently.

In a few applications two protective schemes are applied to lines and they are so connected that both schemes must operate before the opening of circuit-breakers can be initiated. This practice, similar to that adopted with busbars, ensures that a major line will not be opened because of the mal-operation of one scheme. Again, a given line is provided with schemes operating on different principles because two identical schemes could both maloperate for a particular condition.

11.9 THE PROTECTION OF FEEDER TRANSFORMERS

Feeder transformers of several forms are used in networks because they reduce the number of circuit-breakers needed and thus allow financial savings to be made. Two typical arrangements are shown in Fig. 11.20.

These circuits may be protected in various ways. Overcurrent or unit-type schemes are employed and in many cases distance-type schemes are used to protect the feeder. It will be appreciated, however, that the transformers will have their own protective schemes and devices such as Buchholz and thermal relays and therefore intertripping equipment must be provided to open circuit-breakers at the remote ends of feeders.

Because transformers possess a significant series impedance which is effectively lumped at the end of the feeder, a distance protective scheme at the opposite end of the feeder (end A in Fig. 11.20) may be provided with a single, high-speed zone covering the whole length of the feeder and reaching part way into the transformer without any risk of over-reaching to faults on the low voltage side of the transformer.

Such distance schemes could provide cover for both phase to earth and interphase faults but in many cases instantaneous earth fault protection is applied at the input end of the feeder (end A) and the distance schemes are then only required to detect interphase faults.

In interconnected networks where power could flow in either direction in a feeder transformer, then a distance-type scheme could also be provided at the

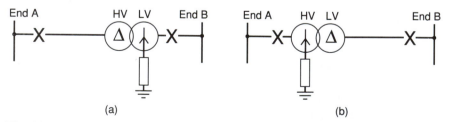

(a)   (b)

Fig. 11.20 Typical arrangements of feeder transformers.

transformer to cover the feeder, the necessary current transformers being mounted within the power transformer. Such a scheme would then be set in the normal way.

11.10 THE PROTECTION OF CABLES

Underground cables, which are used extensively in the lower voltage sections of networks, behave in similar ways to overhead lines and they may be protected by distance-type schemes. The ratios of the series inductive reactances to resistances of cables are lower than those of overhead lines because of the reduced spacings of the phase conductors and therefore the characteristic angles of mho and other relays must be set at the appropriate values in such applications.

The phase displacements between the positive- and zero-sequence impedances of cables may be greater than those of overhead lines and therefore when sound-phase or residual forms of compensation are to be used the phase displacements should be taken into account and the transformers described earlier in section 11.4.3 may not be suitable. When allowance for the phase displacement is made the compensation is described as vectorial compensation.

11.11 THE SETTING OF DISTANCE TYPE SCHEMES

Modern schemes provide ranges of ohmic values at which the various measuring elements may be set to operate. Continuous adjustment is usually possible and the calibrations are quoted in terms of positive-sequence impedance referred to the secondary level, i.e. the secondary levels of the voltage and current transformers. A typical range is $0.25-32\,\Omega$ for a scheme operating from current and voltage transformers with secondary ratings of 1 A and 110 V (line) respectively. In the case of mho elements the above values are the diameters of the characteristic circle, whereas for reactance relays they are the values of reactance below which operation will occur. In addition the characteristic angles of elements may also be set to correspond with those of protected line, i.e. $\tan^{-1}\omega L/R$.

To determine the required settings of elements, it is necessary to calculate the desired reaches along line sections in primary ohms, i.e. for a Zone 1 reach of 80% of a line of length, ℓ, the value would be $0.8\ell\,[R^2+\omega^2 L^2]^{\frac{1}{2}}$ in which R and L are the series resistance and inductance per unit length of the line. The required element settings would then be given by:

$$Z_s = \frac{\text{CT ratio}}{\text{VT ratio}} \times 0.8\ell\,[R^2+\omega^2 L^2]^{\frac{1}{2}}$$

Calculations should be performed using symmetrical component or other techniques to determine the minimum voltages which may be applied to measuring elements during faults of all types at the Zone 1 reach to ensure that the necessary accuracy will be obtained during such conditions. Examples of such calculations are provided in reference [13].

11.12 ACCELERATED CLEARANCE OF FAULTS NEAR THE ENDS OF LINES

There are clearly sections at each end of a line where a fault will be in the Zone 1 reach of one of the distance-type schemes and in the Zone 2 reach of the scheme at the other end. Such a situation, which is illustrated in Fig. 11.21, would cause the circuit-breaker at end B to open rapidly but the circuit-breaker at end A would open after a time delay. This behaviour, as stated earlier in section 11.4, may be unacceptable, this certainly being so when major lines are being protected, and means of accelerating the clearance of such faults must be provided. Several arrangements which are available for this purpose are considered below.

Transfer tripping

Considering the situation shown in Fig. 11.21, operation of a Zone 1 element at end B could cause a signal to be sent to end A to initiate the tripping of the circuit-breaker at that end directly. This arrangement, which is referred to as a 'transfer trip under-reaching scheme' requires high security signalling equipment and coding of signals to avoid maloperation.

An alternative scheme may be used in which the tripping of the circuit-breaker at end A is only initiated on receipt of a signal from end B if at least a Zone 2 or 3 element at end A has detected a fault. This scheme is termed a 'permissive intertrip under-reaching scheme'.

A further alternative which may be used is to detect faults within the Zone 2 reach instantly and send intertripping signals to the other end. Should a Zone 2 element or an extra directional element at the second end have also detected the fault then tripping of the circuit-breaker would be permitted and this process would clearly occur at both ends. These schemes are known as

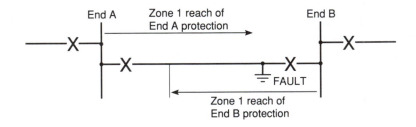

Fig. 11.21 Network configuration used to illustrate tripping arrangements.

'permissive intertripping over-reaching' or 'directional-comparison' schemes. It will be clear that Zone 1 elements should not be needed when these intertripping arrangements are used but such schemes would be very dependent on the signalling channel and equipment. To provide greater security Zone 1 elements are therefore often provided.

Acceleration schemes

A further alternative is to arrange for a signal to be sent when a Zone 1 relay operates at one end of a line. Receipt of this signal at the other end then extends the reach of the elements at that end from Zone 1 to Zone 2 immediately. This scheme, which is termed an 'acceleration scheme', does not operate quite so rapidly as the permissive intertrip schemes because time is required for the measuring elements to operate after their range has been extended, but it is not prone to maloperation.

Blocking schemes

When these schemes are used, reverse-looking directional relays are provided at each end of a protected line to detect faults on circuits connected to it, i.e. external faults. Should these relays operate at one end (B) of the line then blocking signals are sent to the other end of the line (A) to ensure that the circuit-breaker at that end will not trip. In the absence of such a signal, tripping of the circuit-breaker at end A is initiated by the Zone 3, or Zone 2 elements in two-zone schemes, at end A when they detect faults.

It will be appreciated that time must be allowed for the operation of the reverse-looking directional relays and the transmission of the blocking signal before tripping may be initiated by the Zone 2 or 3 relays.

It will be apparent that the schemes described above all have advantages and disadvantages. A factor which must be recognized when schemes are used in which signals have to be sent when faults are present on a protected line is that should the line be used as the signalling channel, then the signals may be attenuated significantly at the fault position.

More detailed information is provided on these schemes in reference [13].

11.13 THE PROTECTION OF TEED LINES

Teed lines with three terminals, which are commonly used in lower voltage networks, can be protected by distance-type schemes at each of the terminals, as shown in Fig. 11.22.

Because each of the sections of such lines will normally be of different lengths it is not possible to set the Zone 1 reaches of the three schemes to the same values. This can be seen from Fig. 11.22 in which the section of line from the tee point to busbar B is much shorter than that to busbar C. In these

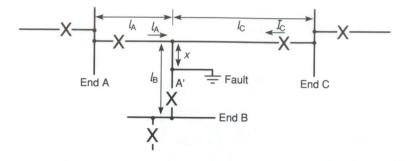

Fig. 11.22 Network containing a teed line.

circumstances, Zone 1 elements at end A would be set to reach to a point A′ near busbar B. These elements would leave a considerable section of the line to busbar C in Zone 2.

A further factor which must be recognized is that the input impedance to a section of line will not be directly proportional to the distance of a fault from the relaying position. This can also be seen from Fig. 11.22 by considering a short circuit on the section of line to busbar B.

If current could be fed to the fault from busbars A and C, i.e. I_A and I_C, then the voltage at busbar A would be:

$$V_A = (R + j\omega L) \{I_A (\ell_A + x) + I_C x\}$$

The apparent input impedance of line A would thus be:

$$Z_{\text{inA}} = \frac{V_A}{I_A} = (R + j\omega L) \left\{ \ell_A + x + \frac{I_C x}{I_A} \right\}_x$$

It can be seen that the magnitude of the apparent input impedance would increase with increase in the infeed from busbar C. As a result, the length of the section of line to busbar B on which faults would be detected by the protection on line A would decrease with increase of the current fed by busbar C, but it would never fall to zero.

Such conditions would cause delayed tripping of one or more of the circuit-breakers for faults in significant sections of the lines because they would be within the Zone 2 reaches. In some applications this performance may be acceptable, but where this is not so, one of the forms of inter-tripping described in the previous section may be provided to accelerate the clearance of faults.

11.14 THE PROTECTION OF SERIES COMPENSATED LINES

The power which can be supplied from the remote end (B) of a line is dependent on the magnitudes and phases of the voltages at both the remote end ($\overline{V}_B$) and at the near end ($\overline{V}_A$) and the series impedance of the line ($\overline{Z}_{AB}$).

For simplicity, the behaviour of a theoretical loss-free, single-phase line with no shunt admittance is considered below. The current (I) in such a line would be given by:

$$\bar{I} = \frac{\bar{V}_A - \bar{V}_B}{\bar{Z}_{AB}} = \frac{\bar{V}_A - \bar{V}_B}{j\,X_{AB}}$$

$$= -j\frac{\bar{V}_A}{X_{AB}} + j\frac{\bar{V}_B}{X_{AB}}$$

If V_B is taken as the reference phasor, then the power output of the line would be:

$$P = Re\left(-j\frac{V_B\,\bar{V}_A}{X_{AB}}\right)$$

This clearly has a value of zero when the voltage V_A is in phase with the voltage V_B and it has a maximum positive value when the voltage at end A leads that at end B by $\pi/2$ rad. In addition, the maximum power is inversely proportional to the inductive reactance of the line.

Similar behaviour occurs on actual three-phase lines and transfer powers would be unacceptably low on many very long lines because of their high inductive reactances and the fact that operation cannot be permitted near the limiting value because of the possibility of instability. To improve the behaviour, it is now common practice to connect capacitors in long lines to effectively reduce the overall reactance between their ends and thus increase their power transfer capabilities.

The ratios of the capacitive reactances to the effective series inductive-reactances of lines used in installations vary, but a typical figure is 0.5. Such degrees of compensation on high voltage lines require that very large voltages be present across the capacitors when their lines are carrying their rated currents and therefore unacceptably high voltages would be produced across the capacitors if they carried high fault currents. It was therefore the practice to limit the voltages across capacitors by connecting arc gaps across them but metal-oxide varistors (MOVs), which have been produced in recent years, are now usually employed. These latter devices, which have resistances which decrease with increase of currents, have the advantage that they are self-extinguishing when normal current conditions are re-established and they provide significant damping to the power systems in which they are installed.

It will be appreciated that the capacitors remain in circuit for short periods after the incidence of faults and are then effectively short-circuited. As a result, protective schemes are presented with two different sets of conditions during the period when a fault is present on a line.

Considering the single-phase line shown in Fig. 11.23(a), in which a capacitor C is installed at the mid-point to provide 50% compensation, i.e. $1/\omega C = 0.5 \ell \omega L$, the steady state input impedance at end A (Z_{in}) for a short circuit at a point on the line at a distance x from end A would vary as shown in Fig. 11.23(b) when the capacitor was in circuit. The variation would, however, be as shown in Fig. 11.23(c) when the capacitor was short circuited. In practice, the situation is further complicated by the fact that the voltages and currents would contain transient components initiated both by the fault and by the short-circuiting of the capacitor.

In the above consideration, the capacitor was taken to be at the mid-point of the line. Whilst such arrangements are used, it is common practice to install capacitors at both ends of a line near the circuit-breakers. Total compensation is typically 70% and such practice has advantages. Firstly a site near the mid-point of a line is not needed and the capacitors can be more readily inspected. A further important factor is that the voltage transformers can be connected on the line sides of the capacitors as shown in Fig. 11.24(a). With this arrangement, the voltage drops across the capacitors do not affect the behaviour of impedance-measuring relays when faults occur on the line although they will do so in the event of faults external to the line, the input

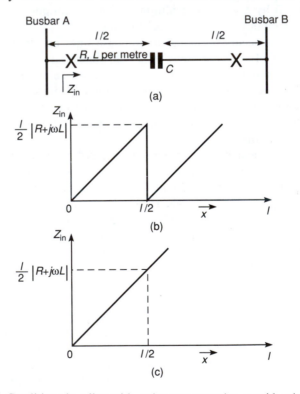

Fig. 11.23 Conditions in a line with series compensation at mid-point.

impedance presented to the relay at end A of the line shown in Fig. 11.24(a) would vary with the distance to a fault on the line or beyond its remote end as shown in Fig 11.24(b) when the remote capacitor is in circuit. The input impedance (Z) would, however, vary as shown in Fig. 11.24(c) when the remote capacitor is short circuited.

Clearly, first and second zone settings which will ensure correct operation for all faults can be selected more readily than is the case when a single capacitor is connected at some point in a line and satisfactory operation can certainly be achieved when accelerating or transfer-tripping features are included. It must be recognized, however, that two sets of transients will be present whenever a capacitor is short circuited. Further details may be found in references [14, 15].

It will be clear from these comments that many factors must be considered when distance-type protective schemes are to be applied to series-compensated lines. To ensure that satisfactory performance will be obtained it is desirable that studies be performed in which systems are accurately modelled on computer-based test benches, capable of dynamically testing proposed schemes.

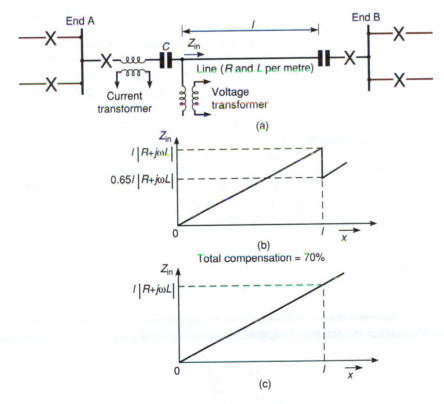

Fig. 11.24 Conditions in a line with series compensation at the two ends.

Newbold and Taylor described studies performed in the above way in a paper presented at an IEE Conference in 1989 [16]. The results of a study of the behaviour of the series-compensated lines of the West Coast, USA were provided and it was shown that distance-type protective schemes could be set so that they would provide satisfactory performance under all conditions.

11.15 THE FUTURE

Distance-type protective schemes have been used successfully for over 50 years to protect overhead lines and cables and it seems probable that they will continue to be used well into the future.

The principles of distance protection have remained essentially the same over the years but the introduction of microprocessors into protective-relay design has introduced new methods of deriving the necessary relaying quantities. The outputs of current and voltage transformers or transducers can be passed through an analogue-to-digital (A/D) converter and thus transformed into a sampled digitized form. Further processing can be employed to implement sophisticated relaying functions [17]. Relaying can thus be regarded as a parameter estimation task whereby samples of the voltage and current signals are processed to obtain quantities such as the r.m.s. value of post-fault voltage and current or the impedance to the fault. Much greater flexibility thus exists in implementing any desired characteristic such as a quadrilateral tripping characteristic of a distance relay.

There are two general approaches to estimating relaying parameters by using digital signal processing techniques.

An estimate can be made based on a model of the waveform, e.g. voltage or current. For conditions encountered during faults it may be postulated that a transducer signal consists of the fundamental, second harmonic, and an exponentially decaying term. Consecutive samples of the signal can then processed to obtain estimates of the various terms. For example, the samples may be processed by an algorithm which implements a Discrete Fourier Transform of the signal and can thus provide the amplitude and phase of the fundamental term.

To illustrate this procedure a voltage signal $v(t)$ of period $2\pi/\omega$ is considered. If this signal is sampled at time intervals T to obtain N samples in any one period $(NT = 2\pi/\omega)$, then the equivalent discrete signal $v(nT)$ is obtained consisting of a succession of samples $v(0), v(T), v(2T). \ldots$

The Discrete Fourier Transform (DFT) of this signal may be obtained by implementing the following algorithm:

$$V(k\Omega_0) = \sum_{n=0}^{N-1} v(nT) e^{-jkn\Omega_0 T} \text{ where } \Omega_0 = \frac{2\pi}{NT}$$

The fundamental-frequency component corresponds to $k = 1$:

$$V_1 = \sum_{n=0}^{\infty} v(nT)\, e^{-j\frac{2\pi n}{N}} = V_1 \underline{|\theta}$$

A similar procedure may be followed for the current signal and an impedance estimate is then obtained from:

$$Z = (V_1 \underline{|\theta})/(I_1 \underline{|\theta})$$

This value is then compared with the required tripping characteristic in the impedance plane and the appropriate action is taken. The number of samples that should be processed before a secure decision is reached is a compromise between speed of operation and immunity to noise. Sophisticated methods are available which can make allowances for the presence of noise in the sampled waveforms so that a rapid estimate is made with the minimum error. An example of this technique is Kalman filtering. This is an example of an adaptive method where after each sample an update of the expected error is made so that estimates improve rapidly [18].

An alternative approach to parameter estimation based on a model waveform, is one based on a model of the system. It may, for example, be assumed that the system may be presented by a series R–L combination. Then the voltage $v(t)$ and the current $i(t)$ seen by the relay are related by [19].

$$v(t) = Ri(t) + \underline{|L}\,\frac{di(t)}{dt}$$

Integrating this equation from t_1 to t_2 gives

$$\int_{t_1}^{t_2} v(t)\, dt = R \int_{t_1}^{t_2} i(t)\, dt + L\,[i(t_2) - i(t_1)] \tag{11.13}$$

These integrals may be evaluated using the trapezoidal rule, i.e.

$$\int_{t}^{t+\Delta t} v(t)dt = \frac{1}{2}\,[v(t+\Delta t) + v(t)]\,\Delta t$$

where Δt is the time interval between consecutive samples.

Implementing equation (11.13) between time intervals $k\Delta t$ to $(k+1)\,\Delta t$ and $(k+1)\,\Delta t$ to $(k+2)\,\Delta t$ and solving for R and L gives:

$$R = \frac{(v_{k+1}-v_k)(i_{k+2}-i_{k+1}) - (v_{k+2}-v_{k+1})(i_{k+2}-i_k)}{(i_{k+1}+i_k)(i_{k+2}-i_{k+1}) - (i_{k+2}+i_{k+1}) + (i_{k+1}-i_k)}$$

$$L = \frac{\Delta t}{2}\left[\frac{(i_{k+1}+i_k)(v_{k+2}+v_{k+1}) - (i_{k+2}+i_{k+1})(v_{k+1}+v_k)}{(i_{k+1}+i_k)(i_{k+2}-i_{k+1}) - (i_{k+2}+i_{k+1})(i_{k+1}-i_k)}\right]$$

where v_k, i_k is the value of the voltage and current sample respectively at time $k \cdot \Delta t$. The estimates R and L can then be used to implement any desired tripping characteristic. The advantage of this formulation is that transient terms are fully accounted for in the model and do not constitute an error. However, it is obvious that the model does not take account of capacitive effects, and hence of travelling-wave phenomena. Another aspect of practical significance is that the algorithm described above is based only on three samples and may therefore be unacceptably sensitive to signal noise. Improvements can be made where the processing window of three samples is made longer so that better noise immunity is obtained [20, 21]. Further details of a number of relaying algorithms may be found in references [22, 23].

Fully algorithmic relay designs based on the principles outlined above are likely to become increasingly available from manufacturers. The fully digital design will allow greater flexibility in algorithm optimization and updating, comprehensive interrogation of the relay from a remote terminal, more sophisticated interlocking arrangements with other relays (e.g. on double-circuit lines), closer coordination with power system control functions via SCADA systems and sophisticated self-test facilities. A recently introduced fully digital relay is described in reference [24].

REFERENCES

1. Ackerman, P. (1922) *Journal of the Engineering Institute of Canada*, Dec. 1922 (also New radial relay protection, *Electrical World*, **81**, *March* 17, 1923, pp. 619–623).
2. Crichton, L. N. (1923) The distance relay for automatically sectionalising electrical networks, *Trans. IEE*, **XLII**, 527–537.
3. McLaughlin, H. A. and Erickson, E. O. (1928) The impedance relay development and application, *Trans. IEE*, July 1928, **47**, 776–784.
4. George, E. E. (1931) Operating experience with reactance type distance relays, *Trans. IEE*, **50**, 288–293.
5. Ellis, N. S. (1982): Distance protection of feeders. In *Power System Protection Manual*, Reyrolle Protection, (ed D. Robertson), Oriel Press, pp. 149–179.
6. Wright, A. (1961) Limitations of distance-type protective equipment when applied to long extremely high voltage power lines, *Proc IEE*, (C), **108**, 271–280.
7. Davison, E. B. and Wright, A. (1963) Some factors affecting the accuracy of distance-type protective equipment under earth fault conditions, *Proc IEE*, **110**, 1678–1688.
8. Micromho-Static distance protection relay, GEC Measurements, *Publication R5406*.
9. Power Swing Blocking Relay, Type LU91, Asea Brown Boveri, (1987), *Publication CH-ES-66.51.11*.
10. Reyrolle Ratio Balance Distance Protection Type XZA Pamphlet 1044 (1951).
11. Type LFZP Optimho Static distance protection relay, GEC Measurements, *Publication R-4056 C*, 18 pages.

12. High-speed distance protection relay 7SL32, Siemens Protective Relays, *Catalog R-1989*, pp 5/21–5/26.

13. Distance protection schemes, In *Protective Relays Application Guide*, GEC Measurements (3rd edn), 1987, Chapter 12.

14. Mathews, C. A. and Wilkinson, S. B. (1980) Series compensated line protection with a directional comparison relay scheme, *2nd IEE Int Conf. on Developments in Power System Protection*, IEE Conf. Publ. No. 185, pp. 215–220.

15. El-Kaleb, M. M. and Cheetham, W. J. (1980) Problems in the protection of series compensated lines, *2nd IEE Int. Conf. on Development in Power System Protection*, IEE Conf. Publ. No. 185.

16. Newbold, A. and Taylor, I. A. (1989) Series compensated line protection: system modelling and relay testing, *4th IEE Int. Conf. on Developments in Power System Protection*, IEE Conf. Publ. No. 302, pp. 182–186.

17. Mann, B. J. and Morrison, I. F. (1971) Digital calculation of impedance for transmission line protection, *Trans. IEE*, **PAS-90**, 270–279.

18. Girgis, A. A. and Brown, R. G. (1981) Application of Kalman filtering in computer relaying, *Trans. IEE*, **PAS-100**, 3387–3397.

19. McInnes, A. D. and Morrison, L. F. (1970) Real time calculation of resistance and reactance for transmission line protection by digital computer, *Elec. Eng. Trans., IE, Australia*, **EE7**, (1), 16–23.

20. Ranjibar, A. M. and Cory, B. J. (1975): Algorithms for distance protection, *Developments in Power System Protection*, IEE Conf. Publ. 125, London, March 1975, pp 276–283.

21. Sanderson, J. V. H. and Wright, A. (1974) Protection scheme for series-compensated transmission lines, *Proc IEEE*, **121**, 1377–1384.

22. Microprocessor relays and protection systems, *IEE Tutorial Course*, (Course coordinator Ms Sachdev), 88EH0269-1-PWR.

23. Phadke, A. G. and Thorpe, J. S. (1988) *Computer Relaying for Power Systems*, RSP.

24. Holweck, C. (1990) Design and performance of PXLN digital protection devices for EHV/UHV power systems, *GEC Alsthom Technical Review*, No. 2, pp. 23–24.

FURTHER READING

Analysis of distance protection, V. Cook, RSP 1986, 188 pages.

Protection aspects of multi-terminal lines, *IEE Power Eng. Soc. Publ*, IEEE Report No 79, TH 0056-2-PWR, 1979, pp. 1–17.

Determination of time settings in a distance protection scheme based on statistical methods, J. De Haas, *CIGRE-report 31–05*.

12

Ultra-high-speed schemes for the protection of long transmission lines

INTRODUCTION

It has been made clear in earlier chapters that unit-type protective schemes which compare the quantities at the two ends of transmission lines can provide the necessary discrimination between faults on the circuits they protect and both healthy conditions and faults on other circuits. Such schemes cannot however operate extremely rapidly because time must be allowed for the interchange of signals between the ends of the protected lines.

These factors do not arise when impedance- or distance-type protective schemes are used and the tripping of circuit-breakers may be initiated for faults within the Zone 1 reaches without delays being introduced during which information is obtained about conditions at the remote ends of the protected lines. To achieve the required measurement accuracies, however, a significant portion of a power-frequency cycle must be monitored after fault inception and therefore circuit-breaker tripping cannot usually be initiated in times much less than 20 ms on systems operating at a frequency of 50 Hz.

In many applications the performance provided by these schemes is acceptable but there has long been a desire to obtain more rapid clearance of faults on long major transmission lines. Such performance would reduce the damage which can be caused by high currents flowing in healthy equipment and in addition it would reduce the possibility of asynchronous operation being initiated as a result of faults on transmission lines. It will be appreciated that the incidence of a fault on a line will reduce the power transmitted by it and should it be connected between two sources then the machines at one end will supply less power whilst those at the other end will supply more. As a result one set of machines will accelerate whilst the others will decelerate and their angular displacements will increase. If this situation were allowed to persist, asynchronous operation would occur and, as stated above, the likelihood of such behaviour is reduced if the time needed to clear faults is reduced.

In recent years the possibilities of producing protective schemes using modern electronic-processing equipment to analyse the travelling waves present on long transmission lines during fault conditions have been examined extensively, because such schemes should enable the most rapid operation possible to be obtained. After the next section, which provides historical material, the principles and operation of schemes which may be used extensively in the future are examined.

12.1 HISTORICAL BACKGROUND

It has long been known that a sudden change in the voltage at a point on a transmission line, resulting from a fault or short-circuit, causes travelling waves of both voltage and current to propagate away from the point in both directions.

When studying such conditions it is usual to assume that the steady state conditions prior to a fault or disturbance continue and that transient travelling waves will also be present during the fault. The situation is usually simulated by considering a source of e.m.f. to be connected suddenly at the fault point and it is this e.m.f. which causes the transient currents and voltages to be present.

Because major transmission lines have relatively low series resistances and low shunt conductances, the travelling-wave components which do not involve the earth have a velocity near that of light, i.e. 3×10^8 m/s and thus waves caused by a fault on a line reach its ends very quickly. For example, the first current and voltage waves resulting from a fault at a point 300 km from the end of a line would arrive at that end 1 ms after the incidence of the fault. It will be appreciated that this would be the first indication at the end of the line of the presence of the fault and if information could be quickly obtained from these waves by protective equipment then the most rapid fault clearance possible could be effected.

Analyses performed over 30 years ago indicated the possibilities of producing protection schemes which analysed travelling waves but the equipment available at that time was not capable of performing the necessary signal processing in real time. Subsequent developments in electronic equipment did, however, provide the opportunity for the development of travelling-wave schemes and in recent years much research effort has been devoted to examining the various detection processes which may be employed.

In the later 1970s a considerable number of papers were published on methods of determining the directions of faults from the first travelling waves reaching the ends of long lines, it being intended to use this information in directional-comparison schemes. Two important papers were published by Chamia and Liberman in the IEE *Transactions* in 1978 [1] and Johns in the *Proceedings* of the IEE [2] in 1980 respectively. Because directional-comparison schemes have the disadvantage that signals must be sent between the ends of lines to ensure correct discrimination, other workers sought to obtain further information from the travelling waves so that schemes could be produced in which fault positions could be determined from the quantities at a single end of a line. Papers published by Crossley and McLaren [3] in 1983 and by Christopoulos *et al.* [4, 5] in the *Proceedings* of the IEE in the years 1988 and 1989, describe such schemes. It is made clear [4] that the necessary information cannot be obtained from the first travelling current and voltage waves to arrive at the end of a line after a fault occurs on it and that information must also be extracted from the next wave which arrives as a result of reflections at the line end and the fault.

Detailed considerations of the behaviour of travelling waves and protective schemes based on them are provided in the remainder of this chapter.

12.2 TRAVELLING WAVES

As stated earlier, travelling waves propagate in both directions away from any point on a transmission line at which the voltage is suddenly changed as a result of a disturbance such as a fault.

The voltage waves progressively change the voltage at other points on the line and they are accompanied by current waves which change the charges on the line capacitances.

12.2.1 Initial waves on d.c. lines when short-circuits occur

The above behaviour can be most simply illustrated by considering an ideal, i.e. loss-free, two-conductor line operating in a d.c. system. Referring to Fig. 12.1(a), the voltage between the conductors would be the same at all points

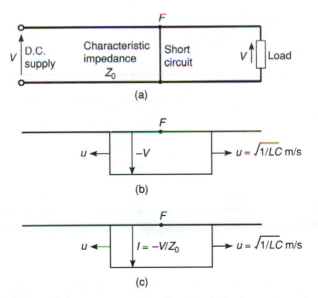

Fig. 12.1 Voltage and current waves on a d.c. line following a short-circuit.

under healthy conditions. Each of the capacitors per unit length (C) would be charged to the system voltage (V) and there would be no voltage drops across the series inductors per unit length (L).

In the event of a short-circuit between the conductors of a line at a point F, travelling voltage waves of magnitude V and negative polarity would be propagated from point F at a velocity of $\sqrt{1/LC}$ m/s, as shown in Fig. 12.1(b). The accompanying current waves would have a magnitude of V/Z_0, where the characteristic impedance Z_0 is equal to $\sqrt{L/C}$ Ω. The polarities of the travelling waves of voltage and current which would reach both ends of the line would all be the same, i.e. negative in the above example.

Should two such lines be connected in series as shown in Fig. 12.2(a) and a short-circuit occurred on one of them, say line P, then the above waves would again be propagated. At both ends of line P the arriving voltage and current waves would again have the same polarities, i.e. negative. As these waves propagated into line Q, the voltage wave would retain its negative polarity but a device detecting the current flowing to the busbar from line Q would regard the current wave as positive. These conditions are illustrated in Fig. 12.2(b). It will therefore be clear that the arrival of voltage and current waves of the same polarity at the end of a line would indicate a fault on the line or beyond its remote end whereas the arrival of waves with opposite polarities would indicate a fault in the opposite direction, i.e. on line P in the above example.

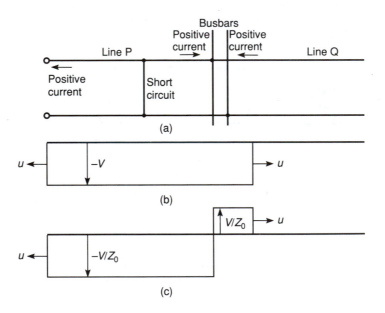

Fig. 12.2 Voltage and current waveforms for conventional positive current reference directions.

It must be recognized that the travelling waves exist in addition to the pre-fault voltage and current and therefore to produce a directional detector the pre-fault, i.e. steady state quantities, must be subtracted from the actual fault currents and voltages to obtain the travelling-wave components.

An important feature of this technique is that it enables correct operation to be obtained under all conditions, unlike other types of directional relays which may not operate when voltages collapse to very low levels during faults.

12.2.2 Initial waves on ideal single-phase a.c. lines when short-circuits occur

As stated earlier, the extra voltages and currents which are superposed on the d.c. lines considered above, in the event of a short-circuit between the conductors, may be determined by connecting a source at the fault position with an e.m.f. equal to but of opposite polarity to that present at the instant of fault occurrence, i.e. $-V$.

The same technique may be used when the behaviour of a.c. lines is being considered, but in such cases the source connected at the position of the short-circuit must provide an e.m.f. equal and opposite to that which would have been present if the fault had not occurred. This situation is shown in Fig. 12.3(a), from which it can be seen that the source e.m.f. would be given by $e = -V_{pk} \sin(\omega t + \alpha)$ if the voltage at the fault position F would have been $v = V_{pk} \sin(\omega t + \alpha)$ had the fault not occurred. In each case the time t is measured from the instant of fault occurrence.

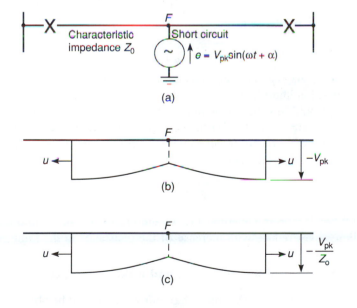

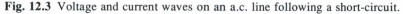

Fig. 12.3 Voltage and current waves on an a.c. line following a short-circuit.

As a result, travelling waves of voltage and current would propagate in both directions from the fault position. Again these waves would travel at the velocity of light on ideal lines and they would not be attenuated as they travelled. Unlike the waves on d.c. lines, however, their amplitudes at given positions would vary with time after the wavefront had passed the various positions but the amplitudes at the wavefronts would be constant. This behaviour is illustrated in Fig. 12.3(b) for a condition in which a short-circuit has taken place at an instant when the voltage at the fault position was positive and of its peak value, i.e. $v = V_{pk}$.

The accompanying current would again be given by $i = e/Z_0$ and it would thus have the same waveform as the voltage. As in the case of d.c. lines, the polarities of the travelling waves of voltage and current which would arrive at the ends of the faulted line would be the same but, as before, they would be of opposite polarities in the event of faults on other lines. The directions of faults could therefore be determined as before.

A factor which must be recognized, however, on an a.c. line is that the initial magnitudes of the travelling waves of voltage and current may be very small or even zero if faults occur at or near instants when the voltage at the fault position is zero. Clearly the directions of faults could not then be determined immediately a fault occurred.

12.2.3 Initial waves on ideal lines when resistive faults occur

The initial value of the travelling voltage waves which would propagate away from a fault position in the event of a fault with a resistance of $R_f \, \Omega$ between the conductors would be given by:

$$V_1 = -\frac{Z_0}{Z_0 + 2R_f} \cdot V_f \qquad (12.1)$$

in which Z_0 is the characteristic impedance of the line and V_f is the voltage at the fault position immediately prior to the incidence of the fault. In the case of a d.c. line V_f would be equal to the line voltage V and for an a.c. line it would be $V_{pk} \sin \alpha$, α being the angle after the voltage zero at which the fault occurred.

The initial values of the accompanying current waves, taking the directions of the waves as the references, would be given by:

$$I_1 = \frac{V_1}{Z_0} = \frac{1}{Z_0 + 2R_f} \cdot V_f \qquad (12.2)$$

It will be clear from equations (12.1) and (12.2) that the amplitudes of the travelling waves reduce with increase of the resistance in the fault path.

12.2.4 The effects of line resistance and leakage conductance

In the sections 12.2.1–12.2.3, the behaviour which would be obtained if faults occurred between the conductors of ideal d.c. or single-phase a.c. lines were considered. The conductors of practical lines possess resistance (R), however, and leakage conductance (G) is present between them. The latter quantity (G) which is usually very low and variable, occurs mainly because of deposits and moisture on the surfaces of the insulators. The presence of resistance and leakage conductance causes the shapes of the fronts of travelling waves to be modified and the amplitudes of the waves attenuate as they travel. These effects tend, however, to be small and this is particularly so when faults occur between the conductors of major three-phase transmission lines which employ several conductors in parallel per phase, i.e. bundle conductors. In such cases, as shown earlier, the directions of interphase faults relative to the ends of lines can be readily detected from the polarities of the initial travelling waves which arrive at the ends of lines after the incidence of such faults.

It must be recognized, however, that in the event of faults which cause currents to flow in the earth, the velocities of the ground components of the travelling waves and also their wavefronts may be affected significantly, the differences being dependent on the resistivities of the ground-return paths. This factor, which must be taken into account, is considered later under section 12.2.6 (page 452).

12.2.5 Later travelling waves resulting from reflections

It will be clear from the preceding sections that the directions of faults relative to the ends of lines can be determined from the polarities of the initial voltage and current waves which reach those ends after faults occur.

The ratio of the amplitudes of the initial incoming voltage and current waves at the end of a line is always equal to the characteristic impedance (Z_0) of the line, which is known, and therefore this parameter does not give any extra information about any fault which is present.

In the case of a d.c. line, the voltages at all points along its length will be almost equal to the voltages at its ends during healthy conditions and should a fault occur at a point on it, the approximate value of the pre-fault voltage at that point (V_f) would be known. Measurement of the magnitude of the first travelling voltage waves to arrive at the line ends (V_1) would therefore enable the fault resistance (R_f) to be assessed from equation (12.1). No information about the position of the fault, other than its direction could, however, be obtained from the initial waves.

In the case of an a.c. line the situation is basically similar. During healthy conditions there may, however, be significant phase displacements between the voltages at various points along the line and therefore the instantaneous voltage at the ends of the line may be considerably different to those at other points. Should the instantaneous voltage at an end, at the time when a fault occurs, be used in equation (12.1) as the value of V_f then the assessment of the fault resistance could be quite inaccurate.

Further information, which could be used in protective schemes to enable

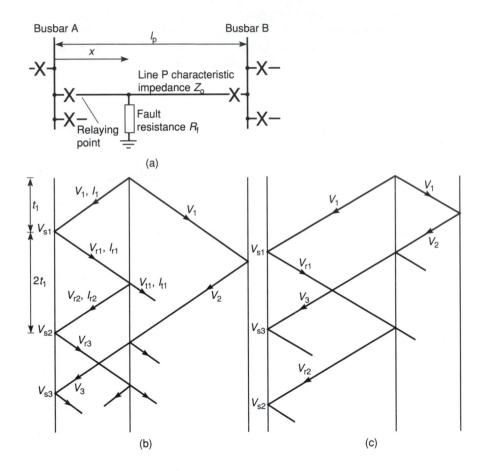

Fig. 12.4 A Bewley lattice diagram for different fault positions.

satisfactory discrimination to be obtained without recourse to signalling between the ends of lines, could be derived from the later travelling waves which are caused by reflections. This is illustrated by making use of the well-known Bewley [4] lattice diagram. To simplify the treatment an ideal single-phase line P connected between busbars A and B as shown in Fig. 12.4(a) is now considered.

The initial values of travelling waves produced by faults on a protected line

In the event of a fault with a resistance R_f at a point a distance x from busbar A, a voltage wave would travel towards busbar A. The initial values of this wave and the accompanying current wave would, as stated earlier in equation (12.1), be given by:

$$V_1 = -\frac{Z_0}{Z_0 + 2R_f} \cdot V_f$$

and

$$I_1 = -\frac{1}{Z_0 + 2R_f} \cdot V_f$$

As there are invariably several connections to a set of busbars, any waves, such as those above, would initiate reflected waves (V_{r1} and I_{r1}) when they arrived at the discontinuity presented by the busbar. This situation is shown in Fig. 12.4(b).

It is well known that the voltage and current reflection coefficients for the waves arriving at busbar A would be:

$$k_{vA} = [R_A - Z_0]/[R_A + Z_0] \tag{12.3}$$

$$k_{iA} = - [R_A - Z_0]/[R_A + Z_0] = - k_{v1} \tag{12.4}$$

In both these equations, R_A is the input resistance presented at busbar A to the incident waves arriving from line P. For the purpose of illustrating the principle of the technique R_A is assumed resistive.

Protective voltage and current transducers on line P near busbar A would not detect the incident and reflected waves separately because their time separation would be very short. They would therefore provide outputs proportional to the sums of the initial values of the two sets of waves, i.e.

$$V_{s1} = V_1 + V_{r1} = V_1 (1 + k_{vA}) = \frac{2R_A V_1}{R_A + Z_0} \tag{12.5}$$

and

$$I_{s1} = I_1 + I_{r1} = I_1(1 + k_{iA}) = \frac{2V_1}{R_A + Z_0} \tag{12.6}$$

The ratio of these equations is

$$\frac{V_{s1}}{I_{s1}} = R_A \tag{12.7}$$

By determining the value of R_A from the above equation and knowing the characteristic impedance of the line (Z_0), the reflection coefficients, k_{vA} and k_{iA}, could be determined from equations (12.3) and (12.4) and then the initial values of the incident at reflected voltages and currents $(V_1, V_{r1}, I_1, I_{r1})$ could be found from equations (12.5) and (12.6).

The first wave reflected from busbar A (V_{r1}, I_{r1}) would return along the line P towards busbar B until it arrived at the fault position where further reflected waves (V_{r2}, I_{r2}) and transmitted waves (V_{t1}, I_{t1}) would be initiated as indicated in the lattice diagram shown in Fig. 12.4(b).

The voltage reflection coefficient k_{vf} at the fault position would be given by:

$$k_{vf} = \frac{[Z_0 R_f/(Z_0 + R_f)] - Z_0}{[Z_0 R_f/(Z_0 + R_f)] + Z_0} \tag{12.8}$$

The reflected voltage wave V_{r2} which would return to busbar A would therefore have an initial value given by:

$$V_{r2} = k_{vf} V_{r1}$$

This wave would in turn reflect from busbar A to initiate a further reflected voltage wave with an initial amplitude V_{r3} given by:

$$V_{r3} = k_{vA} V_{r2}$$

The increment applied at this time to the voltage transducer on line P at its end connected to busbar A would be equal to:

$$V_{s2} = V_{r2} + V_{r3} = k_{vf} (1 + k_{vA}) V_{r1} \tag{12.9}$$

Having already determined the initial value of the reflected voltage wave V_{r1}, the value of the coefficient k_{vf} could be found from equation (12.9) and this would then enable the fault resistance (R_f) to be determined from equation (12.8).

As shown in Fig. 12.4(b), the voltage increment V_{s2} would occur at a time $2t_1$ seconds after the increment V_{s1} produced by the first travelling wave (V_1), indicating that the distance x of the fault from busbar A would be given by:

$$x = ut_1 \quad \text{metres} \tag{12.10}$$

in which u is the propagation velocity of the travelling waves which is known for any particular line.

It will be seen that a considerable amount of information could be obtained by measuring the increments V_{s1}, I_{s1} and V_{s2} and the time interval between the voltage increments V_{s1} and V_{s2}. The results would, however, only be correct provided that the wave reflected from busbar B and transmitted through the fault to produce the increment V_{s3} in Fig. 12.4(b), arrived after the increment V_{s2}. This condition would be met if a fault occurred within half the line length from busbar A, i.e. $x < \ell_p/2$.

For faults on the other half of the line, the increment V_{s3} caused by the reflection from busbar B would, however, occur before the increment V_{s2} caused by the reflection of the wave V_{r1} as shown in Fig. 12.4(c). In these circumstances, if the increment V_{s3} were used in equations (12.1) to (12.7) to determine the fault resistance (R_f) and the distance to the fault (x), the values obtained would be incorrect.

Nevertheless, it will be clear from the above that the positions of faults could be determined using equation (12.10) provided that the increments V_{s2} and I_{s2}, produced by waves travelling between the faults and the busbar A, could be identified.

Several methods of identifying the waves V_{s2} and I_{s2} have been suggested and schemes based on such techniques have been proposed. These are examined later in section 12.3.2.

It will be appreciated that whilst this procedure would enable the positions of faults to be determined, it has the same limitation as existing distance type protective schemes, namely that it could not be set to operate immediately in the event of all faults on a protected line. Provision would have to be made for inevitable errors and it would be necessary to have settings covering two or more zones.

Travelling waves caused by faults external to a protected line

It will be clear that travelling waves may be initiated at many points on an interconnected network when faults or other disturbances occur, and it is essential that any protective scheme which analyses these waves should be able to discriminate correctly.

It has been shown earlier that faults on circuits connected to busbar A in Fig. 12.4(a), other than those on line P, would cause travelling voltage and current waves of opposite polarities to be sensed by a scheme associated with line P, whereas faults on line P would set up waves of the same polarity.

This situation would not obtain, however, for faults on the network which would cause travelling waves to be propagated from busbar B to busbar A. Other checks, such as those described later in section 12.3.2, could be made to ensure correct discrimination between such faults and those on the protected line.

The voltage and current variations at the end of a line when a fault occurs on it

The effects of reflections on the initial values of travelling waves were considered under section 12.2.5. As stated in section 12.2.3, however, the transient voltages and currents at points on a line vary with time after faults occur on it.

When a fault occurs on an a.c. line, therefore, the transient voltages and currents at its end may have step changes at intervals and sinusoidal variations between the steps. For simplicity the behaviour obtained when a short circuit occurs on a loss-free line connected to an infinite source is examined below, the circuit being shown in Fig. 12.5(a).

As stated earlier, a travelling voltage wave (v_1) with an initial value of $-V_{pk} \sin \alpha$ would propagate towards busbar A, the voltage at the fault position at the instant of fault occurrence being $V_{pk} \sin \alpha$. A travelling current wave (i_1) with an initial value of $-V_{pk} \sin \alpha / Z_0$ would accompany the voltage wave. The waves would reach busbar A at a time $t_1 = x/u$, after the incidence of the fault, u being the velocity of light.

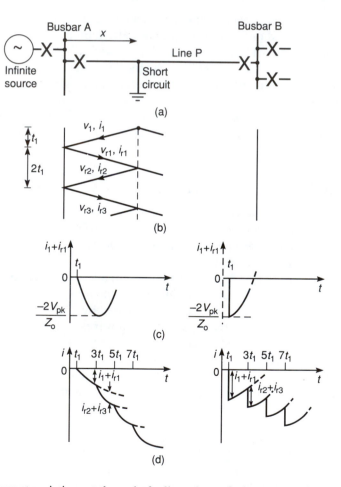

Fig. 12.5 Current variations at the end of a line when a fault occurs on it.

Because of the presence of the infinite source, no change in the level of the voltage of the busbar would occur and therefore a reflected voltage wave (v_{r1}) with an initial value of $V_{pk} \sin \alpha$ would be propagated along the line towards the fault, as shown in Fig. 12.5(b). The accompanying current wave (i_{r1}) would have an initial value of $-V_{pk} \sin \alpha / Z_0$, taking current flow into the busbar from the line to be positive. The current transformer in the line near busbar A would therefore carry a current increment with an initial value of $I_1 + I_{r1}$ which would be equal to $-2 V_{pk} \sin \alpha / Z_0$.

These conditions are illustrated in the Bewley Lattice diagram shown in Fig. 12.5(b). The current in the primary winding of the current transformer would then vary sinusoidally, i.e. $i_1 + i_{r1} = -2 V_{pk}/Z_0 \sin (\omega t + \alpha)$. For faults occurring at instants of voltage zero and at peak voltage, i.e. $\alpha = 0$ and $\pi/2$, the currents would vary as shown in Fig. 12.5(c). The waves v_{r1} and i_{r1} would travel to the fault where they would be reflected, the initial values again being the same as those of the arriving waves, the polarity of the voltage reversing but that of the current remaining the same. The waves v_{r2} and i_{r2} would then travel to busbar A, causing further reflections, i.e. v_{r3} and i_{r3}. The increment in the current on this occasion would again be $-2V_{pk} \sin \alpha / Z_0$. This process would continue indefinitely on a loss-free circuit and currents of the form shown in Fig. 12.5(d) would result.

It will be seen from Fig. 12.5(d) that the polarities of the current increments would always be the same as that of the initial voltage wave arriving at busbar A, i.e. v_1, and that the current would have a finite value after a short time even though a fault occurred at an instant when the voltage at the fault position was zero.

It will also be seen that currents would reach high values very rapidly if faults occurred near the end of a line, i.e. when x and therefore t_1 would have small values. Clearly the build up of current would be slow when faults occurred at more remote positions.

In practice, the travelling waves would attentuate as they travelled because of the line losses and the successive steps would reduce and steady state sinusoidal conditions would eventually be achieved. At this time, the current levels would be controlled by the line series impedance and again the current magnitude would decrease with increase of the distance to a fault, i.e. high currents would flow for close-up faults and lower currents would flow when more remote faults were present.

It will be appreciated that practical busbars are not infinite and that their voltages are affected by the arrival of travelling waves initiated by faults. As a result, the reflected currents do not have the same magnitudes as those of the incoming waves and therefore currents steps of $2V_{pk} \sin \alpha / Z_0$ are not produced. Nevertheless quite rapid build-ups of current are produced when short-circuits occur on a line at points near the busbar and in this connection it is worthy of note that the time interval between the current steps would only be 0.2 ms for a fault at a distance of 30 km from the busbars, i.e. $t_1 = 0.1$ ms for propagation near the velocity of light.

12.2.6 Travelling waves on three-phase lines

In the preceding sections, the propagation of travelling waves has been considered on d.c. and single-phase bases for simplicity. In such situations a line has a single characteristic impedance (Z_0) and velocity of propagation (u) and under ideal loss-free conditions these parameters are given by:

$$Z_0 = \sqrt{\frac{L}{C}} \ \Omega \quad \text{and} \quad u = \sqrt{\frac{1}{LC}} \ \text{m/s}$$

in which L and C are the inductance and capacitance per unit length of the line.

The value of Z_0 is dependent on the diameter of the conductors and the spacing between them. The inductance increases with increase in the spacing and decrease in the conductor diameter, whereas such changes have the opposite effects on the capacitance. As a result, the characteristic impedance behaves similarly to the inductance, i.e. it increases with increase in the spacing and decrease in the conductor diameter. The velocity (u) is not, however, affected by the line spacing and the conductor diameters, and under ideal conditions the velocity of propagation is the same as that of light.

The behaviour of three-phase lines is, however, much more complex. Firstly it must be recognized that some waves propagate only in the line conductors, i.e. say the phase a and b conductors in the event of an interphase fault between phases a and b. Such waves are said to propagate in an aerial mode. In the event of a single-phase to earth fault, however, the current waves flow in a single line conductor, say the phase a conductor, and return in both the earth conductor or conductors which run above the line being supported on the tops of the towers, and also in the ground. These waves are said to propagate in the ground mode.

Aerial-mode propagation

Because the conductors of major transmission lines have very low resistances per unit length and the leakage conductances between the phase conductors are also very low, aerial-mode waves travel at almost the velocity of light and therefore any errors caused by using the velocity of light to assess the positions of faults in the manner described in the previous section would be very small.

The characteristic impedances presented to the waves set up when the various types of interphase faults occur will not all be the same because of the asymmetric positioning of the phase conductors and the fact that conductors are not transposed on major lines because of the practical difficulties that would be entailed in so doing.

Although the phase conductors of a line are each the same, the interphase spacings vary with the configuration employed. As an example, for a horizontal arrangement with equal spacings between the centre and outer conductors, the spacing between the outer phases, say 'a' and 'c', is twice that between the other pairs of phases, 'a' and 'b' and 'b' and 'c'. In this case the characteristic impedance for propagation on the outer phases (Z_{oac}) would be greater than those for the other phases, (Z_{oab}, Z_{obc}), which would both have the same values. Because the inductances and capacitances per unit length are proportional to the logarithms of the spacings and their inverses respectively, the characteristic impedances are also proportional to the logarithms of the spacings. They therefore do not vary as greatly as the spacings. For given lines, they can be readily determined and the appropriate values could be used when the particular interphase faults present have been identified from the initial travelling waves.

Ground-mode propagation

As stated above, voltage and current waves are propagated in parallel along the overhead earth conductors and in the ground when phase to earth faults occur on a line. Because the earth conductors are effectively connected to ground at each tower both sets of waves must propagate at the same velocity.

Clearly, in the absence of the ground, the waves would travel in the faulted phase conductor and the overhead earth conductor and they would thus be of the aerial mode. In the absence of earth conductors, the waves would have to travel in a phase conductor and the ground. The current in the ground tends to follow the route of the line but it is not confined to a particular cross-sectional area and it is affected by the resistivity of the ground. In these circumstances, the inductance per unit length of the conductor-ground loop cannot be calculated simply from basic principles.

To assist in studying telephone and telegraph circuits in which signals were transmitted via a single conductor and returned through the ground, W. Carson produced a model circuit [6] in which the ground was represented by an image conductor, of the same diameter as the aerial conductor, at a depth in the ground given by:

$$d_g = 660\sqrt{\frac{\rho}{f}} \quad \text{m}$$

in which ρ is the average resistivity of the ground and f is the system frequency. For typical resistivities usually encountered, i.e. $100–\Omega$m, the depth of the image conductor in power applications, i.e. at frequencies of 50–60 Hz, is 933 metres.

Using this model, the inductance per metre (L) of a conductor above ground is given approximately by:

$$L = \frac{\mu_0}{\pi} \ln\frac{d_g}{d} \quad \text{H/m}$$

in which d is the effective diameter of the overhead conductor and d_g is an effective height.

This modelling does not enable the capacitance between a conductor and ground to be determined satisfactorily, and for this purpose it is common to assume that the surface of the ground is an equipotential, i.e. at zero volts. The capacitance is then determined assuming an image conductor with a charge equal and opposite to that on the overhead conductor, the depth of the image conductor below the ground surface being equal to the height of the aerial conductor above the ground (h).

Using this technique, the capacitance to ground (C) of a conductor per metre is given by:

$$C = \frac{\pi\varepsilon_0}{\ln 4h/d} \quad \text{F/m}$$

in which d is the effective diameter of the aerial conductor and h is its height above ground.

Because the depth of the image conductor (d_g) in the model used to determine the inductance is greater in practice than the depth of the image conductor in the capacitance model (h), the velocity of propagation of travelling waves which flow in the ground tends to be significantly lower than the velocity of light. As an example, for the line with the configuration shown in Fig. 12.6, the velocity of propagation for waves travelling in a phase conductor and returning through the ground and the overhead earth conductor is 1.8810^8 m/s, i.e. about two thirds of the speed of light. It will be appreciated that such velocities are so far removed from the velocity of light that fault positions calculated using the latter value would be too inaccurate and therefore the actual values for particular lines must be calculated and used.

The values of inductance and capacitance determined in this way must also be used to calculate the ground-mode characteristic impedance.

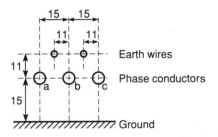

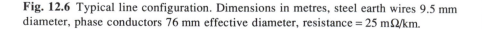

Fig. 12.6 Typical line configuration. Dimensions in metres, steel earth wires 9.5 mm diameter, phase conductors 76 mm effective diameter, resistance = 25 mΩ/km.

Modal signals suitable for protective schemes

The general equation relating the steady state voltage drops and currents in three-phase circuits is of the form shown below:

$$
\begin{bmatrix} V_a \\ V_b \\ V_c \end{bmatrix} = \begin{bmatrix} Z_{aa} & Z_{ab} & Z_{ac} \\ Z_{ba} & Z_{bb} & Z_{bc} \\ Z_{ca} & Z_{cb} & Z_{cc} \end{bmatrix} \cdot \begin{bmatrix} I_a \\ I_b \\ I_c \end{bmatrix}
$$

The use of this equation involves considerable computation because the Z matrix contains nine non-zero elements, i.e. three self impedances and six mutual impedances.

For a completely symmetrical circuit in which the three self impedances (Z_{aa}, Z_{bb}, Z_{cc}) are equal and the six mutual impedances are equal (Z_{ab}, Z_{ac}, Z_{ba}, Z_{bc}, Z_{ca}, Z_{cb}) the above equation can be transformed to diagonalize the impedance matrix so that it only contains three non-zero elements.

A very well-known transformation, dealt with in Appendix 3.2, is that associated with symmetrical components, i.e. positive-sequence, negative-sequence and zero-sequence. For steady state sinusoidal conditions this equation is of the form:

$$
\begin{bmatrix} V_1 \\ V_2 \\ V_0 \end{bmatrix} = \begin{bmatrix} Z_{11} & & \\ & Z_{22} & \\ & & Z_{00} \end{bmatrix} \cdot \begin{bmatrix} I_1 \\ I_2 \\ I_0 \end{bmatrix}
$$

in which V_1, V_2, V_0, I_1, I_2, I_0, Z_{11}, Z_{22} and Z_{00} are the positive-, negative- and zero-sequence voltages, currents and impedances respectively.

A great advantage of this particular equation is that there are no mutual effects between the sequences and as a result the three rows are independent of each other.

The relationships between the positive- and negative-sequence voltages and currents and the actual phase quantities are complex, as shown in Appendix 3. As an example

$$
I_1 = \frac{1}{3} (I_a + a I_b + a^2 I_c)
$$

in which I_a, I_b and I_c are the phase currents and

$$
a = \frac{1}{2} (-1 + j\sqrt{3}) = 1 \left\lfloor \frac{2\pi}{3} \right. \text{ rad}
$$

It will be appreciated that instantaneous voltages and currents cannot be transformed in the above manner. There are, however, a number of transformation matrices containing only real numbers which may be used. A transformation that has been widely employed is:

$$
\begin{bmatrix} V_\alpha \\ V_\beta \\ V_0 \end{bmatrix} = \frac{1}{3} \begin{bmatrix} \frac{3}{2} & 0 & -\frac{3}{2} \\ \frac{1}{2} & -1 & \frac{1}{2} \\ 1 & 1 & 1 \end{bmatrix} \cdot \begin{bmatrix} V_a \\ V_b \\ V_c \end{bmatrix}
$$

The same transformation applies to the currents and the following relationship between the voltages and currents associated with travelling waves is produced.

$$
\begin{bmatrix} V_\alpha \\ V_\beta \\ V_0 \end{bmatrix} = \begin{bmatrix} Z_{\alpha\alpha} & & \\ & Z_{\beta\beta} & \\ & & Z_{00} \end{bmatrix} \cdot \begin{bmatrix} i_\alpha \\ i_\beta \\ i_0 \end{bmatrix} \tag{12.11}
$$

It will be clear that three independent propagation-modes of waves could be used in studies of circuits containing symmetrical overhead lines. Two of these modes, i.e. α and β, would be associated with the phase conductors and they are thus referred to as aerial modes. The current associated with the other mode, i.e. 0, which is the same as the zero-sequence mode, must at least flow partly in the ground and it is therefore referred to as the ground mode. It will be clear that a single characteristic impedance and propagation velocity could be determined for each mode.

It must be recognized, however, that the conductor configurations of major overhead lines are always asymmetric and it is not the practice to transpose the conductors at regular intervals along the lengths of line. As a result, the self and mutual inductances and capacitances per unit length of lines are not the same for each of the phases. The differences are not great, however, because the various quantities are dependent on the logarithms of the spacings. If the actual values were used to produce the elements in the Z matrix of equation (12.11), all of them would have non-zero values. Those on the principal diagonal, i.e. $Z_{\alpha\alpha}$, $Z_{\beta\beta}$ and Z_{00}, would, however, have larger values than the others and limited errors would be produced in determining the voltages and currents, if the elements not on the diagonal of the Z matrix were regarded as zero.

It will be appreciated that the same approximation is usually made when symmetrical-component techniques are used, i.e. it is assumed that the sequences are independent of one another. Various other transformations are now available which allow the errors caused by neglecting the physical asymmetry of overhead lines to be reduced. Some are considered in detail in Appendix 3.3.

12.3 PROTECTIVE SCHEMES WHICH DETECT TRAVELLING WAVES

As indicated in the preceding sections, it is possible to determine the direction of a fault from the first travelling current and voltage waves which arrive at the end of a line and directional-comparison schemes using this technique have been produced. Two of them are considered in the following section.

12.3.1 Directional-comparison schemes

During the 1970s, the possibilities of producing protective schemes based on travelling waves were being studied with the object of obtaining very rapid detection and clearance of faults. Such relays, which were designated as UHSR, i.e. ultra-high-speed relays, were later defined to include schemes which would initiate the opening of circuit-breakers in one quarter of one cycle in 60 Hz systems, i.e. in 4.17 ms. Some of the schemes which have been produced are described below.

RALDA Scheme produced by ASEA

The Bonneville Power Administration in the United States launched a development programme and awarded a contract to ASEA in Sweden, in July 1974, to develop a UHSR system using the travelling-wave approach. The contract specified an operating time of 4 ms or less for close-up faults and 8 ms or less for all other faults.

As a result, after optimizing the relay performance using digital and analogue simulation techniques, a scheme given the designation Ralda was produced. The scheme was described in detail in a paper [1] by Chamia and Liberman which was presented at the IEEE PES Winter Meeting in January and February 1978.

The scheme, represented in block diagram form, is shown in Fig. 12.7. The analogue interfaces were designed to accept signals from either conventional or other types of current and voltage transducers and to provide both isolation and surge immunity.

As stated earlier in section 12.2, it is necessary that the steady state voltages and currents which would have been present in the absence of a fault be subtracted from the voltages and currents which are present during faults so that the transient superimposed qualities are obtained for processing. In the Ralda scheme, this was achieved by including steady state frequency suppressors which had transfer functions which ensured that the initial changes caused by faults were transmitted without significant time delays. Amplifiers were included to provide signals of the levels needed for subsequent processing. These signals were fed to directional detectors I and D and with the conventions adopted a tripping direction signal (T_D) was produced by detector D if the polarities of the voltage and current signals present after the incidence of a fault were different. This corresponded with the voltage and current waves approaching the end of the line being of the same polarity, i.e. the condition for a fault on the protected line, as shown in section 12.2.2. In the event of the polarities of the signals being the same a blocking directional signal (B_D) was produced.

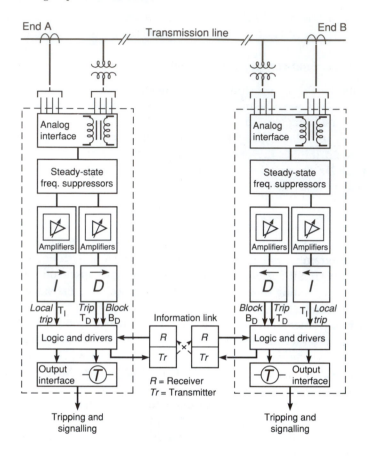

Fig. 12.7 Block diagram of the RALDA scheme. (Reproduced from Chania and Liberman, 1978, *IEEE Trans PAS-97*, with the permission of the IEEE.)

To ensure that the above tripping and blocking signal outputs would not be produced by minor disturbances on a system, for example, by a sudden loss of one of the output voltages of the voltage transducers, the directional detectors *D* were so arranged that the magnitudes of the transient current and voltage inputs should both be above certain set levels before they provided output signals.

Clearly this process had to be performed separately for the quantities obtained from each of the three phases of a protected line and the outputs from the directional detectors *D* were fed to logic circuits. These determined the sequence of the T_D and B_D signals produced after the incidence of a fault to establish whether the first signal was T_D or B_D. This was done to avoid errors which could otherwise have arisen if the signals changed because of the arrival of later reflected waves.

The production of the T_D and B_D signals energized the communication link between the ends and either permissive or blocking signals could be sent and clearly tripping of the circuit-breakers at both ends of the protected line was initiated if T_D signals were present at both ends of the line or it was prevented if a B_D signal was present.

Because of the time delay introduced by the communication link, the time between the incidence of a fault and the initiation of tripping of the circuit-breakers was about 8 ms.

To enable the operating times for close-up faults to be shorter, the directional detectors I were included. These detectors only operated if the magnitude of the transient current was above a level which could only be present for faults within a short distance along the protected line. For such a condition, the current increases rapidly, as explained under section 12.2.5 (page 445), because many waves travel between the fault and the end of a line in a very short time. Operation of a directional detector I produced a T_I signal which directly initiated the tripping of the circuit-breaker at its end of the line without energizing the communication link. This mode of operation was referred to as the independent mode and this led to the designations directional detector I and signal T_I. Its use enabled tripping of the circuit-breaker to be initiated in 4 ms or less.

The other mode of operation was referred to as the dependent mode and this led to the terms directional detector D and signals T_D and B_D, used above.

It will be appreciated that the directional detectors I had to be set so that they would only operate in the event of faults on a section of protected line, i.e. they had to under-reach, whereas the directional detectors D had to operate for all faults on the protected line and in consequence they had to over-reach. It was necessary therefore, when applying the Ralda scheme, to do extensive fault studies to ensure that the directional detectors I and D were set at levels which would ensure correct operation.

The scheme was so arranged that the types of fault present were determined, e.g. phase a to earth, phase b to c, three phase, and the logic circuits then initiated the appropriate circuit-breaker operation, i.e. single-pole or three-pole opening.

The communication equipment incorporated a single-tone transmitter-receiver which operated at a carrier frequency of $72\,kHz \pm 1\,kHz$, its general arrangement being as shown in Fig. 12.8. This was an expensive arrangement with four channel spaces which was suitable for use in schemes for application to short lines where rapid fault clearance was needed. For longer lines where 8–10 ms relaying times would be acceptable in the event of faults near the remote ends, an audio-tone communication equipment operating on a single voice channel was adequate.

During the later stages of testing the prototype equipments, part of the Bonneville Power Administration's 500 kV system was modelled on an analogue transient-system simulator. System voltages and currents for a large

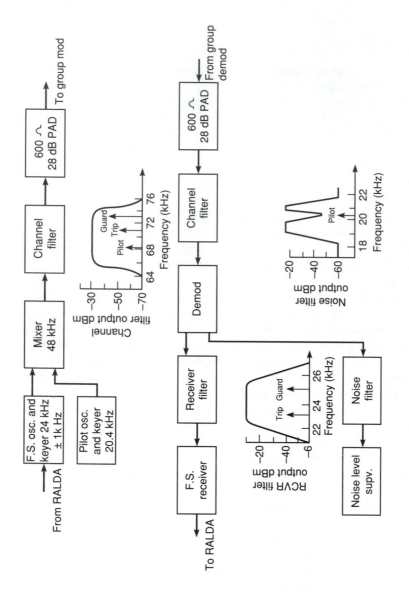

Fig. 12.8 Block diagram of the communication equipment used with the RALDA scheme. (Reproduced from Chania and Liberman, 1978, *IEEE Trans PAS-97*, with the permission of the IEEE.)

number of conditions were recorded and these were later used to provide inputs to the Ralda relays.

Subsequently two Ralda terminals were installed on the 500 kV John Day–Grizzly No. 1 line of the Bonneville Power Administration in April 1976. The setting procedures and one year's operational experience together with the performance of the relays during staged fault tests were summarized in a paper [7] presented by Yee and Esztergalyos at the IEEE PES Winter meeting held in January and February 1978. In the conclusion of the paper it was stated that the collected field data, combined with laboratory test results, prove that the UHSR system is directional and that it has consistent high-speed performance from 2–6 ms. It was further stated that the UHSR system reliability in the forward direction and in the reverse blocking mode is 100%.

Since that time many series-compensated lines in the United States have been protected by Ralda schemes and satisfactory behaviour has been obtained.

Scheme developed by the University of Bath and GEC Measurements

This scheme [2, 8] processed the transient currents and voltages produced at the ends of lines by faults and discrimination was obtained using carrier-blocking type communication equipment. Although it was basically similar to the Ralda scheme, produced by ASEA, the processing techniques were different and it did not detect the directions of faults relative to the ends of lines by directly comparing the polarities of the incoming travelling waves of voltage and current initiated by faults.

It will be clear that if a fault occurs beyond end A of an ideal line as shown in Fig. 12.9(a), i.e. a reverse direction fault, then the resultant change of voltage at the fault position would cause travelling waves of voltage (v_1) and

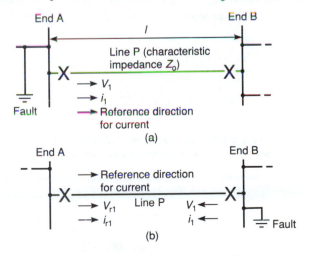

Fig. 12.9 Voltage and current waves for fault at ends A and B.

current (i_1) to be propagated into the protected line P. Taking the current flow from busbar A to busbar B to be positive, the transient current would be given by:

$$i_1 = \frac{v_1}{Z_0}$$

This condition would persist until the waves had travelled to busbar B and then returned to busbar A, the elapsed time being:
Signals S_1 and S_2 are given by:

$$t = \frac{21}{u}$$

$$S_1 = v_1 - i_1 Z_0 \qquad \text{and} \qquad S_2 = v_1 + i_1 Z_0$$

with values of $S_1 = 0$ and $S_2 = 2v_1$ throughout this period.

In the event of a fault on the protected line or beyond its remote end, a travelling voltage wave (v_1) would be propagated towards busbar A as shown in Fig. 12.9(b). The accompanying current wave would be given by:

$$i_1 = \frac{v_1}{Z_0}$$

with the above reference direction for the current.

Because several circuits would be connected to busbar A it would present an impedance R_A to the waves which would not be equal to the characteristic impedance of the line (Z_0). As a result reflections would occur at the busbar and travelling waves v_{r1} and i_{r1} would be propagated back along the line towards busbar B.

As shown under section 12.2.5 (page 445), the incoming and reflected waves would sum at busbar A to produce voltages and currents given by:

$$v_{s1} = v_1 + v_{r1} = \frac{2R_A V_1}{R_A + Z_0}$$

and

$$i_{s1} = i_1 + i_{r1} = -\frac{2V_1}{R_A + Z_0}$$

The voltage and current transducers near busbar A would not detect the initial incoming waves but would give outputs proportional to the above quantities (v_{s1} and i_{s1}).

Under these conditions the signals S_1 and S_2 would have the values:

$$S_1 = v_{s1} - i_{s1} Z_0 = 2v_1$$

and

$$S_2 = v_{s1} + i_{s1} Z_0 = \frac{2v_1 (R_A - Z_0)}{R_A + Z_0}$$

Comparison of these values with those obtained during reverse faults shows that discrimination can be obtained because the magnitude of S_1 is less than that of S_2 for a reverse fault but it is greater for faults on a protected line or beyond it.

Clearly both signals would have zero magnitudes initially for both external and internal fault conditions which occurred at instants of voltage zero but the signals S_1 and S_2 would grow to significant levels for a fault on the protected line before further reflected waves arrived, i.e. after a time of 21/u s.

This detection process was used in the scheme developed at the University of Bath and it was implemented using the α, β, 0 modal quantities referred to under section 12.2.6(page 452).

A block schematic diagram of the equipment required at one end of a line is shown in Fig. 12.10(a). The whole arrangement was initially analogue based and the outputs of the voltage and current transducers were fed to mixing circuits to produce the α, β and 0 components of both sets of quantities.

The asymmetry of the line configuration was neglected and a single characteristic impedance was used for each mode of propagation, i.e. $Z_{\alpha\alpha}$, $Z_{\beta\beta}$, Z_{00} and in fact these impedances were assumed to be purely resistive in the surge-replica circuits. These circuits were fed with the three current components and the quantities $Z_{\alpha\alpha} i_\alpha$, $Z_{\beta\beta} i_\beta$ and $Z_{00} i_0$ were produced for subtraction from and addition to the components of the voltages to obtain quantities such as $v_\alpha - Z_{\alpha\alpha} i_\alpha$ and $v_\alpha + Z_{\alpha\alpha} i_\alpha$. These quantities contained steady state terms as well as extra transient terms caused by faults and therefore the transient components were extracted in the superimposed component circuits. This was achieved by feeding differential amplifiers shown in Fig. 12.10(b), with the $v - Zi$ and $v + Zi$ signals. A time delay of one period of the power system frequency, i.e. $T = 1/f$, was introduced in the connection to the negative input of each of the six differential amplifiers so that the output would be the difference of the input and its value one cycle earlier. This clearly provided the transient components only for the first period after the incidence of a fault and this was considered satisfactory because the fault direction would have been determined within this time. The signals produced by pairs of these circuits, e.g. $S_{\alpha1}$ and $S_{\alpha2}$ were fed to detectors which determined the fault directions indicated by each of the modes and either trip or block outputs were obtained. These outputs were fed to 'OR' gates so that a single trip or block signal was provided by the equipment at an end of a line. In the event of a fault being detected on the line or beyond its remote end, tripping of the local circuit-breaker was initiated unless a blocking signal was obtained from the communication link.

To evaluate the performance of the scheme the conditions which would obtain on the network shown in Fig. 12.11 were determined using a digital simulation technique described in reference [2]. Real-time signals proportional to the system voltages and currents calculated for a wide range of different conditions were supplied to a prototype relay in which the replica surge

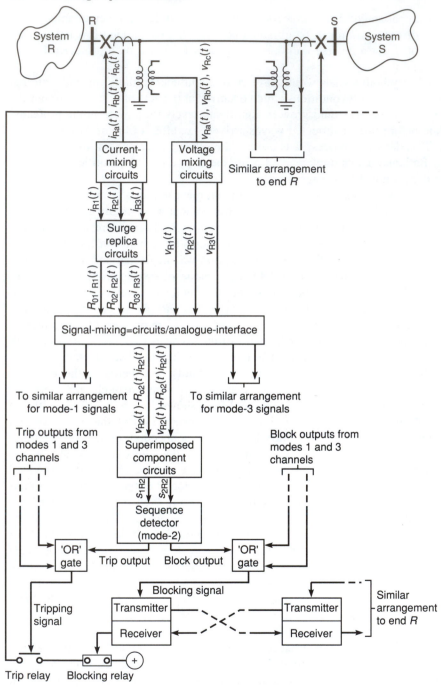

Fig. 12.10 Block diagram of the Bath/GEC scheme. (Reproduced from Johns, 1980, *IEE Proc C*, 127 with the permission of the IEE.)

impedances were set to be proportional to the real component of the actual characteristic impedances, the values of which were $Z_{\alpha\alpha} = Z_{\beta\beta} = 266\ \Omega$ and $Z_{00} = 585\ \Omega$.

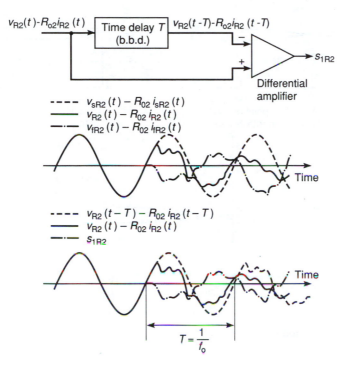

Fig 12.10 (b)

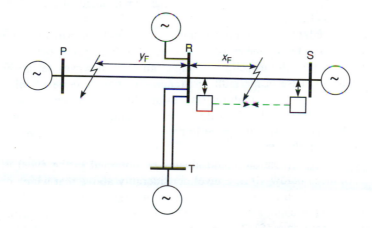

Fig. 12.11 Schematic diagram of network configuration. (Reproduced from Johns, 1980, *IEE Proc C*, 127 with the permission of the IEE.)

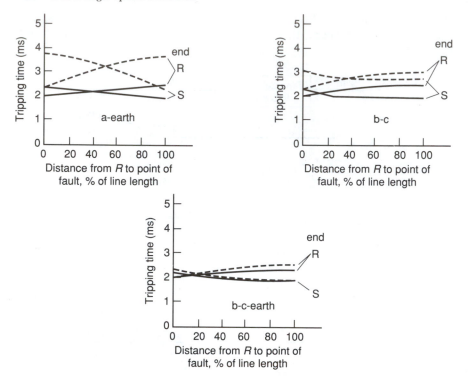

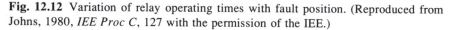

Fig. 12.12 Variation of relay operating times with fault position. (Reproduced from Johns, 1980, *IEE Proc C*, 127 with the permission of the IEE.)

The scheme operated satisfactorily under all conditions and operating times under 4 ms were obtained, the actual times for a range of conditions being shown in Fig. 12.12.

Unlike impedance-measuring schemes, directional-comparison schemes are not directly affected by the resistances present in fault paths, but nevertheless the magnitudes of the travelling voltage and current waves set up by faults reduce with increase in the fault resistance. Tests were done on this scheme to determine the limiting fault resistances at which correct operation could be obtained. Understandably, the values were dependent on the position of the fault on the line and also the point in the cycle at which it occurred.

It was found that the limiting values were in the range 40–600 Ω and therefore correct operation would always be obtained in the event of faults with resistances below 40 Ω, a level considerably above that which could be accepted by a distance-measuring scheme. A high-speed digital directional-comparison relay using a signalling channel based on the above principles is described in reference [9]. Typical operating times of half a power frequency cycle plus signalling delays are possible with this relay. In addition to normal relaying functions, the relay features continuous self-monitoring and diagnostic facilities, and can be interrogated from a remote location via an RS232 serial communication link.

Scheme developed by Asea Brown Boveri

This scheme was described in paper [10] by Vitins which was presented at the IEEE PES Meeting held in February 1980. It employed a correlation technique which had been described in a paper [11] presented at the IEEE PES Meeting held in July 1977.

The algorithm proposed by Vitins determined the trajectories followed by the transient voltages and currents caused by faults (Δv and Δi) in the Δv, $R\Delta i$ plane. This technique is a variation of the directional-sensing processes described earlier, in which the polarities of the voltage and current increments caused by faults are compared. In such schemes, the polarities of the first increments of voltage and current produced at a busbar in the event of a fault either on a protected line or beyond it, i.e. in the forward direction, are the same whereas they are different when faults occur in the reverse direction, i.e. for a fault in the forward direction, Δv and Δi are either both positive or both negative. Such quantities when plotted in the Δv, $R\Delta i$ plane must therefore lie in the first and third quadrants for faults in the forward direction and in the second and fourth quadrants for faults in the reverse direction. They move from the origin at the instant when the first waves arrive at the end of the line and take up positions dependent on the angle in the voltage cycle at which the fault has occurred. When subsequent increments are produced by later waves new positions are taken up and the trajectories follow a clockwise path when faults are in the forward direction and an anticlockwise path when faults are in the reverse direction.

By setting boundaries in the Δv, $R\Delta i$ plane of the form shown in Fig. 12.13, discrimination between forward and reverse faults can be obtained by determining whether the boundaries in the second or fourth quadrants are crossed before the boundaries in the first or third quadrants. In addition, the fault direction can also be determined by detecting the direction of the trajectory. This has the advantage that discrimination is possible at times after the arrival of the initial waves.

The above principle is employed in the relay, designated LR91, produced by Asea Brown Boveri [12]. This scheme, like that described under section 12.3.1 (page 457), incorporates current detectors so that should current levels be so high during a fault that it could not be beyond the end of the line protected, then tripping of the circuit-breaker would be initiated immediately if the directional detector indicated a forward fault. In these circumstances operation of the circuit can be initiated in less than 5 ms. In the event of more remote faults then the directional detection process is used and tripping of the circuit-breakers is only initiated if the equipments at both ends indicate that the faults are in their forward directions.

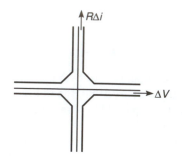

Fig. 12.13 Operational characteristics of the ABB scheme.

The burdens imposed on the voltage and current transducers are very low.

Schemes of this type have been installed since about 1985 and they are now in use in several countries including China where they have been applied to several lines.

12.3.2 Schemes which determine the positions of faults

It was indicated in section 12.2.5 (page 445) that the distance of a fault from a relaying position may be determined by measuring the time interval $(2t_1)$ between the arrival at the end of a line of the first travelling wave caused by a fault and the next wave which results from reflections at the busbar and the fault, i.e. the time between the voltage increments V_{s1} and V_{s2} shown in Fig. 12.4(b).

Because many travelling waves may be present on lines during fault conditions, the increment V_{s2} may not always be the next increment after V_{s1} and therefore, to ensure that fault positions will be determined correctly, it is necessary that protective schemes operating on this basis should identify the increment V_{s2}.

Extensive studies have been conducted on two schemes incorporating different methods of wave identification and these are described in some detail in the following sections.

Scheme developed at the University of Cambridge

This scheme, which was described in a paper [3] by Crossley and McLaren, presented at the IEEE PES Meeting in January and February 1983, employed similar techniques to those described earlier (section 12.3.1 (page 457)).

The signals S_1 and S_2 given by

$$S_1 = v_{s1} - i_{s1} Z_0 \quad \text{and} \quad S_2 = v_{s1} + i_{s1} Z_0$$

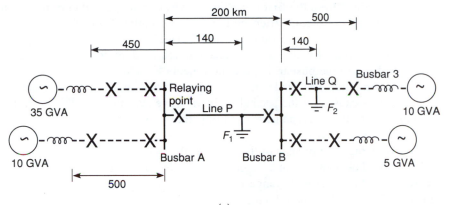

(a)

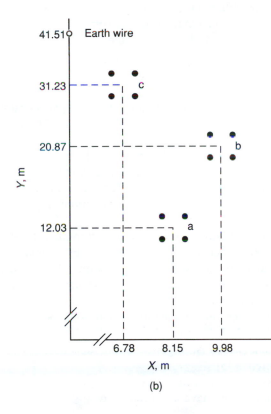

(b)

Fig. 12.14 Schematic diagram of network configuration. (Reproduced from Christopoulos, Thomas and Wright, 1988, *IEE Proc C*, 135 with the permission of the IEE.)

were obtained from the aerial-modal quantities and the directions of faults were determined from the first travelling waves of voltage and current which arrive at the relaying point after the incidence of faults. Thereafter the cross-correlation technique described in the above paper was used to identify waves which have been reflected at the busbars, returned to the fault and then travelled back to the busbar, i.e. the increments V_{s2} and I_{s2}.

The assumption was made that the largest peak in the cross-correlation function was caused by the second incident wave (V_{s2} and I_{s2}). The positions of faults were then determined by timing the interval between the first and second incident waves.

To assess the performance of the scheme, the voltage and current variations which would be obtained on a single-circuit, three-phase line with two infeeds were computed for a wide range of conditions and an off-line program was then used to simulate the relaying processes.

The results obtained indicated that satisfactory performance should be obtainable. In further studies which were done to evaluate a number of ultra-high-speed relay algorithms [13] it was found, however, that the necessary discrimination would not be obtained under some conditions. Enhancements to the original algorithm have been suggested in references [14] and [15].

Scheme developed at the University of Nottingham

This scheme was described in considerable detail in papers published by the Institution of Electrical Engineers in 1988 and 1989 [4, 5]. For simplicity the scheme is considered below on a single-phase basis.

It was shown earlier (section 12.2.5 (page 445)), that a great deal of information can be extracted from the first travelling waves of voltage and current which reach the end of a line after the incidence of a fault on it and the later waves of voltage and current caused after reflections at the busbar and the fault.

On arrival of the first waves of voltage and current, with initial values of V_1 and I_1, reflections occur at busbar A and the increments V_{s1} and I_{s1} are produced, their values being as given in equations (12.5) and (12.6). Their ratio (V_{s1}/I_{s1}), as shown in equation (12.7), is the surge impedance presented by the busbar A (R_A). In practice, because R_A may not be entirely resistive, the reflected wave from the relaying point may be obtained directly from the expression $V_r = (V + IZ)/2$, without the need to calculate the reflection coefficient at the busbar which is, in general, time-dependent. This is explained more fully in reference [5].

As shown in section 12.2.5, the increment V_{s2} caused by reflection of the wave V_{r1} at the fault would have an initial value given by:

$$V_{s2} = k_{vf}(1 + k_{v1}) V_{r1}$$

in which k_{vf} is the voltage reflection coefficient at the fault. The magnitude of this coefficient can thus be found and because it is given by equation (12.8), i.e.

$$k_{vf} = \frac{[Z_0 R_f/(Z_0 - R_f)] - Z_0}{[Z_0 R_f/(Z_0 + R_f)] + Z_0}$$

the value of the fault resistance (R_f) can be determined.

In addition, and most importantly, the time interval between the voltage increments V_{s1} and V_{s2} enables the distance x to the fault to be determined from equation (12.10).

A check must be made, however, to ensure that the second increment which occurs at the busbar has been caused by the reflected voltage V_{r1} being reflected at the fault.

Referring to Fig. 12.4(b), it can be seen that the fault must have occurred at a time t_1 seconds before the arrival of the first travelling wave at the busbar, i.e. half the time between the increments V_{s1} and V_{s2}. A knowledge of the voltage and current conditions at the busbar A end of the line, prior to the fault, would enable the voltage (V_f) at a point distance x along the line at the end of fault incidence to be determined using the well-known long-line equations. Substitution of this voltage in equation (12.1) would allow the value of the fault resistance to be calculated again and obviously this value should correspond closely with that derived earlier.

In the event of a fault near the remote end of a line, an increment V_{s3} would arrive before the increment V_{s2} as shown in Fig. 12.4(c) and should the second value of the fault resistance be determined using the time elapsed between the increments V_{s1} and V_{s3} then the value obtained would not agree with the earlier value. In these circumstances, the process would be repeated later when the increment V_{s2} arrived and the necessary agreement would be obtained.

It was proposed that the scheme should be implemented by processing digital samples of the sums of the two aerial-mode quantities and the ground-mode quantities associated with three-phase lines, i.e.

$$v_{aa} = v_{an} - v_g; \quad v_{ab} = v_{bn} - v_g; \quad v_{ac} = v_{cn} - v_g$$

$$i_{ac} = i_{an} - i_g; \quad i_{ab} = i_{bn} - i_g; \quad i_{ac} = i_{cn} - i_g$$

$$v_g = 1/3 (v_{an} + v_{bn} + v_{cn})$$

$$i_g = 1/3 (i_{an} + i_{bn} + i_{cn}) \tag{12.12}$$

It was further proposed that the values of fault resistances would be determined using the time intervals between the increments present in the aerial-mode phase voltages because the velocity of propagation associated with the aerial mode is constant and near that of light, whereas the propagation velocity of the ground mode is dependent on the resistivity of the ground, which may vary along the length of a line and it may also be affected by climatic conditions.

To evaluate the performance of the proposed scheme its behaviour when applied to transmission lines connected in various power networks was studied. The various voltages and currents at the end of a line at which the scheme was assumed to be mounted were determined for a variety of fault conditions, using a frequency-domain, power-system simulation program provided by GEC Measurements. These quantities, which were sampled at 25 kHz in all numerical computations, were processed as described above.

Studies were based on the network shown in Fig. 12.14(a) in which the lines were assumed to be single circuit and untransposed with the conductor configuration shown in Fig. 12.14(b). The protective scheme being studied was taken to be associated with line P and to be mounted near busbar A. Faults at various points on the network were simulated and the responses of the scheme were determined. In each case, the sampled voltages and currents were added to the corresponding values of the previous half cycle and in this way the presence of a disturbance was indicated by a set of non-zero values and these and the subsequent values were the transient quantities produced by the disturbance. The third set of non-zero quantities were taken to represent the magnitudes of the initial voltage and current increments (V_{s1}, I_{s1}) at the relaying point.

A set of aerial-mode transients caused by a symmetrical, three-phase fault on a line P (of length 200 km) at a point 140 km from busbar A are shown in Fig. 12.15. These were obtained assuming fault resistances of 60 Ω between each of the phases, the fault occurring at an instant of voltage maximum on phase a at the fault position. Because the fault was beyond the mid-point of the line, the initial waves and reflections which travelled between the fault and busbar A arrived at the times t_1 and t_3, whereas the waves which went from the fault to busbar B before returning to busbar 1 at time t_2.

The two sets of estimates obtained of the fault resistances between the phases using the voltage and current increments at times t_1 and t_2 were as follows:

Phase 'a':   76 Ω and − 568 Ω
Phase 'b':   0 Ω and 459 Ω
Phase 'c':   − 60 Ω and − 516 Ω

This lack of agreement indicated that the increments were not those referred to earlier as V_{s1} and V_{s2}. The estimates obtained using the increments at times t_1 and t_3 were:

Phase a: 67 Ω and 48 Ω
Phase b: 67 Ω and 56 Ω
Phase c: 68 Ω and 63 Ω

This agreement indicated that the increments were V_{s1} and V_{s2} and the time

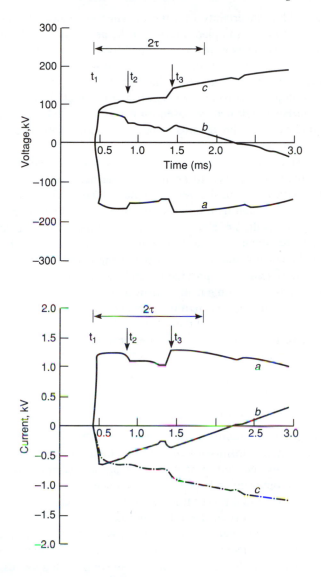

Fig. 12.15 Aerial-mode transients caused by a symmetrical three-phase internal fault 140 km from the relaying point (fault resistance 60 Ω/phase, fault at phase a positive voltage maximum). (Reproduced from Christopoulos, Thomas and Wright, 1988, *IEE Proc C*, 135 with the permission of the IEE.)

interval between t_1 and t_3 corresponded with a fault 146 km from the relaying position, i.e. an error of 6 km.

In this connection it should be noted that the use of a 25 kHz sampling rate introduces an inherent 40 µs uncertainty into the timing estimates and an associated 12 km uncertainty into the distance estimates, this being the distance travelled in one sampled interval at the aerial-mode velocity.

Satisfactory estimates were also obtained for phase to phase and most single phase to earth faults. In a few cases, however, it was not possible to distinguish between earth faults on line P and line Q when using aerial-mode quantities but it was found that satisfactory discrimination could be achieved by incorporating a further check feature based on the ground-mode waves.

Later studies based on circuits containing double-circuit lines also gave satisfactory results.

Having established that the scheme was soundly based, it was felt that features should be included to ensure that the necessary information would always be extracted from the very complex waves which can be present on protected networks and therefore the use of cross-correlation processing was studied. As a result, an algorithm was developed which enabled correct behaviour to be obtained for a wide range of test conditions. Details of this algorithm, the performance obtained when using it and the influence of factors such as window length are provided in reference [5]. Implementation of the basic algorithms in real time using hardware based on a digital signal processor (DSP) confirms that the required processing can be done within a time period not exceeding 2 ms and therefore a relay based upon this principle can have a very fast operating time. To date, however, this scheme has not been applied to an actual line.

12.4 THE APPLICATION OF ULTRA-HIGH-SPEED RELAYS TO SERIES-COMPENSATED LINES

As stated earlier, in section 11.14 (page 431), it is now common practice to install capacitors in transmission lines so that their effective series impedances in steady state are reduced. In this way the phase displacements between the voltages at the opposite ends of lines are reduced for a given power transfer and as a result the rated power transfer over a line may be increased without jeopardizing the stability of the associated network.

To be effective the voltages present across such capacitors at the rated currents of their associated lines must be quite high. The voltages which would appear across them if they carried currents above the rated values during fault conditions would rise to unacceptably high levels and therefore metal oxide varistors are connected across them to ensure that they will be effectively short circuited at such times.

It will be clear therefore that the performance of protective schemes associated with series-compensated lines should not be affected by the presence of the capacitors or their effective removal.

Because the voltages across capacitors cannot be changed suddenly, the initial values of travelling waves are not affected by the capacitors included in series-compensated lines. This can be seen by considering the situation

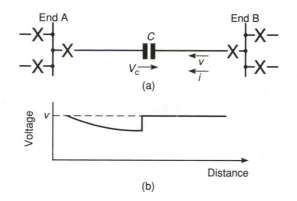

(a)

(b)

Fig. 12.16 Conditions in a series compensated line.

illustrated in Fig. 12.16(a). When the wave of voltage magnitude v reaches the capacitor C, the current $i = v/Z_0$ will begin to flow through the capacitor producing a voltage v_c across it, v_c being given by:

$$v_c = \frac{1}{C} \int_0^t i \, dt = \frac{it}{C} = \frac{v}{Z_0 C} t$$

The initial magnitudes of the waves travelling from the capacitor towards busbar A would therefore be v and i, but at any point behind the wavefronts the voltages and currents would reduce with time as the voltage across the capacitor increased. The form of these variations is shown in Fig. 12.16(b).

In practice the effects produced by the voltage across capacitors during the periods in which ultra-high-speed relays are operating is small. As an example, if a single-phase 100 kV, 50 Hz line of length 150 km and inductive reactance of 0.3 Ω/km was provided with 50% series compensation, the capacitive reactance required would be 22.5 Ω, i.e. $150 \times 0.3 \times 0.5$ Ω. The capacitance would therefore be 141.5 μF.

In the event of a short-circuit at an instant of peak voltage a travelling voltage wave of initial magnitude 141.4 kV would be propagated and for a line with a characteristic impedance of 250 Ω the accompanying current wave would have an initial magnitude of 565 A. This current flowing through the capacitor would cause the voltage across it to rise at a rate of 4 kV/ms, i.e. 2.83% of the initial voltage per millisecond. The time taken for a wave to travel the line length would only be 0.5 ms and therefore the presence of the capacitor would only reduce the voltage surge magnitude by 1.4% in this time. Clearly this would have no effect on schemes which detect the directions of faults from the polarities of the first travelling waves arriving at the relaying positions. Even in schemes which determine fault positions from later travelling waves, the effects would not be significant; for example, the changes in the magnitudes of waves which have traversed the line length three times would be less than 5%.

As stated in section 12.3.1 (page 457), Ralda directional-comparison schemes have been applied to series-compensated lines and to date they have performed satisfactorily.

The suitability of the scheme described in section 12.3.2 (page 468) for application to series-compensated lines has been investigated in some detail. This scheme, as explained earlier in section 12.3.2, discriminates by comparing estimates of the resistances present in faults, one of the estimates being obtained from data stored of the conditions prior to the incidence of a fault and the other from post-fault conditions.

Because the changes in voltage across a capacitor in a short period after fault incidence are small, its presence may be ignored in obtaining the second estimate of fault resistance. During this calculation the time taken for waves to travel between a fault and the end of a line is determined, and should this indicate that the fault is between the capacitor and the relaying point, then the first estimate of resistance obtained from the pre-fault data need not take account of the presence of the capacitor. If, however, a fault appears to be beyond the capacitor then the first estimate must be obtained from a three-section model of the complete line because there may be quite large voltages across a capacitor during healthy conditions. The capacitor forms the centre section of the model and the two end sections represent the lines on each side of the capacitor.

Apart from the above feature, the scheme operates in the same manner as that described earlier for use with uncompensated lines.

Clearly the scheme could not operate satisfactorily if the voltage limiter connected across a compensating capacitor operated within the time interval required for the signals to be processed, but extensive investigations have shown that this situation will not occur. It has also been confirmed that accidental operation of voltage limiters would not generate travelling waves of sufficient magnitude to significantly affect the operation of the scheme. Operation of voltage limiters could occur, however, in the event of faults external to protected lines when the series capacitors are sited near one of the line ends. Various studies have shown, however, that these would not cause the scheme to maloperate.

Further information about the above points and other details are provided in a paper [16] which was presented at the IEEE Meeting held in July 1991.

12.5 THE FUTURE

It is clear that protective schemes based on processing information contained in the travelling waves set up by faults can potentially offer very high-speed operation. Such schemes are particularly useful on long important lines, where a short relaying time is desirable. A number of directional-comparison schemes based on travelling wave, or superimposed quantities are already commercially available [9, 12] and more such schemes are likely to be introduced. The need for communication between the two line ends introduces inherent delays and additional costs.

Distance schemes where the position of the fault is determined are being developed. In order to exploit the inherent speed advantages of such schemes, it is necessary to employ voltage and current transducers of sufficient bandwidth to allow accurate detection of the fast-rising waveforms associated with travelling waves. Although such transducers are available, most existing installations use voltage and current transformers which are designed with 50 or 60 Hz signals in mind. Noise immunity and the response of travelling-wave algorithms when faults occur near a voltage zero are additional factors which suggest that the introduction of such schemes will in practice be slow. The most likely initial application of such techniques will be as part of a combined scheme where, for example, a conventional distance protection scheme is supplemented by an ultra-high-speed algorithm based on travelling waves, to obtain accelerated performance for faults at the remote end of the line.

REFERENCES

1. Chamia, M, and Liberman, S. (1978) Ultra high speed relay for EHV/UHV transmission lines-development design and application, *Trans. IEE*, **PAS-97**, 2104–2112.
2. Johns, A. T. (1980) New ultra-high-speed directional comparison technique for the protection of ehv transmission lines, *Proc. IEE*, **127** (C), 228–239.
3. Crossley, P. A. and McLaren, P. G. (1983) Distance protection based on travelling waves, *Trans. IEE*, **PAS-102**, 2971–2983.
4. Christopoulos, C., Thomas, D. W. P. and Wright, A. (1988): Scheme, based on travelling waves, for the protection of major transmission lines, *Proc. IEE*, **135** (C), 63–73.
5. Christopoulos, C., Thomas, D. W. P. and Wright, A. (1989) Signal processing and discriminating techniques incorporated in a protective scheme based on travelling waves, *Proc. IEE*, **136**, (C), 279–288.
6. Wagner, C. F. and Evans, R. D. (1933) *Symmetrical Components*, McGraw-Hill.
7. Yee, M. T. and Esztergalyos, J. (1978) Ultra high speed relay for ehv/uhv transmission lines–installation staged fault test and operational experience, *Trans. IEE*, **PAS-97**, 1814–1825.
8. Johns, A. T. and Walker, E. P. (1988) Co-operative research into the engineering and design of a new directional comparison scheme, *Proc. IEE*, **135** (C), 334–368.
9. Type LFDC digital directional comparison protection relay, *Publication R-4078*, GEC Measurements.
10. Vitins, M. (1981) A fundamental concept for high speed relaying, *Trans. IEE*, **PAS-100**, 163–168.
11. Vitins, M. (1978) A correlation method for transmission line protection, *Trans. IEE*, **PAS-97**, 1607–1617.

12. Type LR91 ultra-high-speed directional relay, *Publication CH-ES 63-85.11*, Asea Brown Boveri.
13. Cabeza-Resendez, L. Z., Greenwood, A. N. and Lauber, T. S. (1985): Evaluation of ultra-high-speed relay algorithms, *EPRI Report EL-3996*.
14. Rajendra, S. and McLaren, P. G. (1985) Travelling-wave technique applied to the protection of teed circuits: principle of travelling-wave technique, *Trans. IEE*, **PAS-104**, 3544–3550.
15. Shehab-Eldin, E. H. and McLaren, P. G. (1988) Travelling wave distance protection – problem areas and solutions, *IEE Transaction on Power Delivery*, **3**, 894–902.
16. Thomas, D. W. P. and Christopoulos, C. (1991) Ultra-high speed protection of series compensated lines, Paper presented at the IEE/PES summer meeting, San Diego, California, *Paper 91 SM 359-0 PWRD*.

FURTHER READING

McLaren, P. G., Travelling wave and ultra high speed (UHS) relays Chapter 6 in *IEE Tutorial Course-Microprocessor relays and protection systems* (ed M. S. Sachdev), 88 EH 0269-1-PWR.
Bollen, M. H. J. and Jacobs, G. A. P. (1988) Extensive testing of an algorithm for travelling-wave-based directional detection and phase-selection by using TWON-FIL and EMTP, Eindhoven University of Technology, *EUT Report 88-E-206*.
Bollen, M. H. J. and Jacobs, G. A. P. (1989) Implementation of an algorithm for travelling-wave-based directional detection, Eindhoven University of Technology, *EUT Report 89-E-214*.
Johns, A. T., Martin, M. A., Barker, A., Walker, E. P. and Crossley, P. A. (1986) A new approach to E.H.V. direction comparison protection using digital signal processing techniques, *Trans. IEE*, **PWRD-1**, 24–34.
Crossley, P. A., Elson, S. F., Rose, S. J. and Williams, A. (1989) The design of a directional comparison protection for ehv transmission lines, Fourth International Conference on Developments in Power System Protection, *IEE Conf Publ. 302*, pp. 151–155.

Appendix A

The testing and application of power-system protective equipment

The capital costs of power-system equipment are high and this is particularly so for major items such as large turbines, alternators, transformers and overhead lines and cables. They represent investments which provide subsequent income derived from the payments made for the electrical energy supplied to consumers. It is therefore vital that the maximum continuity of supply be maintained to consumers and that power-system equipments should not suffer unavoidable consequential damage when faults occur within networks. Similar considerations also apply to smaller installations such as those in manufacturing plants.

To satisfy these requirements, protective equipments, such as those described in the earlier chapters, must be provided and their not insignificant costs may be considered to be insurance premiums, because they do not directly increase the income from consumers. Having paid the premiums, supply authorities and others understandably expect protective equipments to operate correctly under all conditions. To achieve this goal it is necessary that they and the suppliers of protective equipment ensure that the necessary studies are conducted so that suitable protective schemes are applied and that the appropriate fault settings are chosen for each application. In addition, adequate development, production and site tests must be performed.

The procedures and tests which are undertaken to ensure that the various types of protective equipments now available will meet these requirements are examined in the following sections.

A.1 PERFORMANCE REQUIREMENTS

It will be evident that a knowledge of the performances required of protective equipments is needed both when new schemes are to be developed or when existing schemes or items are to be applied. Such information may be derived from studies of the behaviour of power systems or from the manufacturers of items of equipment and it will affect the selection of the types of protective equipment needed for different applications. As examples, fuselinks may be used to protect relatively low voltage loads or components whereas distance- or impedance-measuring schemes may be applied to major transmission lines and in each case the information required to apply the protection will be different, as will the testing which must be done.

A.2 TESTING OF PROTECTIVE DEVICES AND EQUIPMENT

The testing requirements for the various forms of protection are considered below.

A.2.1 Fuselinks

Fuselinks should ideally operate at currents above their minimum-fusing levels in times slightly shorter than those for which the items or circuits being protected by them can withstand the currents without being damaged. Clearly, therefore, the withstand time/current characteristics of items to be protected must be known and at very high current levels the $I^2 t$, i.e.

$$\int_0^t i^2 \, \mathrm{d}t$$

values which may be withstood must also be known.

Over the years ranges of fuselinks have been produced with known and published characteristics and in many cases fuselinks with characteristics suitable for particular applications can be selected from those already available.

The testing of new designs of fuselinks

Unlike other items of protective equipment, an individual fuselink cannot be fully tested to determine its time/current characteristic because it is destroyed when it operates for the first time. Consequently, when a new fuselink is being developed, a large number of prototypes are produced, care being taken to ensure that they are all as similar as possible. The actual tests done depend on the particular fuselinks being examined, but in general their dimensions are checked to ensure that they are within the tolerances that will be used in the subsequent volume production. The electrical resistance of each fuselink is measured in an ambient temperature, typically in the range 20–25°C and normally, in the case of low voltage fuselinks, the power dissipated by each fuselink at its rated current when it is mounted in a standard test rig is measured. The prototype fuselinks which meet the necessary conditions are then used in the following tests.

The determination of conventional fusing current Conventional fusing current is strictly the current which is needed to cause a fuselink to operate after an infinite time and its determination is clearly impractical. British and International Standards therefore define the conventional current as that current which will cause operation in a given period, say 4 hours, but even the determination of this value would be very time consuming and therefore to simplify the situation, IEC specifications require that all of a specified number of fuselinks mounted in a standard type-test rig should operate in less than the conventional time, for example 4 hours, and not operate in the conventional time when carrying the conventional non-fusing current. The conventional time, which depends on the current rating, is specified in the various standards. For industrial fuses the conventional fusing current is 1.6 × (rated current) and the conventional non-fusing current is 1.25 × (rated current).

The determination of breaking-current capacity Fuselinks must be capable of operating in service under the most onerous conditions which can arise. Their breaking capacities must therefore be determined from tests done under specified conditions in single-phase inductive circuits of low power factor (typically 0.2 for low voltage fuselinks), arranged as shown in Fig. A.1. Equipment must be included to enable test circuits to be closed at any desired points in the voltage cycle so that conditions of varying severity may be produced. The fuselinks are mounted in standard rigs and tests are performed to determine not only their breaking capacities but parameters such as I^2t let through, arc voltages and cut-off currents and when the latter are being found, the circuits must be switched so that arcing commences just prior to an instant when the voltage is at its peak value. During testing the appropriate source voltages specified in the various standards must be provided, for example 10% more than the rated voltage of low voltage fuselinks, and the voltages must be maintained for at least 30 s and in the case of high voltage fuselinks they must be maintained for 60 s.

Type tests must be done on fuselinks over the range of currents between their maximum and minimum breaking capacities and at the lower-current

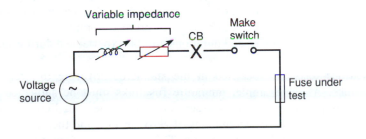

Fig. A.1 Fuse test circuit.

levels the power factors of the test circuits are usually higher than those used at the maximum. This takes account of the situations which arise in service and IEC specifications usually require tests at high current levels to be done at power factors at or below 0.2, whereas tests at lower currents may be done at power factors in the range 0.3 to 0.5.

The maximum breaking-capacity tests are usually conducted in short-circuit testing stations used also for testing switchgear. Such stations have high MVA outputs which enable high currents to be obtained at the required voltage levels. At lower-current levels for which the operating times of fuselinks may be of the order of an hour or more it would be uneconomic to use a major test plant and two-part test methods, which are allowed in IEC specifications, may be used. In such cases, a fuselink is supplied initially from a low voltage source and then when the element is nearing its melting temperature, the test circuit is switched to obtain the current from a short-circuit alternator which is excited to provide the required recovery voltage.

Time/current characteristics   Specific time/current characteristic curves have not been standardized or quoted in standard specifications because this could stifle developments and prevent the introduction of new fuselinks with different characteristics. The present trend is therefore to specify a number of points which form gates through which the actual characteristics of all manufacturers' fuselinks must pass. As an alternative, some of the IEC recommendations on low voltage fuses specify zones in which all characteristics must lie. The zones have been chosen so that all fuselinks of a given rating would operate in a shorter time at any current than any fuselink of twice the rating. This method of specification is likely to be discontinued and in future the gating method will probably be adopted as standard.

Compliance with the gates or zones is checked during the breaking-capacity tests, described above, by supplying prototype fuselinks with the necessary currents and noting the corresponding operating times.

The determination of overload-withstand capability   Some fuselinks must be capable of carrying specified overload currents for particular periods without their subsequent performance being affected. Such a capability is necessary for example when they are to be used to protect motors where current-surges of particular magnitudes and durations will be experienced during starting and run up.

Various tests are specified in the standards covering different types of fuselinks. As an example, miniature fuselinks must be subjected to a 100 cycles, during each of which a current of 1.2 times rated current must flow for 1 hour followed by 15 minutes of zero current, and then finally 1.5 times rated current must flow for one hour. This test must be done on three fuselinks and at the ends of the tests the voltage drops across the fuselinks must not have changed by more than 10%.

Manufacture, quality assurance and inspection

The standards set for the construction of prototype fuses in both materials and manufacturing processes become the standards which must be maintained when bulk production commences. This is essential to ensure that the production fuses will have the performance characteristics indicated by the type tests and because electrical tests cannot be performed on the bulk output.

Detailed specifications and procedures must be prepared for all stages of manufacture from the purchase of materials and components to the final inspection and testing. Many physical checks must be made to ensure that components are of dimensions within the acceptable limits. In the case of high voltage fuselinks, radiographs of completed units are taken to check that the elements are not broken, twisted or damaged in any other way and that they are correctly positioned within their bodies.

As a final check, the resistances of finished fuselinks are usually measured with a digital instrument with an error of 0.5% or less. After correcting for ambient temperature the measured values must comply with the original design limits.

Service checks

In general fuses do not require any regular maintenance but it is desirable that those in important circuits be inspected physically at regular intervals to check that the housing and fuselinks are in good condition. In addition, fuselink resistances can be measured and compared with the original values.

Further information on testing is provided in reference [1] and in National and International Standard Specifications.

A.2.2 Current and voltage transformers

It will be clear from Chapter 2 and the later chapters, which dealt with protective schemes and relays, that separate current transformers are used to supply the various equipments in power systems. As an example, a housing within or adjacent to a circuit-breaker may contain several magnetic cores and secondary windings associated with a common primary winding. One of the secondary windings may supply accurate metering equipment, whilst other secondary windings may separately supply IDMT relays and a current-differential protective scheme.

To enable appropriate current transformers to be supplied the burdens which will be imposed on them by the equipment to be connected to them must be known and in addition the behaviour required of them over the current range which they will encounter must be known.

In many cases transformers conforming to particular classes designated in National and International Standards will be found to be suitable; for example, transformers with a ratio of 500/1 and of Class 5P may be chosen to supply IDMT relays in a particular installation. In these circumstances, the transformers, on completion, are routinely tested to check that they meet the requirements laid down for their class in the relevant standards. Various limits are quoted and bridges and other equipment including sources capable of driving the necessary primary currents, are used.

When the more complex protective schemes, such as those based on current balancing, are being developed matched sets of current transformers are produced for use in conjunctive tests which are performed in the laboratories of manufacturers. In these tests, where the current conditions which could occur in service are produced, the suitability of the current transformers is determined and subsequent production transformers are checked to ensure that they have very similar characteristics. Further information on conjunctive testing is provided later in section A.2.4.

Whereas separate current transformers are used to supply different items of measuring and protective equipment, individual voltage transformers supply several burdens connected in parallel and clearly they must have the necessary volt-ampere outputs. In addition they must provide output voltages within the accuracy ranges required. Each transformer is tested on completion to verify that it complies with the limits specified in the appropriate Standard Specification using equipment such as bridges.

In many applications of both voltage and current transformers it is essential that the polarities of their outputs should be correct. As examples, current transformers feeding current-differential protective schemes must be so connected that operation will not occur during healthy or external fault conditions and current and voltage transformers must be so connected that directional relays supplied by them will operate and restrain correctly. It is therefore necessary that site tests be conducted during commissioning to ensure that the interconnections between transformers and their burdens are correct.

Detailed information on the testing of current and voltage transformers is provided in references [2–4].

A.2.3 The testing of relays

Very large numbers of IDMT and other relays are produced annually for use in power systems. Many of them have adjustable settings to enable them to be applied to any network. As an example, IDMT relays are provided with ranges of current and time settings.

National and International Standard Specifications have been produced in which the operating characteristics of many relays are defined and acceptable performance limits are stated.

Both prototype and production relays are tested by their manufacturers to confirm that their performances comply with the requirements. Because the current and voltage inputs to relays are usually relatively low, typically no more that 100 V and 100 A, there is little difficulty in providing adequate testing facilities.

It is clearly desirable that relays be checked both when they are put into service and at regular intervals thereafter. In the case of electro-mechanical relays, it is necessary that their performances be checked electrically by energizing them from suitable test sets and in addition they should be inspected physically. The electrical performances of electronic relays should also be checked by energizing them with the appropriate signals obtained from test sets.

Further information on this topic can be obtained from reference [5].

A.2.4 The testing of protective schemes

Protective schemes are usually quite complex, and because of their important roles in power systems their performances must be thoroughly examined during their development and subsequent production. Testing is then necessary after schemes are installed and at regular intervals thereafter.

It will be clear that the procedures which must be adopted are dependent on the operating principles and methods of implementation of individual schemes and these cannot be covered in detail in this book. The basic methods and testing equipments which are used are, however, described in the following sections.

Development testing

Because protective schemes are required to operate when the conditions on power systems are abnormal it is essential that their performances be determined during such conditions. Manufacturers have therefore to set up testing facilities which will enable schemes to be subjected to the full range of conditions which they may encounter in service. Over the years, facilities have been developed and extended to meet the changing needs.

The testing of current-differential and phase comparison schemes As stated in Chapter 4, current-differential schemes, as proposed by C. H. Merz and B. Price, were produced and installed on networks in the early years of this century and such schemes have been continually improved and are still being installed. Their performance, as explained earlier, is dependent on the relaying arrangements, the communication links between the ends of the protected units and the current transformers. It therefore became the practice to perform development tests, called conjunctive tests, on prototype schemes fed from current transformers the same as those which were subsequently to be used in service.

Because fault settings are relatively low they can be checked readily by supplying the appropriate currents to the current transformers. To check that stability will be maintained under external fault conditions, it is required that the maximum currents which may flow in service be supplied to the current transformers and that these should contain the transient components with the highest time constants which may be encountered. This latter factor is of great importance as it affects the core fluxes present in current transformers, as explained in Chapter 2, and determines the possible degrees of saturation which may occur.

Initially, manufacturers installed motor-driven-three-phase alternator sets of relatively low volt-ampere outputs which supplied step-down transformers which could supply high currents at low voltages to the test circuits. Impedors with high X/R ratios were available for connection in series with the primary windings of the current transformers and in this way the possible service conditions could be produced. A circuit-breaker was included in the test circuit to enable tests to be initiated and terminated. At that time it was not possible to control the point on the voltage wave at which a circuit-breaker would close and therefore it was the practice to do many stability tests so that the most onerous conditions were likely to have been produced.

These testing arrangements are still used today but features such as the control of the point on the wave at which the circuit closes are now standard and modern recording facilities and instruments are incorporated. A modern medium-current test plant [5] is shown in single line form in Fig. A.2. It can be seen that it incorporates a 500 kVA supply transformer with 11 kV primary windings and secondary windings which can provide line voltages of 880, 660 and 440 V. The inductive reactors can be set to give wide ranges of X/R ratios and current magnitudes, the maximum current being 510 A per phase. This plant feeds relays from built-in current transformers with ratios of 10/1 A and

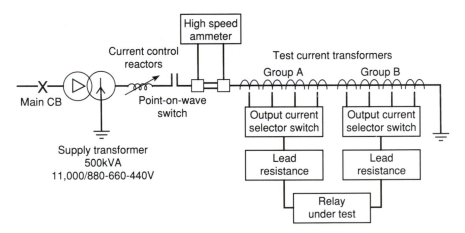

Fig. A.2 Schematic diagram of a medium-current test plant. (Reproduced from *Protective Relays – Application Guide, 3rd edn*, GEC Measurements, 1987 with the permission of GEC Alsthom Protection and Control Ltd.)

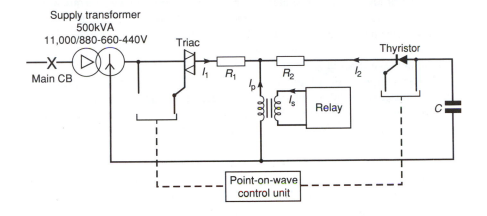

Fig. A.3 Circuit to achieve a high X/R ratio in current transformer primary current. (Reproduced from *Protective Relays – Application Guide, 3rd edn*, GEC Measurements, 1987 with the permission of GEC Alsthom Protection and Control Ltd.)

20/1 A and the effective knee-point voltages of the transformers can be varied between 160 V and 2000 V by connecting one or more of them in series.

The nominal X/R ratio of the reactors is 30, but this value is reduced in practice by the resistances of the connections, the windings of the current transformers and the relays. It will be appreciated that current transformers with the cores and secondary windings to be used in service, but with the appropriate multi-turn primary windings, could be fed by test plants such as that described above to enable completely conjunctive tests to be performed.

The increase in power-system fault levels and X/R ratios in recent years has created conditions which cannot be simulated economically with these types of equipment. Because of this Cavero *et al* [6] commenced investigations at the University of Nottingham into alternative methods of generating the required test currents. This led to development of test plants with the basic arrangement shown in single-phase form in Fig. A.3. In this plant, the transient component of the test current is produced by discharging the capacitor C through the resistor R_2 and the primary winding of the current transformer which supplies the protective scheme or relay. The desired transient current is obtained by previously charging the capacitor to the appropriate voltage and by setting the resistance (R_2) to the necessary value. The steady state component of the current is supplied from a transformer, the desired magnitude being obtained by choosing the appropriate voltage for the secondary winding of the transformer and adjusting the resistance of resistor R_1. This current also flows through the primary winding of the current transformer. Current flow is initiated at the appropriate instant in the cycle of the voltage provided by the secondary winding of the transformer by firing the triac and thyristor.

The technique has been implemented on a three-phase basis and such plants can be used in the development of protective schemes, and again the desired currents can be fed to either current transformers similar to those to be used in service or to built-in current transformers interconnected as described earlier to achieve desired knee-point output voltages.

The most recent development has been the introduction of programmable power-system simulators which can be widely used to test all types of protective equipment. These are described in some detail in the next section.

The testing of impedance-measuring schemes   Unlike the schemes considered in the previous section, impedance-measuring, or distance protection, schemes are supplied with both voltages and currents and in service these are obtained from voltage and current transformers connected near the ends of the lines being protected.

In the past, the power-system currents and voltages which would be present at the end of a given transmission line or cable during faults at different positions on it or beyond it could have been calculated for the range of possible operating conditions. It would, however, have been difficult to then feed the appropriate quantities to the voltage and current transformers connected to a protective scheme during conjunctive testing. Because of this, testing equipments containing so-called 'artificial lines' were produced. Such an equipment is fed from a three-phase voltage source and it contains variable impedors which can be set to represent the effective source impedance at the busbars at the end of the line to which the protective equipment was considered to be connected and also to represent the line and fault impedances and that of the load at the remote end of the line. Prior to a test, the impedances have to be set so that they represent particular network conditions. These have to take account of the source MVA, and voltage, the ratio of source and line impedances (Z_s/Z_L) and other factors such as the X/R ratios of sections of the network.

These equipments are supplied from three-phase sources at voltage levels of 415 V or 660 V (line) at a frequency of 50 Hz (in Britain) and it is usual to be able to obtain an alternative supply from a variable-frequency machine. The maximum currents available are of the order of 120 A (r.m.s.) and these are fed to current transformers with multi-turn primary windings with cores and secondary windings similar to those which are to be used in service with the protective scheme being tested.

All types of faults can be simulated and they can be applied in different positions; for example, on the protected line or beyond it. Point-on-wave control is invariably provided so that faults may be applied at any desired instant in the voltage cycle and adjustable resistors are included so that the presence of fault-arc resistance can be simulated.

As stated in section 11.14 (page 431), programmable power-system simulators have been developed to test distance-measuring schemes including those intended for application to series-compensated transmission lines. In addition, they can be used to test other protective schemes including those in which current comparisons are effected or those in which discrimination is achieved by examining travelling waves.

Simulators may be produced in several ways but basically they contain a computer programmed to calculate the voltages and currents which will be present under a wide range of healthy and fault conditions on a network to which a particular protective scheme is to be applied. Several computer codes are available for this task, a particular example being the Electromagnetic Transient Program (EMTP) [7]. The digital data are converted into analogue quantities which are applied to analogue models of the voltage and current transducers and their outputs are in turn applied to the actual protective scheme. Its behaviour is monitored and its suitability is thereby determined.

Research institutions [8], and manufacturers have constructed such test equipments. As an example, a block diagram of a simulator used by GEC Measurements is shown in Fig. A.4.

The testing of travelling-wave schemes   These schemes are implemented in a similar way to modern distance-type protective schemes and because of their complexity their behaviour should be determined using programmable power-system simulators.

In addition to the tests described above, which determine the electrical performances of relays, additional tests are necessary to ensure that relays are able to withstand a range of environmental conditions. Protective relays must be capable of withstanding vibrations [9], impact and extremes of temperature and humidity. Another class of important tests, particularly for digital relays, is those related to electromagnetic compatibility (EMC). These include high-frequency bursts, impulses, fast transients, static discharges and radiated interference tests [10].

The production testing of schemes

Tests on protective schemes before they are dispatched to site may be conducted using equipments such as those described in the preceding sections but only a limited number of checks should be necessary at this stage. Alternatively, tests conducted using simpler equipment may be adequate to ascertain that the schemes are operating correctly.

Commissioning and subsequent routine testing

Tests are performed before schemes are commissioned using secondary injection and other equipment and at this stage checks should be made to ensure that the correct settings have been selected and that voltage and current transducers have been connected correctly.

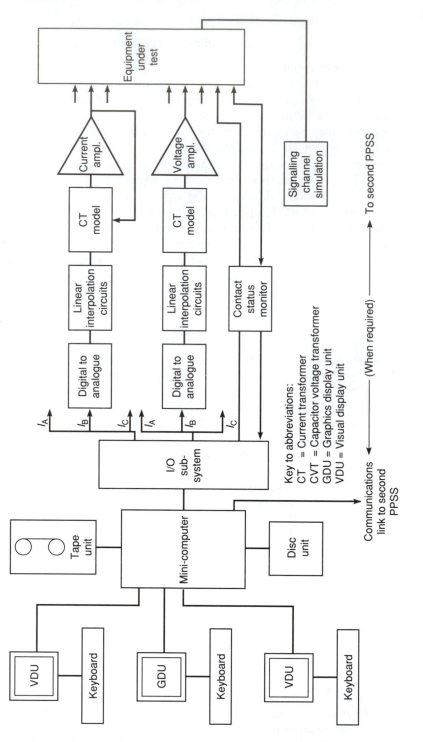

Fig. A.4 Block diagram of a programmable power-system simulator. (Reproduced from *Protective Relays – Application Guide, 3rd edn*, GEC Measurements, 1987 with the permission of GEC Alsthom Protection and Control Ltd.)

Routine checks should be performed at regular intervals after schemes have been commissioned to ensure that they are capable of operating correctly, the nature of the checks being dependent on the way in which the schemes are implemented. An indication of the on-site testing procedures adopted by a major utility may be found in reference [11].

A.3 AUTOMATIC TESTING OF RELAYS AND PROTECTIVE SCHEMES

As indicated above, it is necessary that relays and protective schemes be tested at regular intervals after they are installed and commissioned. In many cases this is done by performing individual tests using secondary-injection equipment and by making physical checks. This was certainly the practice in the past. Equipments are now available, however, to automate the process and they may be used with electro-mechanical relays, such as IDMT elements or modern static relays and schemes.

These equipments provide testing voltage and current waveforms representing power-system quantities which are mathematically synthesized. Digital-to-analogue converters and power amplifiers are incorporated to produce the levels of voltage and current required to cause the relays under test to operate. The relay outputs are monitored and quantities such as operating times are recorded.

The use of such equipments enables tests to be conducted quickly even on quite complex schemes, such as distance protection. Further information on such equipment is provided in reference [12].

It was the practice in the past for manufacturers to provide some built-in test features in relatively complex schemes. As an example, the early phase-comparison protective schemes applied to transmission lines incorporated switches which enabled high-frequency signals to be sent between the line ends to check that correct operation would be obtained during both internal- and external fault conditions. These switches could be operated manually or automatically at regular intervals, say every 6 hours.

This practice has continued and many modern systems incorporate facilities for regular testing. Comprehensive self-test facilities can be easily incorporated in digital relays where hardware and software are designed to allow continuous testing of the functioning of various parts of the relay, including the accuracy of computation and memory integrity. Self-checking is performed as a background task, without interfering with the operation of the relay, and it is claimed that identification of malfunctions is possible in most cases. When malfunctioning is detected an alarm is raised to alert the operator to the problem.

REFERENCES

1. Wright, A. and Newbery, P. G. (1982) *Electric Fuses*, Peter Peregrinus Ltd.
2. *BS 3938 : 1973 British Standard Specification for Current Transformers.*
3. ANSI/IEE C57.13.1-1981, *IEE Guide for Field Testing of Relaying Current Transformers.*
4. *BS 3941 : 1975 Specification for Voltage Transformers.*
5. GEC Measurements, *Protective Relays Application Guide* (3rd edn), 1987, Chapter 23.
6. Paull, C. J., Wright, A. and Cavero, L. P. (1976) Programmable testing equipment for power-system protective equipment, *Proc. IEE*, **123**, 343–349.
7. *Electromagnetic Transients Program-Reference Manual* (*EMTP Theory book*), Prepared by H. W. Dommel for the Bonneville Power Administration, August 1986.
8. Redfern, M. A., Aggarwal, R. K. and Husseini, A. H. (1990) A laboratory power system simulator for evaluating the performance of modern protective relays, *25th Universities Power Eng Conf*, Aberdeen 12–14 Sept 1990, pp 167–170.
9. *BS 142 : 1989, Electrical Protection Relays*, Section 1.5, Vibration, shock, bump and seismic testing.
10. *IEC 255-22: Electrical disturbance tests for measuring relays, Parts 1 to 4.*
11. Laycock, W. J. (1989) On site testing of protection and automatic switching equipment in 275 and 400 kV substations, *Proc. 4th Int Conf on Development in Power System Protection*, Edinburgh 11–13 April 1989, IEE Conf. Publ. No 302, pp 27–29.
12. Webb, A. C. and Webb, M. (1988) Automated testing of power system protection relays, *Power Eng J.*, November, pp 291–296.

Appendix B

Percentage and per-unit quantities

Whilst the performances of power-system networks can be determined using the actual ohmic impedances of the components within the networks, the various transformations caused by the presence of power transformers have to be taken into account. These transformations could be eliminated in a model network used to represent a particular power system if one voltage level was chosen as the reference and all the impedances were converted to the equivalent values associated with this voltage level. A similar effect is obtained somewhat more simply by using percentage or per-unit values in calculations associated with power systems.

A given percentage value for an impedor means that the voltage drop across it when it is carrying the rated current of the section of the circuit in which it is connected is the percentage of the rated voltage of that section of the circuit. As an example, if a resistor of $20\,\Omega$ is connected in a circuit for which the rated voltage and current are 1000 V and 10 A respectively, then the resistor has a percentage value of 20%, i.e. it drops 200 V when carrying 10 A in a circuit rated at 1000 V. It may alternatively be given a per-unit value of 0.2, i.e. 200/1000.

Voltages and currents are also assigned percentage or per-unit values based on the rated voltages or currents of circuits. As an example, if the voltage between two points in a circuit rated at 100 V is 50 V, it is referred to as 50% or 0.5 per unit. Similarly, if a current of 20 A flows in a circuit rated at 50 A then the current is either 40% or 0.4 per unit. The term per-unit is usually abbreviated to pu, e.g. $I = 0.4$ pu.

It is well known that impedances can be referred across transformers, i.e. an impedor in the secondary circuit of a transformer may be represented by an equivalent impedor in the primary circuit. This is illustrated in Fig. B.1(a) in which an ideal transformer with N_p and N_s primary and secondary winding turns has an impedor $\overline{Z}_s$ in its secondary circuit which carries a current $\overline{I}_s$. The impedor has a per-unit value of $|\overline{I}_s \overline{Z}_s|/V_s$, V_s being the rated voltage of the secondary circuit. In the equivalent circuit, shown in Fig. B.1(b), the impedor $\overline{Z}_s$ is replaced by an impedor $\overline{Z}_s'$ in the primary circuit. The current in this circuit is $\overline{I}_p = -\overline{I}_s N_s/N_p$ and the drop across the impedor $\overline{Z}_s'$ is $\overline{I}_p \overline{Z}_s' = -\overline{I}_s N_s \overline{Z}_s'/N_p$. The rated voltage of the primary circuit is $V_p = -V_s N_p/N_s$ and

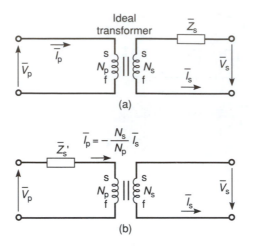

Fig. B.1 Equivalent circuit and parameters of a transformer.

therefore the per-unit value of the equivalent impedor ($\overline{Z}_s'$) is given by:

$$\frac{|\overline{I}_p \overline{Z}_s'|}{V_p} = \frac{|\overline{I}_s|}{V_s} \ |\overline{Z}_s'| \left(\frac{N_s}{N_p}\right)^2 \text{pu}$$

This has the same per-unit value as the actual secondary impedor, $|\overline{I}_s \overline{Z}_s|/V_s$ when

$$\overline{Z}_s' = \left(\frac{N_p}{N_s}\right)^2 \overline{Z}_s$$

i.e. the value used when impedance is transferred across a transformer.

It will also be appreciated that if a particular per-unit voltage is applied to the primary winding of an ideal transformer then the secondary output voltage has the same per-unit value. As an example, if the primary voltage has a value of 0.5 pu then the secondary voltage has a value of 0.5 pu. Similar relationships exist for the currents.

Figure B.2(a) shows a simple circuit containing a transformer and Fig. B.2(b) shows the equivalent circuit in which per-unit values are used.

The technique is also applied to three-phase circuits where further simplifications can be made. For example, if the phase voltages of a circuit are each of a particular per-unit value then the line voltages also have the same per-unit values.

Actual voltages, currents and impedances associated with circuits during steady state conditions are usually expressed in complex or polar form, for example $\bar{I}_p$ could be $(50 - j\,50)$A or $50\sqrt{2}\,\lfloor -\pi/4$ A and $\bar{Z}_s$ could be $20(1 + j\sqrt{3})\,\Omega$ or $40\,\lfloor \pi/3\ \Omega$. If these quantities were associated with a circuit in which one per-unit current and impedance were 100 A and 200Ω respectively then they would be expressed as:

$$\bar{I}_p = (0.5 - j\,0.5)\ \text{pu} \qquad \text{and} \qquad \bar{Z}_s = 0.1\,(1 + j\sqrt{3})\ \text{pu}$$

An important factor which must be taken into account when percentage or per-unit quantities are to be used is the volt-ampere ratings of the component parts of the circuit to be studied. This will be evident from the following example, which is based on a star-connected three-phase circuit in which there is an impedor Z in one of the phases.

The full load or rated current (I_r) of each phase of the circuit is given by:

$$I_r = \frac{VA}{\sqrt{3}\ V_L}$$

in which VA is the rated volt-amperes of the three-phase circuit and V_L is the rated line (phase-to-phase) voltage.

The voltage drop across the impedor Z when carrying the rated current of the circuit would be:

$$I_r Z = \frac{VAZ}{\sqrt{3}\ V_L}$$

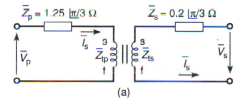

(a)

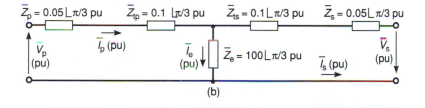

(b)

Fig. B.2 Transformer (a) and its equivalent circuit in per-unit quantities (b). Primary rated voltage and current: 500 V, 20 A. Primary winding impedance: $Z_{tp} = 2.5\,\lfloor \pi/3\ \Omega$. Exciting current at 500 V: $0.2\,\lfloor -\pi/3$ A. Secondary rated voltage and current: 200 V, 50 A. Secondary winding impedance: $Z_{ts} = 0.4\,\lfloor \pi/3\ \Omega$.

The per-unit impedance of Z would therefore be given by:

$$Z_{\mathrm{pu}} = \frac{I_{\mathrm{p}} Z}{V_{\mathrm{ph}}} = \frac{VAZ}{\sqrt{3}\, V_{\mathrm{L}} \frac{V_{\mathrm{L}}}{\sqrt{3}}} = \frac{VA}{V_{\mathrm{L}}^2} \cdot Z \qquad (\mathrm{B.1})$$

in which V_{ph} is the rated phase voltage of the circuit.

It will be clear that the per-unit value of an impedor Z in a single-phase circuit is given by:

$$Z_{\mathrm{pu}} = \frac{VA}{V^2} \cdot Z \qquad (\mathrm{B.2})$$

in which V is the rated voltage of the circuit.

When determining the behaviour of circuits which contain sections or items with different VA ratings it is necessary to use a common or reference volt-ampere base throughout. This procedure is used in the following example which should also illustrate the other points referred to above.

Example   For simplicity, the single-phase circuit shown in Fig. B.3(a) is examined. A transformer with a step-down ratio of 2 : 1 is included and this feeds three load circuits with different current ratings. The impedors $\bar{Z}_{\mathrm{tp}}$ and $\bar{Z}_{\mathrm{ts}}$ represent the impedances of the primary and secondary windings of the

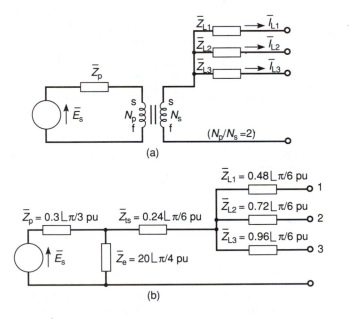

(a)

(b)

Fig. B.3 Example of a network (a) and its per-unit equivalent (b).

transformer and $\bar{Z}_e$ is its exciting impedance. The source, which provides an e.m.f. E_s, has an impedance $\bar{Z}_s$ and the connections between the source and the transformer have an impedance $\bar{Z}_c$. The impedance $\bar{Z}_p$ shown in Fig. B.3(a) is therefore given by:

$$\bar{Z}_p = (\bar{Z}_s + \bar{Z}_c + \bar{Z}_{tp})\Omega$$

The outgoing circuits have impedances of $\bar{Z}_{L1}, \bar{Z}_{L2}$ and $\bar{Z}_{L3}$ as shown.

The rated voltages of the primary and secondary circuits are 1000 V and 500 V respectively and the rated currents are as follows:

> Rated current of load circuit 1 $(I_{L1}) = 50$ A
> Rated current of load circuit 2 $(I_{L2}) = 40$ A
> Rated current of load circuit 3 $(I_{L3}) = 30$ A
> Rated current of primary circuit $= (50 + 40 + 30)/2 = 60$ A.

The impedances in the circuit are:

$$\bar{Z}_{L1} = 2 \lfloor \pi/6 \ \Omega; \bar{Z}_{L2} = 3 \lfloor \pi/6 \ \Omega$$

$$\bar{Z}_{L3} = 4 \lfloor \pi/6 \ \Omega; \bar{Z}_p = 5 \lfloor \pi/3 \ \Omega$$

$$\bar{Z}_e = 1000/3 \lfloor \pi/4 \ \Omega; \bar{Z}_{ts} = 1 \lfloor \pi/6 \ \Omega$$

The rated *VA* associated with the primary circuit is $1000 \times 60 = 60$ kVA and this is used as the reference base below.

Voltage drop across Z_p at 1000 V 60 kVA $= 60 \times 5 \lfloor \pi/3$

$$\therefore \bar{Z}_p = \frac{300}{1000} \lfloor \pi/3 \ \text{pu} = 0.3 \lfloor \pi/3 \ \text{pu}$$

Alternatively

$$\bar{Z}_p = \frac{VA}{V^2} Z = \frac{60 \times 1000}{1000^2} \cdot 5 \lfloor \pi/3 = 0.3 \lfloor \pi/3 \ \text{pu}$$

For load circuit 1 the rated current $= 50$ A and at this current the voltage drop on $\bar{Z}_{L1} = 100 \lfloor \pi/6 \ $ V.

$$\therefore \bar{Z}_{L1} = \frac{100}{500} \lfloor \pi/6 \ \text{pu at } 500 \times 50 = 25 \text{ kVA}$$

With reference to the 60 kVA base,

$$\bar{Z}_{L1} = \frac{60}{25} \cdot 0.2 \lfloor \pi/6 = 0.48 \lfloor \pi/6 \ \text{pu}$$

Alternatively,

$$\bar{Z}_{L1} = \frac{VA}{|V|^2} \bar{Z}_{L1} = \frac{60 \times 10^3}{500^2} \times 2 \lfloor \pi/6 = 0.48 \lfloor \pi/6 \ \text{pu}$$

Similarly,

$$\bar{Z}_{L2} = \frac{60 \times 10^3}{500^2} \times 3 \lfloor \pi/6 = 0.72 \lfloor \pi/6 \text{ pu}$$

$$\bar{Z}_{L3} = \frac{60 \times 10^3}{500^2} \times 4 \lfloor \pi/6 = 0.96 \lfloor \pi/6 \text{ pu}$$

$$\bar{Z}_e = \frac{60 \times 10^3}{1000^2} \times \frac{1000}{3} \lfloor \pi/4 = 20 \lfloor \pi/4 \text{ pu}$$

and

$$\bar{Z}_{ts} = \frac{60 \times 10^3}{500^2} \times 1 \lfloor \pi/6 = 0.24 \lfloor \pi/6 \text{ pu}$$

The equivalent circuit, using the 60 kVA base, is shown in Fig. B.3(b).

This circuit could be used more simply than the one shown in Fig. B.3(a) to determine the behaviour over widely ranging conditions.

As an example, should load terminals 2 and 3 be open-circuited and the load terminal 1 be short circuited to earth when the source e.m.f. (E_s) was 1000 V, i.e. 1 pu, and the small effect of the exciting impedance (Z_e) was neglected, then the short-circuit current (I_{sc}) would be given by:

$$\bar{I}_{sc} = \frac{1}{\bar{Z}_p + \bar{Z}_{ts} + \bar{Z}_{L1}} = \frac{1}{0.3 \lfloor \pi/3 + 0.72 \lfloor \pi/6} \text{ pu}$$

$$= 1.018 \lfloor -0.676 \text{ pu}$$

This is based on a 60 kVA base, i.e. 60 A in the primary circuit and 120 A on the secondary side of the transformer.

The actual current in the primary circuit would therefore be 61.1 A and the current in the short circuit would be 122.2 A.

Appendix C

Transformations of three-phase quantities

The steady-state voltage/current relationships in three-phase circuits with three phase conductors and a neutral conductor are of the general form:

$$\begin{bmatrix} \bar{V}_{an} \\ \bar{V}_{bn} \\ \bar{V}_{cn} \end{bmatrix} = \begin{bmatrix} \bar{Z}_{aa} & \bar{Z}_{ab} & \bar{Z}_{ac} \\ \bar{Z}_{ba} & \bar{Z}_{bb} & \bar{Z}_{bc} \\ \bar{Z}_{ca} & \bar{Z}_{cb} & \bar{Z}_{cc} \end{bmatrix} \cdot \begin{bmatrix} \bar{I}_a \\ \bar{I}_b \\ \bar{I}_c \end{bmatrix} \tag{C.1}$$

The current in the neutral conductor (I_n) is given by:

$$\bar{I}_n = -(\bar{I}_a + \bar{I}_b + \bar{I}_c)$$

There are basically three degrees of freedom and each phase voltage is dependent on the three phase currents. It will be clear that both the amount of computation involved in determining the behaviour of power systems and the processing in protective schemes can be reduced if equation (C.1) is transformed so that the elements in the Z matrix, other than those in the principal diagonal, all have a value of zero. In these circumstances, three independent equations are provided by the rows of the matrix equation.

It will be evident that the voltage, current and impedance matrices in equation (C.1) must all be so transformed that the two sides of the resulting equation are equal. The basic technique when dealing with steady-state r.m.s. quantities is outlined in the following section.

C.1 VOLTAGE, CURRENT AND IMPEDANCE TRANSFORMATION

Using a matrix $[\bar{C}]$, the relationship between the original and transformed current matrices may be expressed as:

$$[\bar{I}] = [\bar{C}][\bar{I}'] \tag{C.2}$$

in which $[\bar{I}]$ and $[\bar{I}']$ are the original and transformed matrices respectively.

To ensure that the volt-amperes are not changed the original and transformed quantities must be the same and because voltage-amperes are given by $[VA] = [\bar{V}_t][\bar{I}^*]$, in which $[\bar{V}_t]$ is the transpose of the voltage matrix and $[\bar{I}^*]$ contains elements which are the complex conjugates of the elements in the current matrix, the following relationship must be maintained:

$$[\bar{V}_t'][\bar{I}'^*] = [\bar{V}_t][\bar{I}^*] \tag{C.3}$$

in which $[\bar{V}_t']$ is the transpose of the transformed voltage matrix.

The relationship in equation (C.2) may be expressed as:

$$[\bar{I}^*] = [\bar{C}^*][\bar{I}'^*]$$

and thus equation (C.3) may be expressed as:

$$[\bar{V}_t'][\bar{I}'^*] = [\bar{V}_t][\bar{C}^*][\bar{I}'^*]$$

and therefore:

$$[\bar{V}_t'] = [\bar{V}_t][\bar{C}^*]$$

This equation may be rearranged to give:

$$[\bar{V}'] = [\bar{C}_t^*][\bar{V}] \tag{C.4}$$

Equation (C.1) may be expressed as:

$$[\bar{V}] = [\bar{Z}][\bar{I}]$$

and therefore equation (C.4) may be written in the form:

$$[\bar{V}'] = [\bar{C}_t^*][\bar{Z}][\bar{I}] = [\bar{C}_t^*][\bar{Z}][\bar{C}][\bar{I}']$$

$$= [\bar{Z}'][\bar{I}']$$

It is clear that the transformed impedance matrix $[\bar{Z}']$ is given by:

$$[\bar{Z}'] = [\bar{C}_t^*][\bar{Z}][\bar{C}] \tag{C.5}$$

When the matrix $[\bar{C}]$ is square and has an inverse $[\bar{C}]^{-1}$ then the following equations may all be used.

$$[\bar{I}] = [\bar{C}][\bar{I}'] \qquad\qquad [\bar{I}'] = [\bar{C}]^{-1}[\bar{I}]$$

$$[\bar{V}] = [\bar{C}_t^*]^{-1}[\bar{V}'] \qquad\qquad [\bar{V}'] = [\bar{C}_t^*][\bar{V}]$$

$$[\bar{Z}] = [\bar{C}_t^*]^{-1}[\bar{Z}'][\bar{C}]^{-1} \qquad\qquad [\bar{Z}'] = [\bar{C}_t^*][\bar{Z}][\bar{C}]$$

When the matrix $[C]$ is also orthogonal then

$$[\bar{C}][\bar{C}_t^*] = [\bar{C}_t^*][\bar{C}] = 1$$

i.e.

$$[\bar{C}_t^*] = [\bar{C}]^{-1} \qquad \text{and} \qquad [\bar{C}_t^*]^{-1} = \bar{C}$$

Using these relationships, the following equations are obtained:

$$[\bar{I}] = [\bar{C}][\bar{I}'] \qquad\qquad [\bar{I}'] = [\bar{C}_t^*][\bar{I}]$$

$$[\overline{V}] = [\overline{C}][\overline{V}'] \qquad\qquad [\overline{V}'] = [\overline{C}_t^*][\overline{V}]$$

$$[\overline{Z}] = [\overline{C}][\overline{Z}'][\overline{C}_t^*] \qquad\qquad [\overline{Z}'] = [\overline{C}_t^*][\overline{Z}][\overline{C}]$$

It should be noted that $[\overline{C}_t^*]$ can usually be determined more rapidly than $[\overline{C}]^{-1}$.

C.2 TRANSFORMATION INTO SYMMETRICAL COMPONENTS

Symmetrical components are a special form of transformation in which steady state three-phase voltage and current phasors are replaced by three sets of phasors designated as being of positive-, negative- and zero-sequence.

As an example, an unbalanced set of current phasors ($\overline{I}_a$, $\overline{I}_b$ and $\overline{I}_c$) is replaced by the nine current phasors shown in Fig. C.1, the relationships being that:

$$\overline{I}_a = \overline{I}_{1a} + \overline{I}_{2a} + \overline{I}_{0a}$$

$$\overline{I}_b = \overline{I}_{1b} + \overline{I}_{2b} + \overline{I}_{0b}$$

$$\overline{I}_c = \overline{I}_{1c} + \overline{I}_{2c} + \overline{I}_{0c} \qquad (C.7)$$

Again there are only three degrees of freedom and therefore the relationship can be expressed in the form:

$$\begin{bmatrix} \overline{I}_a \\ \overline{I}_b \\ \overline{I}_c \end{bmatrix} = \begin{bmatrix} 1 & 1 & 1 \\ a^2 & a & 1 \\ a & a^2 & 1 \end{bmatrix} \cdot \begin{bmatrix} \overline{I}_{1a} \\ \overline{I}_{2a} \\ \overline{I}_{0a} \end{bmatrix} \qquad (C.8)$$

in which the operator $a = -0.5 + j\sqrt{3}/2$ and $a^2 = -0.5 - j\sqrt{3}/2$ or $a = 1\underline{|2\pi/3}$ rad and $a^2 = 1\underline{|4\pi/3}$ rad.

The transformation matrix $[\overline{C}]$ is thus:

$$[\overline{C}] = \begin{bmatrix} 1 & 1 & 1 \\ a^2 & a & 1 \\ a & a^2 & 1 \end{bmatrix} \qquad (C.9)$$

The matrix $[\overline{C}_t^*]$ is therefore:

$$[\overline{C}_t^*] = \begin{bmatrix} 1 & a & a^2 \\ 1 & a^2 & a \\ 1 & 1 & 1 \end{bmatrix}$$

It must be recognized, as stated above, that when sequence quantities are used, the three phase voltages are replaced by nine sequence voltage and currents.

The volt-amperes associated with the actual phase voltages and currents are given by:

$$\overline{VA} = \overline{V}_{an}\overline{I}_a^* + \overline{V}_{bn}\overline{I}_b^* + \overline{V}_{cn}\overline{I}_c^* = [\overline{V}_t][\overline{I}^*]$$

and the voltage-amperes associated with the sequence quantities are given by:

$$\overline{VA} = 3\overline{V}_{1a}\overline{I}_{1a}^* + 3\overline{V}_{2a}\overline{I}_{2a}^* + 3\overline{V}_{0a}\overline{I}_{0a}^* = 3\,[\overline{V}_t'][\overline{I}'^*]$$

With this transformation therefore:

$$[\overline{V}_t][I^*] = 3\,[\overline{V}_t'][\overline{I}'^*] = [\overline{V}_t][\overline{C}^*][\overline{I}'^*]$$

and as a result:

$$3\,[\overline{V}_t'] = [\overline{V}_t][\overline{C}^*]$$

or

$$[\overline{V}'] = 1/3\,[\overline{C}_t^*][\overline{V}] \tag{C.10}$$

The full transposed voltage matrix $[V']$ is therefore:

$$\begin{bmatrix} \overline{V}_{1a} \\ \overline{V}_{2a} \\ \overline{V}_{0a} \end{bmatrix} = 1/3 \begin{bmatrix} 1 & a & a^2 \\ 1 & a^2 & a \\ 1 & 1 & 1 \end{bmatrix} \begin{bmatrix} \overline{V}_{an} \\ \overline{V}_{bn} \\ \overline{V}_{cn} \end{bmatrix} \tag{C.11}$$

In a circuit which is physically symmetrical, the elements on the principal diagonal of the impedance matrix, i.e. the impedances $\overline{Z}_{aa}$, $\overline{Z}_{bb}$, $\overline{Z}_{cc}$ are each equal to self impedance $(\overline{Z}_s)$ and the other elements, which are all the same, are equal to the mutual impedance $(\overline{Z}_m)$. The actual impedance matrix is therefore given by:

$$[\overline{Z}] = \begin{bmatrix} \overline{Z}_s & \overline{Z}_m & \overline{Z}_m \\ \overline{Z}_m & \overline{Z}_s & \overline{Z}_m \\ \overline{Z}_m & \overline{Z}_m & \overline{Z}_s \end{bmatrix}$$

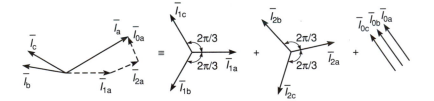

Fig. C.1 Transformation into symmetrical components

$$\begin{aligned} |\overline{I}_{1a}| &= |\overline{I}_{1b}| = |\overline{I}_{1c}| \\ |\overline{I}_{2a}| &= |\overline{I}_{2b}| = |\overline{I}_{2c}| \\ |\overline{I}_{0a}| &= |\overline{I}_{0b}| = |\overline{I}_{0c}| \end{aligned}$$

The sequence-impedance matrix is:

$$[\overline{Z}'] = 1/3\,[\overline{C}_t^*][\overline{Z}][\overline{C}]$$

$$= 1/3\begin{bmatrix} 1 & a & a^2 \\ 1 & a^2 & a \\ 1 & 1 & 1 \end{bmatrix}\begin{bmatrix} \overline{Z}_s & \overline{Z}_m & \overline{Z}_m \\ \overline{Z}_m & \overline{Z}_s & \overline{Z}_m \\ \overline{Z}_m & \overline{Z}_m & \overline{Z}_s \end{bmatrix} \cdot \begin{bmatrix} 1 & 1 & 1 \\ a^2 & a & 1 \\ a & a^2 & 1 \end{bmatrix}$$

i.e

$$= \begin{bmatrix} \overline{Z}_s - \overline{Z}_m & 0 & 0 \\ 0 & \overline{Z}_s - \overline{Z}_m & 0 \\ 0 & 0 & \overline{Z}_s + 2\overline{Z}_m \end{bmatrix} \tag{C.12}$$

Hence $\overline{Z}_1 = \overline{Z}_2 = \overline{Z}_s - \overline{Z}_m$ and $\overline{Z}_0 = \overline{Z}_s + 2\overline{Z}_m$.

It will be clear that there are no mutual impedances in the transformed matrix and the sequences are thus independent of each other when circuits are physically symmetrical, i.e.

$$\overline{V}_{1a} = \overline{Z}_1 \overline{I}_{1a}, \quad \overline{V}_{2a} = \overline{Z}_2 \overline{I}_{2a} \quad \text{and} \quad \overline{V}_{0a} = \overline{Z}_0 \overline{I}_{0a}$$

In practice, three-phase transformers and overhead transmission and distribution lines are physically asymmetric, i.e. $\overline{Z}_{aa}$, $\overline{Z}_{bb}$, $\overline{Z}_{cc}$ are not all equal nor are the mutual impedances such as $\overline{Z}_{ab}$ and $\overline{Z}_{ac}$. In these circumstances, the transformed sequence impedance matrix does not only have non-zero elements on the principal diagonal and the sequences are not therefore independent of each other.

Methods of determining the behaviour of three-phase power systems during unbalanced conditions, using symmetrical component techniques, are described later in Appendix D.

C.3 OTHER TRANSFORMATIONS

In many situations, such as the treatment of transient signals, the sequence transformations described in the previous section cannot be used. It is however useful in such cases to be able to decouple phases and the approach outlined below may be used.

The phase voltages $[\overline{V}^{(p)}]$ and currents $[\overline{I}^{(p)}]$ are related by the following expression:

$$\frac{d}{dx}[\overline{V}^{(p)}] = -[\overline{Z}^{(p)}][\overline{I}^{(p)}]$$

$$\frac{d}{dx}[\overline{I}^{(p)}] = -[\overline{Y}^{(p)}][\overline{V}^{(p)}]$$

where $[\overline{Z}^{(p)}]$ and $[\overline{Y}^{(p)}]$ are the system impedance and admittance matrices

respectively. In general they contain components which are frequency dependent and have real and reactive parts.

These equations may be combined to obtain:

$$\frac{d^2}{dx^2}[\overline{V}^{(p)}] = [\overline{P}][\overline{V}^{(p)}]$$

where
$$[\overline{P}] = [\overline{Z}^{(p)}][\overline{Y}^{(p)}]$$

Transformation matrices $[\overline{S}]$ and $[\overline{Q}]$ are sought so that component voltages $[V^{(c)}]$ and currents $[I^{(c)}]$ can be obtained.

$$[\overline{V}^{(p)}] = [\overline{S}][\overline{V}^{(c)}]$$

$$[\overline{I}^{(p)}] = [\overline{Q}][\overline{I}^{(c)}]$$

Expressing the phase quantities in terms of the component quantities gives:

$$\frac{d^2}{dx^2}[\overline{V}^{(c)}] = [\overline{S}]^{-1}[\overline{P}][\overline{S}][\overline{V}^{(c)}] = [\gamma^2][\overline{V}^{(c)}]$$

The three component quantities are decoupled provided the matrix $[\gamma^2]$ is diagonal.

Matrices $[\overline{S}]$ and $[\overline{Q}]$ are selected to diagonalize $[\gamma^2]$.

For lines in which the conductors are symmetrically positioned, diagonaliz-
ation is straight forward and the transformation matrices are:

$$[\overline{S}] = [\overline{Q}] = \begin{bmatrix} 1 & 1 & 1 \\ 1 & 0 & -2 \\ 1 & -1 & 1 \end{bmatrix}$$

This is known as Clarke's or $(\alpha, \beta, 0)$ transformation.

In practice the conductors of lines are not arranged symmetrically. Diago-
nalization is again possible but the transformation matrix varies with frequency. As an example, for a 400 kV, 50 Hz line in which the conductors are
in a vertical configuration and for a ground resistivity of 100 Ωm the $[\overline{S}]$
matrix is

$$\begin{bmatrix} 0.66 & -0.553 + j\,0.1 & -0.27 - j\,0.08 \\ 0.58 - j\,0.006 & 0.23 - j\,0.11 & 0.824 \\ 0.48 - j\,0.02 & 0.787 & -0.47 + j\,0.13 \end{bmatrix}$$

In transient calculations two approximations are necessary. Firstly, an $[\overline{S}]$
matrix at a particular frequency is selected typically at 1 kHz. Secondly, the
imaginary part of the components of the $[\overline{S}]$ matrix is neglected. For the 400
kV line considered above, the approximate matrix at 1 kHz is:

$$[\tilde{S}] \cong \begin{bmatrix} 0.68 & -0.56 & -0.2 \\ 0.57 & 0.36 & 0.79 \\ 0.46 & 0.75 & -0.59 \end{bmatrix}$$

The component or modal voltages and currents are then related by the expression

$$[V^{(c)}] = [Z^{(c)}][I^{(c)}]$$

where

$$[\overline{Z}^{(c)} = \begin{bmatrix} 542.4 & 1.53 & 0.1 \\ -3.2 & 293.2 & -0.7 \\ -0.2 & 0.1 & 246.6 \end{bmatrix}$$

It will be seen that because an approximate S matrix was used the off-diagonal elements are small but not zero and hence complete decoupling has not been achieved. In practice the contribution from the off-diagonal elements can be neglected.

Further details of transformation matrices may be found in references [1–3].

REFERENCES

1. Wilson Long, R. and Gelopulos, D. (1982) Component transformations – eigenvalue analysis succinctly defines their relationships, *Trans, IEE* **PAS-101**, 4055–4060.
2. Wasley, R. G. and Selvavinayagamoorthy, S. (1974) Approximate frequency response values for transmission-line transient analysis, *Proc. IEE*, **121**, 281–286.
3. Wedepohl, L. M. and Mohqwed S. E. T. (1969) Multiconductor transmission lines – theory of natural modes and Fourier integral applied to transient analysis *Proc. IEE* **116**, (C), 1553–1563.

Appendix D

The determination of power-system behaviour using symmetrical components

As indicated in Section C.2 of Appendix C, the sequences are only independent of each other when circuits are physically symmetrical. Although items of plant such as alternators are completely symmetrical, this is not the case for overhead lines or three-phase transformers but, nevertheless, the effects of asymmetry are usually neglected to simplify the calculations of the behaviour which will occur in power systems during both normal and abnormal conditions. This practice is acceptable because it is not normally necessary to determine fault-current levels with great accuracy.

In all networks, current flows are affected by the various impedances which are present and therefore the impedances presented to each of the current sequences must be known. In the following section the sequence impedances of items of equipment used in power systems are considered.

D.1 SEQUENCE IMPEDANCES

Because of the differences in the constructions and modes of operation of rotating machines and other static equipment, their effective impedances are examined separately below.

D.1.1 Three-phase synchronous machines

These machines are always physically symmetrical in that they contain three a.c. windings, each of the same number of turns, distributed similarly around their peripheries. A set of steady-state positive-sequence currents in these windings sets up a magnetic field of constant magnitude which rotates relative to the windings at synchronous speed (n_s) in the same direction as the main magnetic field set up by the field winding. An extra positive-sequence set of e.m.f.s is thereby induced in the windings, their magnitudes being proportional to the winding currents. This effect is normally allowed for by assigning a reactance to each winding, i.e. a positive-sequence reactance, which together with the winding leakage reactance is termed the positive-sequence synchron-

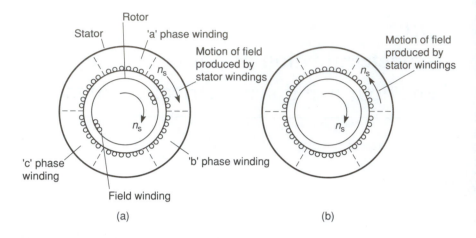

Fig. D.1 Field set up by stator currents rotates at n_s as shown for positive (a) and negative (b) sequence currents.

ous reactance (X_1). The three windings each present the same resistance (R_1) to the currents and as a result a machine can be assigned a positive-sequence synchronous impedance $(\bar{Z}_1)$. In practice this impedance is made relatively high to limit the currents which may flow in the events of faults, but, of course, the resistance is kept to a minimum to ensure that high efficiencies are obtained. As a result $X_1 \gg R_1$.

It will be appreciated that the magnetic field set up by positive-sequence currents does not cut the field circuit of a machine. This is illustrated in Fig. D.1(a).

A set of steady-state negative-sequence currents in the windings of a machine sets up a magnetic field which rotates relative to the windings at synchronous speed (n_s) in the opposite direction to that produced by the positive-sequence currents. This field, shown in Fig. D.1(b), cuts the three-phase windings and induces e.m.f.s in them but it is of reverse or negative sequence. This field rotates at twice synchronous speed relative to the field windings of the machine and the core material associated with them, causing hysteresis effects and eddy-currents. As a result, the negative-sequence impedance $(\bar{Z}_2)$ is not equal to the positive-sequence impedance $(\bar{Z}_1)$.

A set of zero-sequence currents in the windings of a machine sets up three time-varying magnetic fields which are stationary with respect to the windings. They induce equal e.m.f.s of the same phase in each of the windings, i.e. a set of zero-sequence e.m.f.s with a magnitude dependent on the zero-sequence current. These zero-sequence fields move at synchronous speed relative to the field windings and the core material causing power losses within them. A machine has therefore a zero-sequence impedance $(\bar{Z}_0)$, which is different from the other two sequence impedances.

For a large turbo-alternator, typical values for the steady-state positive-sequence reactance (synchronous reactance) range between 2 and 2.5 pu. However, immediately after the incidence of a fault, the machine reactance is much lower, typically 0.1–0.2 pu (subtransient) reactance. The subtransient phase lasts for approximately one or two cycles and it is followed by the transient phase when typical reactance values range from 0.15–0.28 pu. Steady state is reached after typically four or five cycles. Typical values for negative- and zero-sequence reactances range from 0.13–0.25 pu and 0.05–0.13 pu respectively.

D.1.2 Transformers

When positive-sequence currents flow in the primary and secondary windings of a transformer, magnetic fields are set up in leakage paths and positive-sequence e.m.f.s are induced in both sets of windings. This effect may be allowed for by assigning a positive-sequence leakage reactance to the primary and secondary windings (X_{1p} and X_{1s}). The currents flow through the resistances of the windings and therefore the windings possess positive-sequence impedances $\bar{Z}_{1p}$ and $\bar{Z}_{1s}$.

Similar effects are produced by negative-sequence currents and the impedances which may be assigned have the same values as those associated with the positive-sequence currents, i.e. $\bar{Z}_{2p} = \bar{Z}_{1p}$ and $\bar{Z}_{2s} = \bar{Z}_{1s}$.

The passage of zero-sequence currents also causes magnetic fields to be set up in leakage paths, but because these fluxes are in phase with each other they may not follow the same paths as those set up by the other sequence currents and as a result the zero-sequence impedances, $\bar{Z}_{0p}$ and $\bar{Z}_{0s}$ may not have the same values as the other sequence impedances.

Typical per-unit positive- and negative-sequence impedances for transformers rated at 10 MVA and above range between 0.05 and 0.2, depending on their insulation levels, construction and winding connections.

The zero-sequence impedance depends on the type of the core used and winding connections as described in section D.2.2. In general it has a value which is smaller than the other sequence impedances as explained in more detail in reference [1].

D.1.3 Overhead lines and cables

A similar situation to that with transformers exists with overhead lines and cables, namely that if their physical asymmetry is neglected then positive-, negative- and zero-sequence impedances may be assigned to them, the positive- and negative-sequence values being the same whereas the zero-sequence value is different. In the case of lines and cables, this occurs because the positive- and negative-sequence currents flow only in the phase conductors but the zero-sequence currents flow in the phase conductors and return in the earth or neutral conductor and ground. This causes not only the reactances to differ but also the resistances of the paths taken by the currents. As an example, the positive and negative sequence impedances for a 400 kV line in the British system, consisting of bundles of four conductors, is $\overline{Z}_1 = \overline{Z}_2 - 0.02 + j0.3\,\Omega/\text{km}$, and the zero-sequence impedance is $\overline{Z}_0 = 0.1 + j0.8\ \Omega/\text{km}$.

D.1.4 The effects of physical asymmetry

Because physical asymmetry is present in parts of power networks, the mutual impedances that are present cause interactions between the sequences. These can clearly be taken into account but they are generally neglected in the interests of simplifying the determination of the behaviour of networks during both normal and fault conditions and in the following sections the techniques which are usually employed are described.

D.2 SEQUENCE NETWORKS

An actual three-phase network is represented by three separate sequence networks which have to be interconnected appropriately to represent particular conditions.

As an example, the network shown in Fig. D.2(a), which is represented in single-line form for simplicity, would be replaced by the three networks shown in Fig. D.2(b).

Alternators produce only positive-sequence e.m.f.s, i.e. E_{A1} and E_{B1}, and therefore no e.m.f.s are present in the negative- and zero-sequence networks.

Other important factors which must be taken into account are considered below.

D.2.1 Earthing of neutral points

In Fig. D.2(a), the star points of the generators were shown to be solidly earthed and therefore they were shown to be connected directly to the zero-voltage conductor in each of the sequence networks.

In practice alternators are normally earthed through resistors to limit the currents which will flow in the event of faults to earth. It will be appreciated that the sum of the three positive-sequence currents $(\overline{I}_{a1}, \overline{I}_{b1}$ and $\overline{I}_{c1})$ is zero and that no current flows through an earthing resistor as a result of the flow of these currents and no voltage drop is caused across such resistors. They are therefore not included in the positive-sequence network. Similar considerations apply when negative-sequence currents flow. When zero-sequence currents flow, however, the current in an earthing resistor (R_g) is $3I_0$ because the three-sequence currents are of the same phase and the voltage drop across the resistor is $3I_0 R_g$. To allow for this situation, a resistor of $3R_g$ must be included in the zero-sequence network to allow for the fact that the current flow is only I_0. This representation is shown in Fig. D.3.

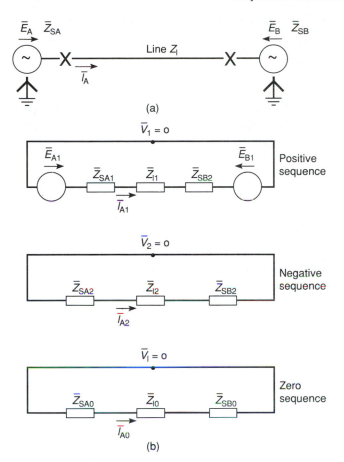

(a)

Positive sequence

Negative sequence

Zero sequence

(b)

Fig. D.2 A network (a) and its three phase-sequence networks (b).

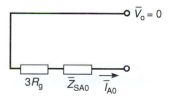

Fig. D.3 Zero-sequence network for an alternator earthed through a resistor R_g.

D.2.2 Transformer connections

The three-phase windings of power transformers may be connected in star or delta and indeed it is common practice to have the high voltage windings connected in star because this reduces the cost of insulation, as stated in Chapter 6, and the low voltage windings are often connected in delta. There are also situations where transformers have three sets of three-phase windings, a common arrangement then being to have the two higher sets of voltage windings connected in star and the other set of windings connected in delta.

Clearly positive- and negative-sequence currents can flow in both star- and delta-connected windings through the sequence impedances of the primary and secondary windings. It should be noted that the exciting impedances are usually regarded as infinite in the interests of simplicity, i.e. the exciting currents are assumed to be zero, and therefore ideal transformation is assumed, namely $I_p N_p + i_s N_s = 0$. A transformer may therefore be represented in both the positive- and negative-sequence networks by two percentage or per-unit impedances in series, i.e. $\bar{Z}_{1p}$ and $\bar{Z}_{1s}$ and $\bar{Z}_{2p}$ and $\bar{Z}_{2p}$ as shown in Fig. D.4(a).

Clearly zero-sequence currents cannot flow from a circuit into a delta-connected set of transformer windings because there is no path by which they may leave. A delta-connected set of windings therefore presents an infinite impedance to incoming zero-sequence currents. A set of zero-sequence currents can however circulate in a delta-connected winding and such currents would en-

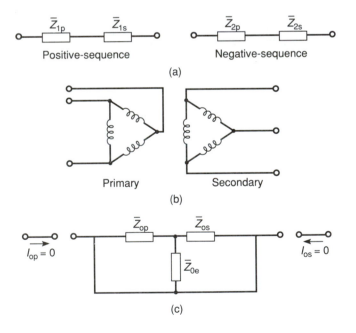

Fig. D.4 Sequence networks of a transformer.(The quantities shown are per-unit.)

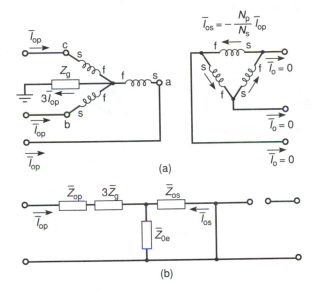

Fig. D.5 A star-delta connected transformer (a) and its zero-sequence network (b).

counter the zero-sequence leakage impedance of the winding ($\bar{Z}_0$). In the unlikely event of a transformer having two delta-connected windings as shown in Fig. D.4(b), the zero-sequence network would be as shown in Fig. D.4(c).

Zero-sequence currents can clearly flow into a star-connected set of windings provided that the star-point is earthed, either directly or via an impedor ($\bar{Z}_g$). As stated above, earthing impedors must be represented in zero-sequence networks by $3\bar{Z}_g$ because the actual impedor carries the three zero-sequence currents.

When a transformer is connected in star-delta, zero-sequence currents circulate around the delta-connected windings when zero-sequence currents flow in the star-connected windings. They do not, however, flow into the lines connected to the delta-connected windings. Because of this, the impedance presented to the zero-sequence currents is that of the leakage impedances of the two sets of windings, i.e. $\bar{Z}_{0p}$ and $\bar{Z}_{0s}$. This situation is shown in Fig. D.5(a) and (b).

If both windings of a two-winding transformer are connected in star, then zero-sequence currents will flow in each of the windings if their star points are both earthed and the magnitudes of the currents in those circumstances will be dependent on the leakage impedances of the windings and the impedances in the ground paths. This situation is illustrated in Fig. D.6. If the star point of only one winding is connected to earth then the zero-sequence current in it will be limited by the zero-sequence exciting impedance of the transformer (Z_{0e}). This situation is shown in Fig. D.7.

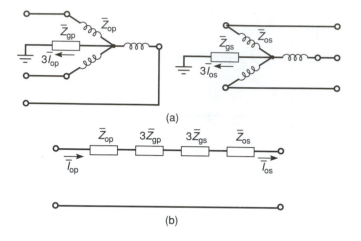

Fig. D.6 A star-star connected transformer with both star points earthed, and its zero-sequence network.(All quantities are per-unit.)

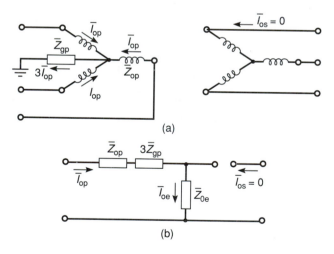

Fig. D.7 A star-star connected transformer with only one star point earthed (a), and its zero-sequence network (b).

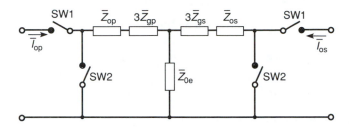

Fig. D.8 General form of zero-sequence network for two-winding transformers.(All quantities are per-unit.)

When dealing with two-winding transformers their zero-sequence circuits can be readily determined from Fig. D.8, in which each switch 1 is closed if its associated winding is star connected and each switch 2 is closed if its associated winding is delta connected.

Similar circuits may be used for three-winding transformers, an example based on a star-delta-star transformer being shown in Fig. D.9.

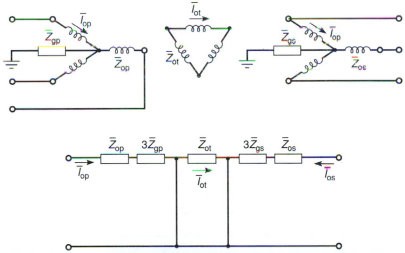

All quantities are shown in per unit

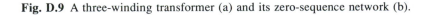

Fig. D.9 A three-winding transformer (a) and its zero-sequence network (b).

D.3 THE INTERCONNECTION OF SEQUENCE NETWORKS

To enable the actual operating conditions of power systems to be determined, the appropriate sequence networks must be interconnected in particular ways. These are described below for a range of possible conditions.

D.3.1 Balanced three-phase normal or fault conditions

Under conditions associated with either normal operation or when three-phase faults are present, the phase voltages and currents are balanced. As a result they have no negative- or zero-sequence components and the behaviour may be determined from the positive-sequence network, an example being shown in Fig. D.10. It will be clear that the phase currents and voltages at points in the network can be readily determined using complex quantities.

D.3.2 Conditions when a phase-to-phase fault is present

When an interphase fault is present, the current leaving one of the faulted phases at the fault position is equal to the current entering the other faulted phase. This situation is illustrated in Fig. D.11(a) for a fault between phases 'a' and 'b'. The fault current $\bar{I}_f$ is equal to the current $\bar{I}_a$ and equal to $-\bar{I}_b$. As a result $\bar{I}_a = -\bar{I}_b$, and of course the current $\bar{I}_c$ is zero. These currents, which are shown in Fig. D.11(b), may be transformed into the positive- and negative-sequence currents shown in Fig. D.11(c). It will be seen that the positive- and negative-sequence currents of the healthy phase (I_{1c}, I_{2c}) are in antiphase and have equal magnitudes given by:

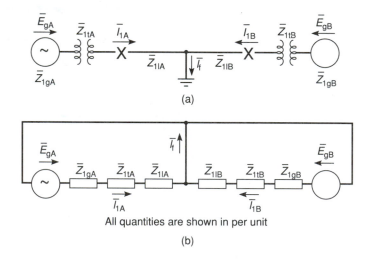

(a)

All quantities are shown in per unit

(b)

Fig. D.10 A network (a) and its sequence network following a balanced fault (b). The excitation impedance Z_{oe} is neglected.

$$\left|\,\overline{I}_{1c}\,\right| = \left|\,\overline{I}_{2c}\,\right| = \frac{\left|\,\overline{I}_{a}\,\right|}{\sqrt{3}} - \frac{\left|\,\overline{I}_{f}\,\right|}{\sqrt{3}}$$

or the complex form:

$$\overline{I}_{1c} = -\overline{I}_{2c} = \frac{j\,\overline{I}_{f}}{\sqrt{3}} \tag{D.1}$$

The positive- and negative-sequence networks representing an actual circuit must be interconnected so that the healthy-phase sequence currents at the fault position are of equal magnitude but of opposite polarities. This is illustrated in the example shown in Fig. D.11(d) which is based on the circuit shown in Fig. D.10(a). It will be appreciated that the actual fault current can be determined using equation (D.1) and the actual phase and voltage currents at other points in the network can be found by transforming the sequence quantities at

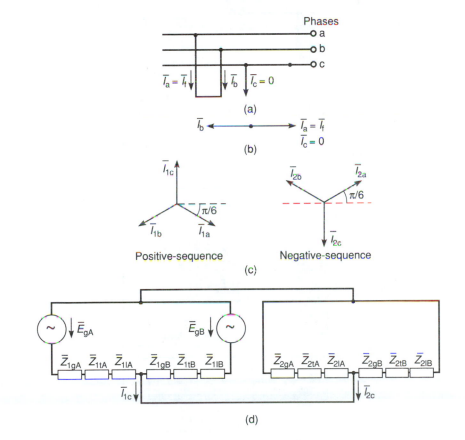

(a)

(b)

(c)

(d)

Fig. D.11 Conditions and sequence network for an interphase fault. (All quantities are per-unit.)

the corresponding points. A numerical example is included later in section D.4.

D.3.3 Conditions when a single phase to earth fault is present

When a single phase to earth fault is present, a current leaves the faulted conductor and flows in earthed conductors and/or the ground. For the fault on phase a shown in Fig. D.12(a), the fault current $\bar{I}_f$ is equal to the current $\bar{I}_a$ and the currents $\bar{I}_b$ and $\bar{I}_c$ are zero. These currents, which are shown in Fig. D.12(b), may be transformed into the positive-, negative- and zero-sequence currents shown in Fig. D.12(c). It will be seen that the three sequence-currents associ-

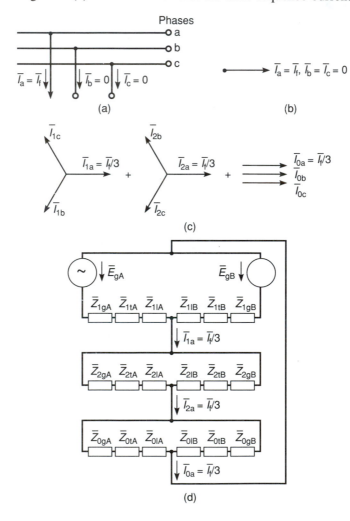

Fig. D.12 Conditions and sequence network for a phase-to-earth fault.

ated with phase a are in phase with each other and have the same magnitude,
i.e.

$$\bar{I}_{1a} = \bar{I}_{2a} = \bar{I}_{0a} = \bar{I}_f/3 \tag{D.2}$$

The three sequence networks representing a circuit must therefore be connected in series, as in the example shown in Fig. D.12(d). Again the actual currents at any point in a network can be determined and the fault current can be calculated using equation (D.2).

A numerical example is provided in section 4.4 and this indicates how factors such as transformer connections, methods of earthing and the presence of impedance in fault paths may be taken into account.

D.3.4 Impedances used in sequence networks

It will be clear that steady-state conditions are not established in a power system immediately after a disturbance, such as the incidence of a fault, occurs, and if it is necessary, it is now possible with the aid of modern computing equipment to determine the instantaneous values of the various currents and voltages which will result from a particular disturbance.

In many cases, however, such detailed information is not necessary. As an example, when choosing the fault settings of protective equipment, a knowledge of the minimum currents likely to flow during fault conditions may be adequate. When considering the stability of a protective scheme during external fault conditions, however, the maximum currents which may flow at such times must be known.

The higher currents which flow immediately after fault incidence are determined by using the sub-transient values of the sequence impedances of the sources which are present in the circuit being considered. The lower currents which are present shortly after a fault occurs may be determined by using the transient values of the sequence impedances and the eventual steady state conditions can be determined by using the synchronous impedances. These various values were referred to earlier in section D.1.

D.4 NUMERICAL EXAMPLE

A small network is shown schematically in Fig. D.13a. The parameters of each item of plant referred to the same base and neglecting the resistances are given below,

Generator G_A $Z_{A1} = j0.3$ pu, $Z_{A2} = j0.2$ pu, $Z_{A0} = j0.05$ pu

Generator G_B $Z_{B1} = j0.25$ pu, $Z_{B2} = j0.15$ pu, $Z_{B0} = j0.03$ pu

Transformer T_A $Z_{tA1} = Z_{tA2} = Z_{tA0} - j0.12$ pu

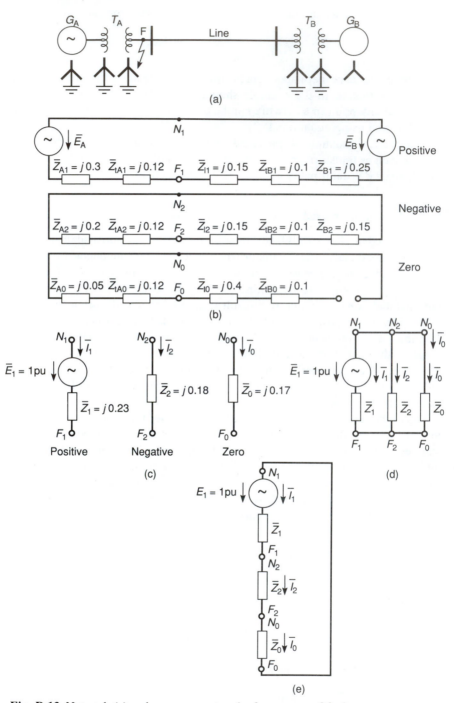

Fig. D.13 Network (a) and sequence networks for a range of faults.

Transformer T_B $\qquad$ $Z_{tB1} = Z_{tB2} = Z_{tB0} = j0.1$ pu

Line $\qquad$ $Z_{l1} = Z_{l2} = j0.15$ pu, $Z_{l0} = j0.4$ pu

It is further assumed that the two generators produce voltages of the same phases and of magnitude 1 p.u. The currents for faults near the secondary terminals of transformer T_A are determined below.

The positive-, negative- and zero-sequence networks are shown in Fig. D.13b. These networks may be further simplified as shown in Fig. D.13c.

D.4.1 Short-circuit between phases b and c and earth

For this fault the three sequence networks are combined as shown in Fig. D.13d. The sequence currents are then

$$\overline{I}_1 = \frac{\overline{V}}{\overline{Z}_1 + \dfrac{\overline{Z}_2\overline{Z}_0}{\overline{Z}_2 + \overline{Z}_0}} = \frac{1}{j0.23 + j\dfrac{0.18 \times 0.17}{0.18 + 0.17}} = -j3.15 \text{ pu}$$

Similarly, $\overline{I}_2 = j1.53$ pu and $\overline{I}_0 = j1.62$ pu.

The phase currents at the fault may be found by substituting the above values and are:

$$\overline{I}_a = \overline{I}_1 + \overline{I}_2 + \overline{I}_0 = 0$$

$$\overline{I}_b - \overline{I}_0 + a^2\overline{I}_1 + a\overline{I}_2 = -4.05 + j2.43 \text{ pu}$$

$$\overline{I}_c = \overline{I}_0 + a\overline{I}_1 + a^2\overline{I}_2 = 4.05 + j2.43 \text{ pu}$$

D.4.2 Short-circuit between phase a and earth

For this fault the three sequence networks are connected as shown in Fig. D.13(e). Hence $\overline{I}_1 = \overline{I}_2 = \overline{I}_0 = \dfrac{1}{j(0.23 + 0.18 + 0.17)} = -j1.72 \text{ pu}$

The phase currents may be found by substituting the above values and are

$$\overline{I}_a = 3\overline{I}_0 = -j5.17 \text{ pu}$$

$$\overline{I}_b = \overline{I}_c = 0$$

REFERENCE

1. Roeper, R. (1985) *Short-Circuit Currents in Three-Phase Systems*, (2nd edn), Siemens and A. G. and J. Wiley and Sons.

CONCLUDING REMARKS

It was stated in the Preface that all aspects of the protection of power systems cannot now be covered in a single book. It is hoped, however, that most of the important areas have been dealt with in this volume and that the information provided will enable readers to cope with any situations which may arise.

Topics such as the protection of the recently introduced co-generation schemes and combined heat and power plants have not been covered. The same basic requirements clearly apply to such installations and they could be protected in the ways described earlier. Because of the relatively small sizes of their sites, however, their protective equipment can be more integrated.

Recent advances in information technology and communications permit information to be conveyed in analogue or digital form from the various items of plant to a central point where it can be processed to detect faults or other abnormal conditions. Clearly, comparisons could be made of the currents entering and leaving particular items as in the current-differential schemes described in Chapter 5. In addition, the processing associated with the protection of the system could be combined with that required for control purposes. It must be recognized, however, that any failure of the central processing equipment could have very serious consequences and measures must be taken to duplicate vital circuits to ensure that the essential protection and control is maintained at all times.

The degree of coordination and integration of protection and control and the speed of implementation of such schemes in the future will be determined by the need to ensure a very high degree of security for protective schemes installed in major systems.

Arthur Wright and
Christos Christopoulos

Index